Ticks of the Southern Cone of America

Ticks of the Southern Cone of America

Diagnosis, Distribution, and Hosts with Taxonomy, Ecology and Sanitary Importance

Santiago Nava

Estación Experimental Agropecuaria Rafaela, Instituto Nacional de Tecnología Agropecuaria, and Consejo Nacional de Investigaciones Científica y Técnicas, Santa Fe, Argentina

José M. Venzal

Facultad de Veterinaria, CENUR Litoral Norte, Universidad de la República, Uruguay

Daniel González-Acuña

Facultad de Ciencias Veterinarias, Universidad de Concepción, Chile

Thiago F. Martins

Faculdade de Medicina Veterinária e Zootecnia, Universidade de São Paulo, Brazil

Alberto A. Guglielmone

Estación Experimental Agropecuaria Rafaela, Instituto Nacional de Tecnología Agropecuaria, and Consejo Nacional de Investigaciones Científica y Técnicas, Santa Fe, Argentina

ACADEMIC PRESS

An imprint of Elsevier

elsevier.com

Academic Press is an imprint of Elsevier
125 London Wall, London EC2Y 5AS, United Kingdom
525 B Street, Suite 1800, San Diego, CA 92101-4495, United States
50 Hampshire Street, 5th Floor, Cambridge, MA 02139, United States
The Boulevard, Langford Lane, Kidlington, Oxford OX5 1GB, United Kingdom

Notices
Knowledge and best practice in this field are constantly changing. As new research and experience broaden
our understanding, changes in research methods, professional practices, or medical treatment may become
necessary.

Practitioners and researchers must always rely on their own experience and knowledge in evaluating and
using any information, methods, compounds, or experiments described herein. In using such information or
methods they should be mindful of their own safety and the safety of others, including parties for whom they
have a professional responsibility.

To the fullest extent of the law, neither the Publisher nor the authors, contributors, or editors, assume any
liability for any injury and/or damage to persons or property as a matter of products liability, negligence or
otherwise, or from any use or operation of any methods, products, instructions, or ideas contained in the
material herein.

British Library Cataloguing-in-Publication Data
A catalogue record for this book is available from the British Library

Library of Congress Cataloging-in-Publication Data
A catalog record for this book is available from the Library of Congress

ISBN: 978-0-12-811075-1

For Information on all Academic Press publications
visit our website at https://www.elsevier.com/books-and-journals

Working together
to grow libraries in
developing countries

www.elsevier.com • www.bookaid.org

Publisher: Sara Tenney
Acquisition Editor: Linda Versteeg-Buschman
Editorial Project Manager: Halima Williams
Production Project Manager: Laura Jackson
Designer: Mark Rogers

Typeset by MPS Limited, Chennai, India

Dedication

This study is dedicated to Juan José Boero (Argentina),
Isaías Tagle (Chile), Luis Enrique Migone (Paraguay),
and Varela Calzada (Uruguay) for their valuable
scientific contributions to different parasitological
fields in the Southern Cone of America.

Contents

Biographies

Santiago Nava is a Senior Scientist in Parasitology at the Estación Experimental Agropecuaria Rafaela, Instituto Nacional de Tecnología Agropecuaria, and Scientist at the Consejo Nacional de Investigaciones Científica y Técnicas, Argentina. E-mail: nava.santiago@inta.gob.ar

Santiago Nava is a young Argentinean researcher whose interest is focused on systematics, ecology and control of ticks, and epidemiology of tick-borne pathogens. Within this subject area he coordinates different projects on ticks with medical and veterinary importance. The principal contribution of Dr. Nava has been in the study of Neotropical ticks, and he also has an active collaboration with scientists from more than 20 countries from America, Europe, and Africa.

José M. Venzal is Professor of Veterinary Parasitology at the Laboratorio de vectores y enfermedades transmitidas, Facultad de Veterinaria, CENUR Litoral Norte, Universidad de la República, Uruguay. E-mail: jvenzal@unorte.edu.uy

José M. Venzal is a young Uruguayan researcher on ticks and tick-borne diseases with a great expertise in systematic and phylogeny of Argasidae, a field mastered for few persons worldwide. The contribution of Prof. Venzal has been fundamental to describe several new Neotropical species of ticks (argasids and ixodids) but also for tick ecology and epidemiology of ticks-transmitted diseases. He received the Scopus Award Uruguay 2011 in the area of Agricultural Sciences, Editorial Elsevier, and the National Academy of Veterinary Medicine of Uruguay Award 2008.

Daniel González-Acuña is full Professor in Zoology and Ornithology at the Facultad de Ciencias Veterinarias, Universidad de Concepción, Chillán, Chile. E-mail: danigonz@udec.cl
 Daniel González-Acuña is a Doctor in Veterinary Medicine at the Institute of Wildlife of the Veterinary Medicine Superior School of Hannover, Germany. As Full Professor of the Faculty of Veterinary at the Universidad de Concepción, he teaches General Zoology for students of Veterinary Medicine and Agriculture Zoology for students of Agronomy. Prof. González-Acuña is specialized in parasites of Chile's wild fauna; within this endeavor he coordinated several research projects about ticks from the wild fauna in Chile. His contribution resulted in the description of several new species of ticks found in Chile and neighboring countries, and relevant biogeography information.

Thiago F. Martins is a Postdoc at the Department of Preventive Veterinary Medicine and Animal Health, Faculty of Veterinary Medicine, University of São Paulo, Brazil. E-mail: thiagodogo@hotmail.com
 Thiago F. Martins is a Brazilian Postdoc student with a prolific contribution on tick taxonomy and ecology. Dr. Martins made a very important contribution for the diagnoses of Argentinian and Brazilian nymphs from the genus *Amblyomma*, the most numerically important tick genus in South America. He is an active collaborator with many tick-research groups in Brazil and elsewhere providing his expertise for tick identification.

Alberto A. Guglielmone is a Senior Scientist in Parasitology at the Estación Experimental Agropecuaria Rafaela, Instituto Nacional de Tecnología Agropecuaria, and Superior Scientist at the Consejo Nacional de Investigaciones Científica y Técnicas, Argentina. E-mail: guglielmone.alberto@inta.gob.ar
 Alberto A. Guglielmone is a senior Argentinean scientist with a long career on ticks and cattle tick-borne diseases, started in 1975, but currently dealing with biogeography of Ixodidae. Dr. Guglielmone contributed with more than 300 scientific articles and several books, and collaborated with scientists from more than 30 countries. He also received several awards and is a member of the Academia Nacional de Agronomía y Veterinaria (Argentina).

Preface

Ticks (Acari: Ixodida) are becoming increasingly relevant as a sanitary problem for human, domestic, and wild animals worldwide. Significant scientific information about ticks of the Southern Cone of America (Argentina, Chile, Paraguay, and Uruguay) has been obtained through time, but no publication condensing this achievement is available. We have undertaken the task to collect the relevant data for diagnosis, distribution, and hosts with comments on taxonomy, ecology, and sanitary problems for each species of tick of the two families (Argasidae and Ixodidae) established in the Southern Cone of America believing that our summary will be of value for people interested in ticks and tick-borne diseases. Data for all species treated here were obtained from a search of the world tick literature that concluded on March 31, 2016.

Acknowledgments

We acknowledge:
Dmitry A. Apanaskevich and *Lorenza Beati*, U.S. National Tick Collection, Institute for Coastal Plain Science, Georgia Southern University, Statesboro, GA, USA; *Darci M. Barros-Battesti*, Laboratório de Parasitologia, Instituto Butantan, São Paulo, Brazil; *Jason Dunlop*, Museum für Naturkunde, Leibniz Institute for Research on Evolution and Biodiversity at the Humboldt-University Berlin, Berlin, Germany; *Agustín Estrada-Peña*, Facultad de Veterinaria, Universidad de Zaragoza, Zaragoza, Spain; *Marcelo B. Labruna*, Department of Preventive Veterinary Medicine and Animal Health, Faculty of Veterinary Medicine, University of São Paulo, São Paulo, Brazil; *Atilio J. Mangold*, Estación Experimental Agropecuaria Rafaela, Instituto Nacional de Tecnología Agropecuaria, Rafaela, Argentina; *João R. Martins*, Laboratório de Parasitologia, Instituto de Pesquisas Veterinárias Desidério Finamor, Eldorado do Sul, Brazil; *Valeria C. Onofrio*, Laboratório de Parasitologia, Instituto Butantan, São Paulo, Brazil; *Trevor N. Petney*, Department of Ecology and Parasitology, Institute of Zoology, Karlsruhe Institute of Technology, Karlsruhe, Germany; *Richard G. Robbins*, Walter Reed Biosystematics Unit, Department of Entomology Smithsonian Institution, Suitland, MD, USA; and *Matias P.J. Szabó*, Faculdade de Medicina Veterinaria, Universidade Federal de Uberlandia, Uberlandia, Brazil, for help in bibliography search, discussions about several tick species, information about type specimens, hosts, and distribution, and assistance for obtaining DNA sequences and construction of phylogenetic trees.

Marta E. Sánchez, Estación Experimental Agropecuaria Rafaela, Instituto Nacional de Tecnología Agropecuaria, Rafaela, Argentina, who significantly contributed to obtain many scientific articles used in this study.

Maria I. Camargo-Mathias, Department of Biology, Institute of Bioscience, São Paulo State University, Rio Claro, Brazil; *Pablo H. Nunes*, Latin American Institute of Life and Natural Sciences, Federal University of Latin American Integration, Foz do Iguaçu, Brazil; *Diego G. Ramirez*, Laboratório de Parasitologia, Instituto Butantan, São Paulo, Brazil; *Fredy A. Rivera-Páez*, Department of Biological Sciences, Faculty of Exact and Natural Sciences, University of Caldas, Manizales, Colombia; and *Patricia L. Sarmiento*, Servicio de Microscopía Electrónica del Museo de La Plata,

Universidad Nacional de La Plata, La Plata, Argentina, for their expertise with and assistance in scanning electron microscopy. *Karen Ardiles Villegas* and *Maria S. de la Fuente*, Departamento de Ciencias Pecuarias, Facultad de Ciencias Veterinarias, Universidad de Concepción, Chillán, Chile; *Oscar Castro, Oscar Correa, Carlos G. de Souza, María Laura Félix*, and *Paula Lado*, Facultad de Veterinaria, Universidad de la República, Uruguay; *Mariano Mastropaolo*, Cátedra de Parasitología y Enfermedades Parasitarias, Facultad de Ciencias Veterinarias, Universidad Nacional del Litoral, Esperanza, Argentina; *Lucila Moreno Salas*, Departamento de Zoología, Facultad de Ciencias Naturales y Oceanográficas, Universidad de Concepción, Concepción, Chile; *Sebastián Muñoz-Leal*, Departamento de Medicina Veterinária Preventiva e Saúde Animal, Faculdade de Medicina Veterinária e Zootecnia, Universidade de São Paulo, São Paulo, Brazil; and *Patrick Sebastian, María N. Saracho Bottero, Evelina L. Tarragona, Oscar Warnke*, and *Mario Wuattier*, Estación Experimental Agropecuaria Rafaela, Instituto Nacional de Tecnología Agropecuaria, Rafaela, Argentina, who contributed in many field and laboratory works useful for the information contained in this book.

The *Instituto Nacional de Tecnología Agropecuaria*, Argentina; *Asociacion Cooperadora de la Estación Experimental Agropecuaria*, Rafaela, Argentina; *Ministerio de Ciencia, Tecnología e Innovación Productiva*, Argentina; and the *Consejo Nacional de Investigaciones, Científicas y Técnicas*, Argentina, for providing institutional and financial support.

Introduction

Ticks are blood-feeding ectoparasites of vertebrates except fishes recognized as a worldwide nuisance by their capacities to cause dermatosis, anemia, toxemia including paralysis, impaired weight gain, while facilitating the occurrence of myiasis and secondary bacterial infections. Nevertheless, the most important problem associated with tick infestation is the competence to transmit pathogenic agents such as virus, bacteria, protozoa, and nematodes to animals and humans, being along mosquitos the most important arthropod vectors of infectious diseases.[1] Most tick species are parasites of wild animals, but reduction of primeval-host habitats by anthropic activities and increase of live animal trade may result in extinction of some tick species or changes in host profile in others as an adaptation to new conditions. As a consequence of the above scenario, studies on tick taxonomy, ecology, effects on animal production and health, including transmission of pathogenic organisms, resulted in a voluminous literature sometimes difficult to track.

The broadest definition for the Southern Cone of America includes Argentina, Chile, Paraguay, and Uruguay, covering a vast territory from north of Antarctica to the Tropic of Capricorn with the Andes Mountains separating Argentina and Chile as a major barrier for gene flow. This geography contains a great variety of wildlife, animal production systems under extensive pasture conditions and landscapes attracting a considerable numbers of tourists regularly. Ticks and tick-borne diseases of cattle have been historically of great concern for meat industry, graziers, scientists, and governments in Argentina, Uruguay, and Paraguay, but the recognition of local ticks as transmitters of human diseases added an additional worry and a notorious increase of regional research on ticks and related problems.

Most studies in the past focused on cattle ticks and cattle diseases transmitted by *Rhipicephalus microplus* with other tick species mostly recognized by scanty records of hosts and localities with eventual descriptions of new species as *Ixodes nuttalli* and *I. abrocomae* by Lahille,[2,3] *I. longiscutatus* by Boero,[4] and *I. neuquenensis* by Ringuelet.[5] The books of Lahille[6] "Contribution a l'étude des ixodides de la République Argentine" and Boero[7] "Las garrapatas de la República Argentina" remained as main references for species different to *R. microplus* for decades and, indeed, the work of Boero is still a sound tool for identification of several species of ticks. Thereafter, studies of *R. microplus* and associated diseases continued, sometimes with considerable success as the development of vaccines against cattle babesiosis, but some workers became interested in ticks of the genus *Amblyomma* in Argentina,[8–10] while Alcaíno[11] in Chile alerted about the presence of the dog tick *R. sanguineus* sensu lato which become a major tick problem throughout the region a few years later. A major shift resulted from the studies of Conti et al.[12] and Ripoll et al.[13] incriminating ticks as vector of rickettsial diseases to humans in Uruguay and Argentina, respectively, which led to diagnosis of rickettsial and others tick-transmitted diseases in humans as well as their agents in certain or potential tick vectors.[14–16]

Tick ecology studies (apart of *R. microplus*) increased the knowledge on hosts, distribution, and effects of abiotic factors for several species of sanitary importance as shown by the works of Guglielmone et al.,[17] González-Acuña et al.,[18] Nava et al.,[19] Venzal et al.,[20] Debárbora et al.,[21] Tarragona et al.,[22] just to cite a few examples. Another major contribution resulted from the combination of morphological and molecular studies, and consequently the description of new species for the Southern Cone of America, namely *Amblyomma boeroi, A. hadanii, A. tonelliae, lahillei, O. microlophi, O. quilinensis, O. rioplatensis,* and *O. xerophylus* from 2008 to 2014.

The paragraphs above show that there is a considerable amount of information about the ticks of Argentina, Chile, Paraguay, and Uruguay. Therefore we attempt to present here a synopsis of that information that may be of value for students, taxonomists, biologists, human and animal health professionals, and the general public.

The work comprises four chapters and a conclusion section. The first chapter contains general considerations about tick classification, external tick anatomy including a glossary, and biological cycles. The second contains information for the species of the family Ixodidae established in the region, and the third includes data for the species of the family Argasidae. The fourth chapter contains morphological dichotomous keys for tick identification (genera and species).

Keys, figures (if available), and diagnosis for male, female, and nymph but not for larvae are presented for Ixodidae. Precise descriptions of the larva of several species of ixodids are lacking being impossible to construct reliable morphological keys; therefore, we decided to exclude taxonomic information and figures for this parasitic tick stage. The situation of Argasidae is almost the opposite to Ixodidae because larvae are of great importance for species identification and many species are known exclusively for this stage; therefore, for each taxon we present texts and figures of male and female (if available) and larva, but no figures of nymphs are included because we consider their taxonomic value of relative usefulness.

For each species we include the name with full reference to the original description; geographic distribution in the Southern Cone of America and elsewhere if proper; biogeographic information referring exclusively to Argentina, Chile, Paraguay, or Uruguay are presented afterwards, followed by data about hosts (class, order, family, binomial scientific name), brief notes on ecology and sanitary importance. Next part of the presentation of each species contains diagnosis and figures with notes on taxonomic information (if applicable), DNA sequences with relevance for identification of the species and eventually condensed phylogenetic trees. Synonyms are not fully considered in this study because a recent work[23] already covers this issue, and those readers interested in tick names should consult it.

Basic geographic distribution of ticks were obtained from Guglielmone and Nava[24,25] for Argentina, González-Acuña and Guglielmone[26] for Chile, Nava et al.[27] for Paraguay, Venzal et al.[28] for Uruguay, and Guglielmone et al.[29] for all those countries plus data contained in several studies published from 2003 to early 2016. Biogeographic information for the Southern Cone of America was based on Morrone[30] who divided this territory into three areas (Neotropical, South America Transition Zone, and Andean) which are further divided into biogeographic provinces as presented in Fig. 1.

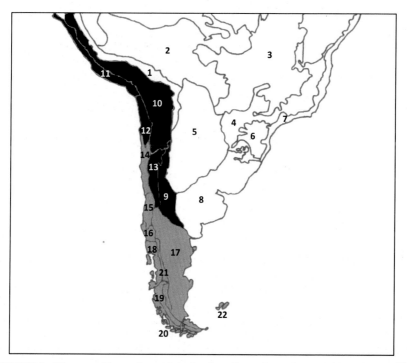

FIGURE 1 Biogeographic provinces of the Southern Cone of America outlined by Morrone.[30] *Neotropical Region*: (1) Yungas; (2) Pantanal; (3) Cerrado; (4) Parana Forest; (5) Chaco; (6) *Araucaria angustifolia* Forest; (7) Brazilian Atlantic Forest; (8) Pampa. *South American Transition Zone*: (9) Monte; (10) Puna; (11) Coastal Peruvian Desert; (12) Atacama; (13) Prepuna. *Andean Region*: (14) Coquimbo; (15) Santiago; (16) Maule; (17) Central Patagonia; (18) Valdivian Forest; (19) Magellanic Forest; (20) Magellanic Paramo; (21) Subandean Patagonia; (22) Malvinas Islands.

Locality records have been obtained partly from the articles cited above, publications with exhaustive information about some species of ticks (i.e., Muñoz-Leal and González-Acuña[31] for *I. uriae*), and partly from a compilation of tick records from the Neotropical Zoogeographical Region contained in about 2000 scientific studies constructed by Alberto A. Guglielmone being available upon request. Information about localities and hosts presented exclusively in dissertations or scientific events were ignored.

The nomenclature for hosts name was based on Wilson and Reeder,[32] Bárquez et al.,[33] Weksler et al.,[34] D'Elía et al.,[35] and Voss and Jansa[36] for Mammalia, Guglielmone and Nava[37] and Cano et al.[38] for Reptilia and Amphibia, and Clements[39] for Aves, but some names were changed to adopt new classifications (data not shown). Aves are now considered a subgroup of Reptilia or Sauropsida for some authors, but we maintain Aves as a class for convenience reasons. Name of live organisms and their phylogenetic position is changing continuously, therefore readers are advice to check if host names including in our perusal correspond to actual valid denomination. A host was included in the corresponding list only when the parasitic tick stage was clearly stated; parasitism of captive or laboratory animals are excluded from the analysis. Records for species whose nymph and larva are undescribed or poorly described where included if the specific diagnosis was based on convincing molecular analysis or from molted engorged specimens.

It should be noted that host list for each species refers only to records from Neotropical Mexico to southern Argentina, Chile, and their islands. Important records and record from the Nearctic or other biogeographic areas are sometimes discussed but never included in the host list.

The morphological diagnosis was based on the original descriptions plus figures and texts provided by several authors[7,40–51] and examination of tick specimens deposited in collections of the Instituto Nacional de Tecnología Agropecuaria, Estación Experimental Agropecuaria Rafaela, Rafaela, Argentina; Facultad de Ciencias Veterinarias, Universidad Nacional del Litoral, Esperanza, Argentina; Facultad de Ciencias Veterinarias, Universidad de Concepción, Chillán, Chile; Facultad de Veterinaria, Universidad de la República, Salto, Uruguay; Faculdade de Medicina Veterinária e Zootecnia, Universidade de São Paulo, São Paulo, Brazil; and Institute for Coastal Plain Science, Georgia Southern University, Statesboro, GA, United States.

Body size of ticks is relevant for the identification of species from the genus *Amblyomma* and larva of most argasids established in the Southern Cone of America. Whenever possible, mean numbers and ranges of important morphological parameters are included for males, females, and nymphs of *Amblyomma* and also for Argasidae. We include measures without range for species belonging to other genera of Ixodidae obtained from the descriptions or redescritptions of each taxon. We found no reliable information for the body size of males (equivalent to the dimension of the scutum), and scutum and body size of females of *A. argentinae*, *A. dubitatum*, *A. ovale*, *A. parvitarsum*, *A. pseudoconcolor*, *A. tigrinum*, and *A. triste*, and male, female, and nymph of *A. neumanni*. Therefore, we measured the body length and breadth of 10 males, and the body length and breadth and scutum length and breadth of 10 unfed females of the species named above from specimens deposited in the Collection of Estación Experimental Agropecuaria Rafaela, and 10 laboratory-reared nymphs of *A. neumanni*, to be included in the diagnosis of these taxa. Whenever possible the descriptions of nymphs and female ticks were based on unfed specimens. Body sizes of ticks of the genus *Amblyomma* were separated as large, medium, and small, considering the relationship among the species of the Southern Cone of America, not as universal criterion for all species of ticks worldwide. We classified the body size of male and females as small when less than 3 mm long, medium size when 3−4 mm long, and large when more than 4 mm in length, and nymphs were categorized as small when less than 1.3 mm long, medium size when 1.4−1.6 mm long, and large when more than 1.6 mm in length.

DNA information was obtained from the Gen Bank and phylogenetic trees were constructed by Santiago Nava. The sections of taxonomical notes, ecology, and sanitary importance include comments with references about phylogenetic positions and taxonomical conflicts, a summary of ecological knowledge, and brief comments of diseases and health problems.

Finally, the great majority of the species found in the Southern Cone of America are exclusive for this geography or with distribution that encompasses other Neotropical countries with some taxa reaching southern Nearctic localities, but four species, *I. auritulus*, *I. uriae*, *R. microplus*, and *R. sanguineus* sensu lato, have distributions that largely exceed the Neotropical and Nearctic Regions. The information about these species refers solely to Argentina, Chile, Paraguay, and Uruguay, with additional data for the Neotropics.

REFERENCES

1. Sonenshine DE, Roe RM. Overview. Ticks, people and animals. In: 2nd ed. Sonenshine DE, Roe RM, editors. *Tick biology*, vol. 1. Oxford: Oxford University Press; 2014. p. 3−16.
2. Lahille F. Sobre dos *Ixodes* de la República Argentina y la medición de las garrapatas. *Bol Minist Agric Repúb Argent* 1913;**16**:278−89.
3. Lahille F. Descripción de un nuevo ixódido chileno. *Rev Chil Hist Nat* 1916;**20**:107−8.
4. Boero JJ. Notas ixodológicas. (I). *Ixodes longiscutatum*, nueva especie. *Rev Asoc Méd Argent* 1944;**58**:353−4.
5. Ringuelet R. La supuesta presencia de *Ixodes brunneus* Koch en la Argentina y descripción de una nueva garrapata *Ixodes neuquenensis* nov. sp. *Notas Mus La Plata* 1947;**12**:207−16.
6. Lahille F. Contribution à l'étude des ixodidés de la Répulique Argentine. *An Minist Agric Repúb Argent* 1905;**2**:1−166.
7. Boero JJ. *Las garrapatas de la República Argentina (Acarina: Ixodoidea)*. Buenos Aires: Universidad de Buenos Aires; 1957.

8. Ivancovich JC. Nuevas subspecies de garrapatas del género *Amblyomma* (Ixodidae). *Rev Invest Agropec Ser 4 Patol Anim* 1973;**10**:55−64.

9. Ivancovich JC. Reclasificación de algunas especies de garrapatas del género *Amblyomma* (Ixodidae) en la Argentina. *Rev Invest Agropec* 1980;**15**:673−82.

10. Guglielmone AA, Hadani A, Mangold A, De Haan L, Bermúdez A. Garrapatas (Ixodoidea-Ixodidae) del ganado bovino en la provincia de Salta: especies y carga en 5 zonas ecológicas. *Rev Med Vet (Buenos Aires)* 1981;**62**:194−205.

11. Alcaíno H. Antecedentes sobre la garrapata café del perro (*Rhipicephalus sanguineus*). *Monogr Med Vet* 1985;**7**:48−55.

12. Conti IA, Rubio I, Somma RE, Pérez G. Cutaneous − ganglionar rickettsiosis due to *Rickettsia conorii* in Uruguay. *Rev Inst Med Trop São Paulo* 1980;**32**:313−18.

13. Ripoll CM, Remondegui CE, Ordoñez G, Arazamendi R, Fusaro H, Hyman, et al. Evidence of rickettsial spotted fever and ehrlichial infections in a subtropical territory of Jujuy, Argentina. *Am J Trop Med Hyg* 1999;**61**:350−4.

14. Abarca K, López J, Acosta-Jamett G, Lepe P, Soares JF, Labruna MB. A third *Amblyomma* species and the first tick-borne rickettsia in Chile. *J Med Entomol* 2012;**49**:219−22.

15. Venzal JM, Estrada-Peña A, Portillo A, Mangold AJ, Castro O, De Souza CG, et al. *Rickettsia parkeri*: a rickettsial pathogen transmitted by ticks in endemic areas for spotted fever rickettsiosis in southern Uruguay. *Rev Inst Med Trop São Paulo* 2012;**54**:131−4.

16. Romer Y, Nava S, Govedic F, Cicuttín G, Denison AM, Singleton J, et al. *Rickettsia parkeri* rickettsiosis in different ecological regions of Argentina and its association with *Amblyomma tigrinum* as a potential vector. *Am J Trop Med Hyg* 2014;**91**:1156−60.

17. Guglielmone AA, Mangold AJ, Luciani CE, Viñabal AE. *Amblyomma tigrinum* (Acari: Ixodidae) in relation to phytogeography of central-northern Argentina with notes on hosts and seasonal distribution. *Exp Appl Acarol* 2000;**24**:983−9.

18. González-Acuña D, Venzal JM, Keirans JE, Robbins RG, Ippi S, Guglielmone AA. New host and locality records for the *Ixodes auritulus* (Acari: Ixodidae) species group, with a review of host relationships and distribution in the Neotropical Zoogeographic Region. *Exp Appl Acarol* 2005;**37**:147−56.

19. Nava S, Szabó MPJ, Mangold AJ, Guglielmone AA. Distribution, hosts, 16S rDNA sequences and phylogenetic position of the Neotropical tick *Amblyomma parvum* (Acari: Ixodidae). *Ann Trop Med Parasitol* 2008;**102**:409−25.

20. Venzal JM, Estrada-Peña A, Castro O, De Souza CG, Félix ML, Nava S, et al. *Amblyomma triste* Koch, 1844 (Acari: Ixodidae): hosts and seasonality of the vector of *Rickettsia parkeri* in Uruguay. *Vet Parasitol* 2008;**155**:104−9.

21. Debárbora VN, Mangold AJ, Oscherov EB, Guglielmone AA, Nava S. Study of the life cycle of *Amblyomma dubitatum* (Acari: Ixodidae) based on field and laboratory data. *Exp Appl Acarol* 2014;**63**:93−105.

22. Tarragona EL, Mangold AJ, Mastropaolo M, Guglielmone AA, Nava S. Ecology and genetic variation of *Amblyomma tonelliae* in Argentina. *Med Vet Entomol* 2005;**29**:297−304.

23. Guglielmone AA, Nava S. Names for Ixodidae (Acari: Ixodoidea): valid, synonyms, *incertae sedis*, *nomina dubia*, *nomina nuda*, *lapsus*, incorrect and suppressed names − with notes on confusions and misidentifications. *Zootaxa* 2014;**3767**:1−256.

24. Guglielmone AA, Nava S. Las garrapatas de la familia Argasidae y de los géneros *Dermacentor*, *Haemaphysalis*, *Ixodes* y *Rhipicephalus* (Ixodidae) de la Argentina: distribución y hospedadores. *Rev Invest Agropec* 2005;**34**(2):123−41.

Introduction xxiii

25. Guglielmone AA, Nava S. Las garrapatas argentinas del género *Amblyomma* (Acari: Ixodidae): distribución y hospedadores. *Rev Invests Agropec* 2006;**35**(3):135−55.
26. González-Acuña D, Guglielmone AA. Ticks (Acari: Ixodoidea: Argasidae, Ixodidae) of Chile. *Exp Appl Acarol* 2005;**35**:147−63.
27. Nava S, Lareschi M, Rebollo C, Benítez Usher C, Beati L, Robbins RG, et al. The ticks (Acari: Ixodida: Argasidae, Ixodidae) of Paraguay. *Ann Trop Med Parasitol* 2007;**101**:255−70.
28. Venzal JM, Castro O, Cabrera PA, De Souza CG, Guglielmone AA. Las garrapatas de Uruguay: especies, hospedadores, distribución e importancia sanitaria. *Veterinaria (Montevideo)* 2003;**38**:17−28.
29. Guglielmone AA, Estrada-Peña A, Keirans JE, Robbins RG. *Ticks (Acari: Ixodida) of the neotropical zoogeographic region. Special publication of the international consortium on ticks and tick-borne diseases-2.* Houten (The Netherlands): Atalanta; 2003.
30. Morrone JJ. Biogeographic areas and transition zones of Latin America and the Caribbean Islands based on panbiogeographic and cladistic analyses of the entomofauna. *Annu Rev Entomol* 2006;**51**:467−94.
31. Muñoz-Leal S, González-Acuña D. The tick *Ixodes uriae* (Acari: Ixodidae): hosts, geographical distribution, and vector roles. *Ticks Tick-borne Dis* 2015;**6**:843−68.
32. Wilson DE, Reeder DM. *Mammals species of the world: a taxonomic and geographic reference.* 3rd ed. Baltimore (MD): The Johns Hopkins University Press; 2005.
33. Bárquez RM, Díaz MM, Ojeda RA. *Mamíferos de Argentina: sistemática y distribución.* Tucumán: Sociedad Argentina para el Estudio de los Mamíferos; 2006.
34. Weksler M, Percequillo AR, Voss RS. Ten new genera of oryzomyne rodents (Cricetidae: Sigmodontinae). *Am Mus Novit* 2006;**3537**:1−29.
35. D'Elía G, Pardiñas UFJ, Jayat P, Salazar-Bravo J. Systematics of *Necromys* (Rodentia, Cricetidae, Sigmodontinae): species limits and groups, with comments on historical biogeography. *J Mammal* 2008;**89**:778−90.
36. Voss RS, Jansa SA. Phylogenetic relationships and classification of didelphid marsupials, an extant radiation of New World metatherian mammals. *Bull Am Mus Nat Hist* 2009;**322**:177.
37. Guglielmone AA, Nava S Hosts of *Amblyomma dissimile* Koch, 1844 and *Amblyomma rotundatum* Koch, 1844. *Zootaxa* 2010;**2541**:27−49.
38. Cano PD, Ball HA, Carpinetto MF, Peña GD. Reptile checklist of Río Pilcomayo National Park, Formosa, Argentina. *Check List* 2015;**11**(3) (article 1658) 13 p.
39. Clements JF. *The Clements checklist of birds of the world.* 6th ed. Ithaca (NY): Cornell University Press; 2007.
40. Robinson LE. *Ticks. A monograph of the Ixodoidea. Part IV. The genus Amblyomma.* London: Cambridge University Press; 1926.
41. Cooley RA. The genera *Boophilus, Rhipicephalus,* and *Haemaphysalis* (Ixodoidea) of the New World. *Natl Inst Health Bull* 1946;**187**:54.
42. Guglielmone AA, Viñabal AE. Claves morfológicas dicotómicas e información ecológica para la identificación de las garrapatas del género *Amblyomma* Koch, 1844 en la Argentina. *Rev Invest Agropec* 1994;**25**(1):39−67.
43. Yunker CE, Keirans JE, Clifford CM, Easton ER. *Dermacentor* ticks (Acari: Ixodoidea: Ixodidae) of the New World: a scanning electron microscopy atlas. *Proc Entomol Soc Wash* 1986;**88**:609−27.
44. Durden LA, Keirans JE. Nymphs of the genus *Ixodes* (Acari: Ixodidae) of the United States: taxonomy, identification key, distribution, hosts, and medical/ veterinary importance. *Thomas Say Publ Entomol Monogr* 1996;**9**:95.

Estrada-Peña A, Venzal JM, Mangold AJ, Cafrune MM, Guglielmone AA. The *Amblyomma maculatum* Koch, 1844 (Acari: Ixodidae: Amblyomminae) tick group: diagnostic characters, description of the larva of *A. parvitarsum* Neumann, 1901, 16S rDNA sequences, distribution and hosts. *Syst Parasitol* 2005;**60**:99−112.

46. Estrada-Peña A, Venzal AJ, Nava S, Mangold A, Guglielmone AA, Labruna MB, et al. Reinstatement of *Rhipicephalus* (*Boophilus*) *australis* Fuller, the Australian cattle tick (Acari: Ixodidae) with redescription of the adult and larval stages. *J Med Entomol* 2012;**49**:794−802.

47. Barros-Battesti DM, Arzua M, Bechara GH. *Carrapatos de importância médico-veterinária da região neotropical: um guia ilustrado para identificação de espécies*. São Paulo: Vox/ICTTD-3/Butantan; 2006.

48. Onofrio VC, Barros-Battesti DM, Labruna MB, Faccini JLH. Diagnoses of and illustrated key to the species of *Ixodes* Latreille, 1795 (Acari: Ixodidae) from Brazil. *Syst Parasitol* 2009;**72**:143−57.

49. Martins TF, Onofrio VC, Barros-Battesti DM, Labruna MB. Nymphs of the genus *Amblyomma* (Acari: Ixodidae) of Brazil: descriptions, redescriptions, and identification key. *Ticks Tick-borne Dis* 2010;**1**:75−99.

50. Martins TF, Labruna MB, Mangold AJ, Cafrune MM, Guglielmone AA, Nava S. Taxonomic key to nymphs of the genus *Amblyomma* (Acari: Ixodidae) in Argentina, with description and redescription of the nymphal stage of four *Amblyomma* species. *Ticks Tick-borne Dis* 2014;**5**:753−70.

51. Nava S, Beati L, Labruna MB, Cáceres AG, Mangold AJ, Guglielmone AA. Reassessment of the taxonomic status of *Amblyomma cajennense* (Fabricius, 1787) with the description of three new species, *Amblyomma tonelliae* n. sp., *Amblyomma interandinum* n. sp. and *Amblyomma patinoi* n. sp., and reinstatement of *Amblyomma mixtum* Koch, 1844 and *Amblyomma sculptum* Berlese, 1888 (Ixodida: Ixodidae). *Ticks Tick-borne Dis* 2014;**5**:252−76.

Chapter 1

Tick Classification, External Tick Anatomy with a Glossary, and Biological Cycles

In this chapter a few excerpts are given for tick classification and biological cycles of Argasidae and Ixodidae with a few remarks about the third family of Ixodoidea, the Nuttalliellidae. Nevertheless, most text will be used to show the most important external morphological features of Ixodidae and Argasidae including a glossary for those structures with relevance for diagnosis of genera and species.

TICK CLASSIFICATION

Ticks belong to the Phylum Arthropoda, Class Arachnida, Subclass Acari, Superorder Parasitiformes, Order Ixodida, Superfamily Ixodoidea.[1] This superfamily contains the families Ixodidae, Argasidae, and Nuttalliellidae (monotypic). Last compilation of Argasidae by Guglielmone et al.[2] listed 193 species worldwide but this number jumped to 208 by the end of 2015 as new species have been described.[3–10] Guglielmone and Nava[11] registered 711 species of Ixodidae but due to description of new taxa, new synonyms, and reinstatement of one species, the actual number is 722 after the studies of Takada[12] (for synonymization of *Haemaphysalis ias* with *Haemaphysalis cornigera*), Apanaskevich et al.,[13,14] Estrada-Peña et al.,[15] Hornok et al.,[16] Nava et al.,[17,18] Apanaskevich and Apanaskevich,[19–22] and Krawczak et al.[23] Argasidae are generally considered to be formed by 5 genera, Ixodidae by 14, and Nuttalliellidae by 1 as presented in Table 1.1.

The Ixodidae are usually further divided into two groups based on morphological and biological characters: **Prostriata** containing all the species of *Ixodes* and **Metastriata** formed by the remainder genera. Prostriata is characterized morphologically by species with anal groove anterior to the anus and venter of males mostly covered by flat plates, while Metastriata contains species with anal groove posterior to the anus or indistinct and venter of males is never mostly covered by flat plates. See "Biological Cycles" section for biological differences between Metastriata and Prostriata groups.

Ticks of the Southern Cone of America. DOI: http://dx.doi.org/10.1016/B978-0-12-811075-1.00001-7

TABLE 1.1 List of Genera for the Three Families of Ixodida

Argasidae[a]	Ixodidae	Nuttalliellidae
Argas Latreille, 1795	*Amblyomma* Koch, 1844	*Nuttalliella* Bedford, 1931
Ornithodoros Koch, 1844	*Anomalohimalaya* Hoogstraal, Kaiser and Mitchell, 1970	
Otobius Banks, 1912	*Bothriocroton* Keirans, King and Sharrad, 1994	
Antricola Cooley and Kohls, 1942	*Compluriscutula*[b] Poinar and Buckley, 2008	
Nothoaspis Keirans and Clifford, 1975	*Cornupalpatum*[b] Poinar and Brown, 2003	
	Cosmiomma Schulze, 1920	
	Dermacentor Koch, 1844	
	Haemaphysalis Koch, 1844	
	Hyalomma Koch, 1844	
	Ixodes Latreille, 1795	
	Margaropus Karsch, 1879	
	Nosomma Schulze, 1920	
	Rhipicentor Nuttall and Warburton, 1908	
	Rhipicephalus Koch, 1844	

[a]The classification of Argasidae is based on Guglielmone et al.[2] but see also Estrada-Peña et al.[c] (2010). Some authors include the genus *Carios* Latreille, 1796, as proposed by Klompen[d] and Klompen & Oliver[e] based mainly on larvae chaetotaxy but others found biological and morphological inconsistencies in this scheme and disperse the species of *Carios* among the genera *Antricola*, *Nothoaspis*, *Argas*, and *Ornithodoros*.[2]
[b]Extinct genera.
[c]Estrada-Peña A, Mangold AJ, Nava S, Venzal JM., Labruna MB, Guglielmone AA. A review of the systematics of the family Argasidae (Ixodida). *Acarologia* 2010;**50**:317−33.
[d]Klompen JSH. Comparative morphology of argasid larvae (Acari: Ixodida: Argasidae), with notes on phylogenetic relationships. *Ann Entomol Soc Am* 1992;**85**:541−60.
[e]Klompen JSH, Oliver JH. Systematic relationships in the soft ticks (Acari: Ixodida: Argasidae). *Syst Entomol* 1993;**18**:313−31.

MAIN EXTERNAL MORPHOLOGICAL CHARACTERS OF THE FAMILIES OF IXODIDA

Ixodidae and Argasidae

The principal characteristics of the external anatomy of both soft and hard ticks (Figs. 1.1−1.6) are defined in the following glossary. It was constructed

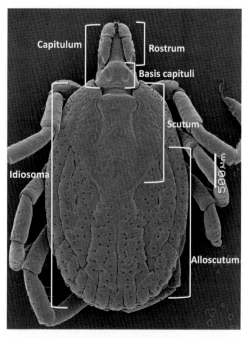

FIGURE 1.1   Major body divisions of hard ticks (Ixodidae).

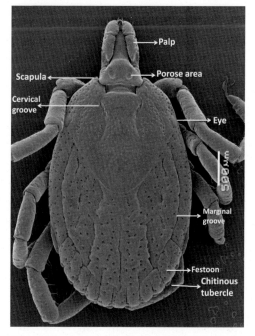

FIGURE 1.2   External anatomy of ticks of the family Ixodidae. Dorsal view of a female.

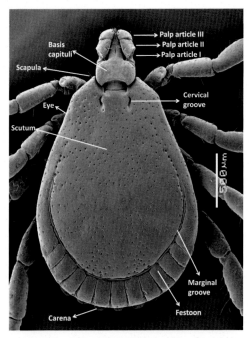

FIGURE 1.3   External anatomy of ticks of the family Ixodidae. Dorsal view of a male.

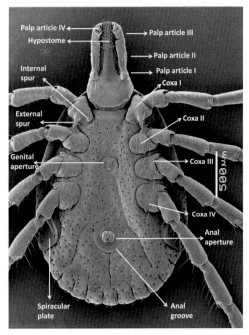

FIGURE 1.4   External anatomy of ticks of the family Ixodidae. Ventral view of a female.

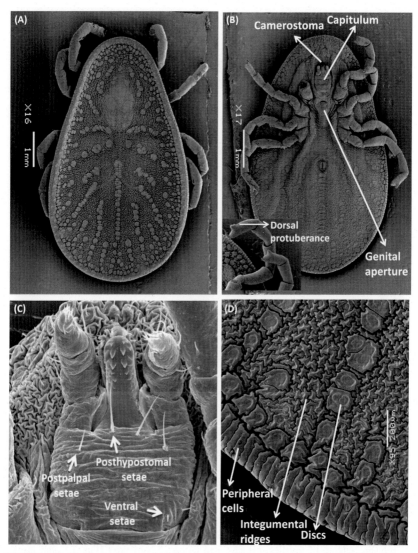

FIGURE 1.5 External anatomy of adult soft tick of the genus *Argas*: (A) dorsal view, (B) ventral view, (C) ventral view of capitulum, (D) details of dorsal integument. The subapical dorsal protuberance of the legs is shown in (B).

based on the terminology and figures from several sources[1,24−27] plus specific definitions referenced in the corresponding description, focusing on morphological characters employed in the diagnosis of species (see chapters: "Genera and Species of Ixodidae" and "Genera and Species of Argasidae") and keys (see chapter: "Morphological Keys for Genera and Species of Ixodidae and Argasidae"). This glossary does not constitute an exhaustive

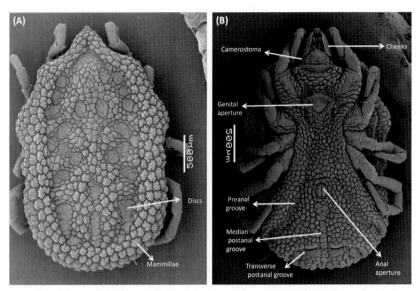

FIGURE 1.6 External anatomy of adult soft tick of the genus *Ornithodoros*: (A) dorsal view and (B) ventral view.

list of the morphological characters of ticks with taxonomic value, but it is aimed to define those characters useful for taxonomic determination of the tick species established in Argentina, Chile, Paraguay, and Uruguay encompassing three genera of Argasidae (*Argas*, *Ornithodoros*, and *Otobius*) and five genera of Ixodidae (*Amblyomma*, *Dermacentor*, *Haemaphysalis*, *Ixodes*, and *Rhipicephalus*). Therefore when referring to characters specific to some genera in the Southern Cone of America, it does not exclude its presence in tick species of those genera established elsewhere (Table 1.2).

Glossary

Alloscutum (=notum) the dorsal extensible surface that surrounds laterally and posteriorly the scutum of females, nymphs, and larvae of hard ticks (Fig. 1.1) but almost imperceptible in males of Ixodidae. It can be glabrous or with setae in different numbers and sizes.

Anal groove a depression surrounding the anus anteriorly, genus *Ixodes*, or posteriorly, remaining genera of hard ticks (Figs. 1.4 and 1.24). Also present in several Argasidae (Fig. 1.6).

Auriculae lateral projection at the sides of the venter of the basis capituli characteristic of the genus *Ixodes* (Fig. 1.7).

Basis capituli the basal portion of the capitulum on which the mouthparts are inserted. Dorsally the shapes of the basis capituli are generally classified as hexagonal, rectangular, subrectangular, triangular, and subtriangular (Fig. 1.8).

TABLE 1.2 Main Morphological Differences Between Adult of Argasidae and Ixodidae

Ixodidae	Argasidae
With dorsal scutum	Without dorsal scutum
Evident sexual dimorphism	Not evident sexual dimorphism (with few exceptions)
Capitulum visible dorsally	Capitulum not visible dorsally
Coxae with spurs	Coxae without spurs
Festoon present or absent	Festoon never present
Porose area of female present (with one exception)	Porose area of female never present
Spiracular plates present	Spiracular plates absent

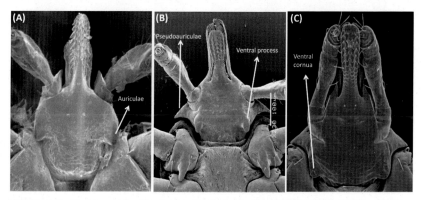

FIGURE 1.7   Basis capituli of hard ticks: (A) Auriculae present in the genus *Ixodes*, (B) pseudoauriculae and ventral process present in nymphs of the genus *Amblyomma*, (C) ventral cornua present in nymphs of the genus *Amblyomma*.

Camerostoma a depression on the ventral surface of soft ticks in which the capitulum of adults and nymphs lies (Figs. 1.5 and 1.6).

Capitulum (=gnathosoma) moveable structure formed by the basis capituli and the rostrum or mouthparts (palps, chelicerae, and hypostome). The capitulum is located at the anterior end of the body in all the parasitic stages of hard ticks and in larvae of soft ticks, and it is visible from a dorsal view (Fig. 1.1). In nymphs and adults of Argasidae the capitulum is inserted in the camerostoma bearing cheeks (=flaps) in several species, hidden from dorsal view (Figs. 1.5 and 1.6). The walls of the camerostoma may be projected anteriorly forming a hood (=anterior projection).

Carena (=distal sclerotized ventral plates) as defined in Guglielmone et al.[28]: chitinous plates that may surpass posteriorly the festoons of males of some species of *Amblyomma*. Carena are present or absent, incised or not incised (Fig. 1.9).

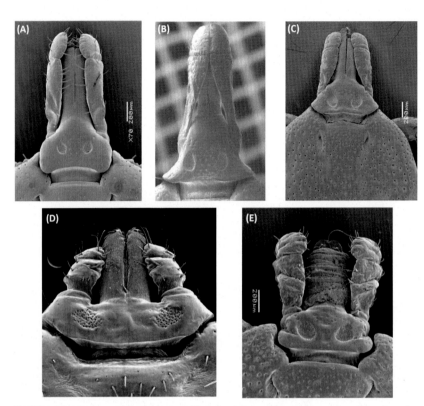

FIGURE 1.8 Shapes of the basis capituli in hard ticks: (A) subrectangular, (B) triangular, (C) subtriangular, (D) hexagonal, (E) rectangular.

Caudal appendage small projection presents at the middle of posterior margin of the body in males of some species of the genus *Rhipicephalus* (Fig. 1.10).

Cervical groove paired depressions in the central anterior part of the scutum of males, females, and nymphs of Ixodidae (Figs. 1.2 and 1.3).

Chelicerae a pair of moveable structures with digits located between the palps (Fig. 1.2) used to pierce the skin of hosts.

Cornua dorsal paired projections at the postero-lateral margins of the basis capituli. They can be long, short, or absent (Fig. 1.11).

Coxae first segment of the legs that allow the articulation between legs and body. The coxae of most of the species of hard ticks have spurs, a single spur, or a pair. These spurs are short or long, narrow or triangular, sharp or blunt (Fig. 1.12). When the coxae have a pair of spurs, they can be equal or unequal in size and shape. These characters are of importance for specific diagnosis.

Discs circular or subcircular depressions present on the dorsal and ventral surface of nymphs and adults of soft ticks (Figs. 1.5 and 1.6). They represent sites of muscle insertion.

Dorsal shield a small plate located centrally on the dorsal surface of the body of larvae of soft ticks (Fig. 1.13). The shapes of the dorsal shield have taxonomic value, usually being oval, pyriform, or triangular.

FIGURE 1.9 Carena (=distal sclerotized ventral plates): (A) present, not incised, dorsal view; (B) present, not incised, ventral view; (C) present, incised, dorsal view; (D) present, incised, ventral view; (E) absent, dorsal view; (F) absent, ventral view.

FIGURE 1.10 Male caudal appendage in hard ticks (*Rhipicephalus microplus*).

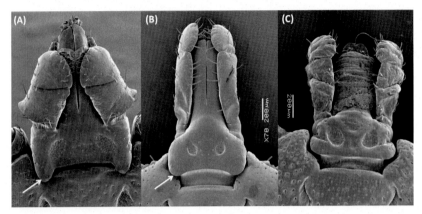

FIGURE 1.11   Cornua of hard ticks: (A) long (marked with an *arrow*), (B) short, (C) absent.

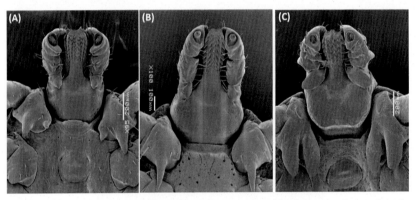

FIGURE 1.12   Spurs on coxa I of hard ticks: (A) spurs unequal in size: external spur triangular and blunt, internal spur triangular and short, (B) spurs unequal in size: external spur long, narrow and sharp, internal spur indistinct, (C) spurs equal in size: spurs long, triangular, and blunt.

Eyes simple circular structures located on the lateral margins of the scutum of hard ticks, but absent in species of the genera *Ixodes* and *Haemaphysalis*. Shapes of eyes are classified as flat, bulging, or orbited (Fig. 1.14). Some species of argasids have eyes which, in the case of Southern Cone of America soft ticks, are only of diagnostic value for larva.

Festoons small areas divided by grooves at the posterior margin of the body of some genera of Ixodidae (Figs. 1.2 and 1.3). In some species of *Amblyomma* there are chitinous tubercles on the postero-ventral side of the festoons (Fig. 1.15). This feature is distorted in fed females and nymphs because the tubercles become positioned in the posterior margin as festoons disappeared with engorgement.

Genital aperture the opening of the reproductive organ is located on the ventral surface; it varies in position and shape among genera and species. The outline of the genital aperture is shaped by the disposition of two lateral lips and has specific taxonomic

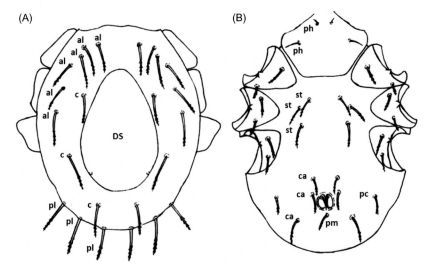

FIGURE 1.13   Setae pattern of larvae of soft ticks: (A) dorsal view: (al) anterolateral setae, (c) central setae, (pl) postero-lateral setae; (B) ventral view: (pm) postero-median seta, (st) sternal setae, (ca) circumanal setae, (pc) postcoxal setae, (ph) posthypostomal setae. Dorsal shield (DS) is also showed.

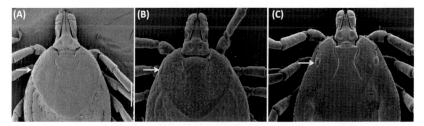

FIGURE 1.14   Eyes of hard ticks: (A) flat, (B) bulging, (C) orbited.

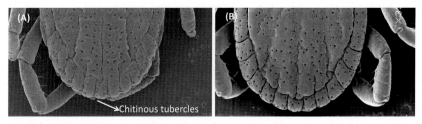

FIGURE 1.15   Chitinous tubercles on the postero-ventral side of the festoons of females of hard ticks (*Amblyomma*): (A) present and (B) absent.

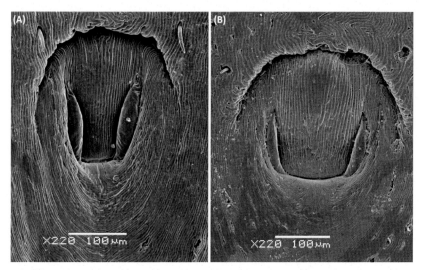

FIGURE 1.16   Genital aperture of hard ticks: (A) "V" shaped and (B) "U" shaped.

value for females of Ixodidae. There are two principal variants in female genital aperture: "U" shaped and "V" shaped (Fig. 1.16).

Goblet cells small round structures which open in the spiracular plate (Fig. 1.17).

Haller's organ a sensorial structure presents on the dorsal surface of the tarsus of leg I. The Haller's organ is composed by an anterior pit and a posterior capsule, containing sensory hairs whose anatomical characteristics may vary among genera and species (Fig. 1.18).

Hypostome a single, central, and toothed structure located in the capitulum which serves to penetrate the skin of the host. The hypostome may have different shapes but usually being blunt, pointed, spatulated, or notched (Fig. 1.19). Teeth are present on the ventral surface of hypostome arranged in columns, and this character has taxonomic value with a specific notation. For example, if there are three columns of teeth on either side of the midline of hypostome, the notation is 3/3. Rows of teeth may also be of some taxonomical value.

Idiosoma the tagmosis of ticks comprises two major divisions, an anterior part called capitulum or gnathosoma, and a posterior part, the idiosoma. In hard ticks the idiosoma is externally divided in scutum and alloscutum (Fig. 1.1).

Integumental ridges striations irregular in shape, and variable in length and width, which are characteristics of the integument of adults and nymphs of several species of the genus *Argas* (Fig. 1.5).

Lateral suture (=marginal suture) in the body of nymphs and adults of the genus *Argas*, the lateral suture forms a division between the dorsal and ventral surface of the body.

Legs the legs of ticks are formed by seven segments: coxa, trochanter, femur, genu, tibia, tarsus, and ambulacrum. The tarsus of adults of Argasidae may bear a subapical dorsal protuberance at distal of Haller's organ, as well as dorsal humps (Fig. 1.5); these

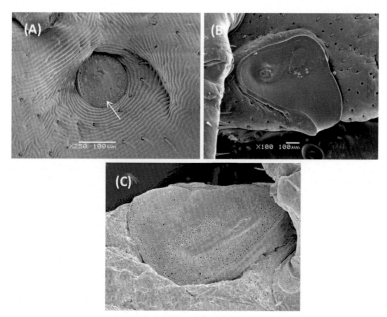

FIGURE 1.17   Spiracular plates of hard ticks: (A) rounded, (B) comma-shaped, (C) oval. The *white arrow* in (A) indicates the goblets cells.

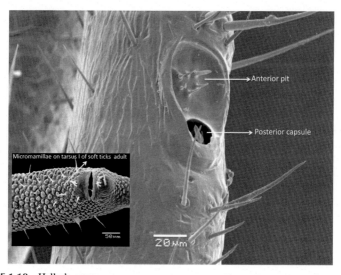

FIGURE 1.18   Haller's organ.

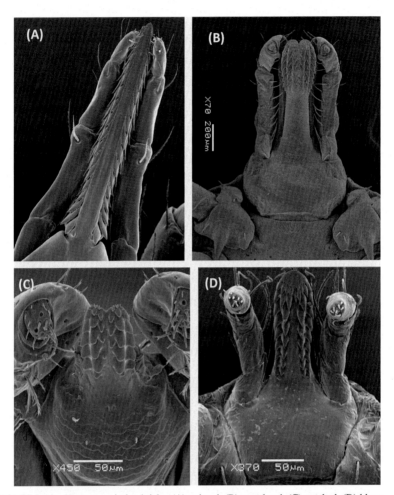

FIGURE 1.19   Hypostome in hard ticks: (A) pointed, (B) spatulated, (C) notched, (D) blunt.

humps or dorsal protuberance may be also present on other segments of the legs, and they are of diagnostic importance.

Mammillae elevations of various shapes on the integument of nymphs and adults of several species of *Ornithodoros* (Fig. 1.6) sometimes used equivocally for other genus of soft ticks.

Marginal groove depression on the lateral and posterior margins of the body of hard ticks. It can be complete, incomplete, or absent (Fig. 1.20), and of great value for the identification of males of the genus *Amblyomma*.

Palps a pair of moveable structures inserted in the capitulum and located laterally to the hypostome and chelicerae. Palps are formed by four segments called articles (Fig. 1.4). Article I is closest to the basis capituli. In Ixodidae article II and III varies in length and shape, and article IV is small and situated in an anterior depression of article III.

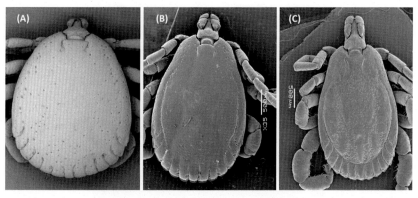

FIGURE 1.20   Marginal grooves in males of hard ticks (*Amblyomma*): (A) absent, (B) incomplete, (C) complete.

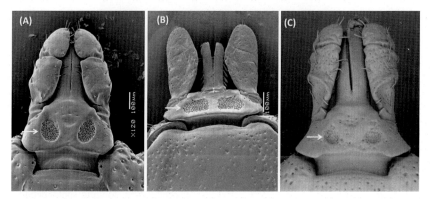

FIGURE 1.21   Porose areas of females of hard ticks: (A) oval, longitudinally elongated, (B) oval, transversally elongated, (C) rounded.

Porose area   paired depressions with a cluster of pores present on the dorsal surface of the basis capituli of females of hard ticks. They may be superficial or deep, with different sizes and shapes but the most common outline is oval or rounded (Fig. 1.21).

Pseudoauriculae   lateral projections of the dorsum of basis capituli that resembles auriculae when the basis capituli is observed ventrally. This structure is present in some nymphs of the genus *Amblyomma* (Fig. 1.7).

Pseudoscutum   division at the end of the anterior third of the scutum of males that gives it a superficial appearance of the dorsum of a female tick (Fig. 1.22). This character is present in the males of some species of the genera *Amblyomma* and *Ixodes*.

Scapula   anterior projections of the scutum in hard ticks which are next to the basis capituli (Figs. 1.2 and 1.3).

Scutum   a single dorsal shield on the anterior dorsal surface of ixodid females, nymphs, and larvae (Figs. 1.1 and 1.2), which covers almost all the dorsal surface of males (Fig. 1.3). The pattern of punctuations (density and deepness) and the relationship-wide length of the scutum have taxonomic value.

FIGURE 1.22   Pseudoscutum. Structure present in males of some species of the genera *Ixodes* and *Amblyomma*.

Scutum ornamentation the pattern of coloration of the scutum in adult ticks of several species of the genus *Amblyomma* established in the Southern Cone of America. The terms used to identify the elements that characterize the ornamentation pattern of the scutum in ornamented *Amblyomma* species are: central area, limiting spots, cervical spot, first lateral spot, second lateral spot, third lateral spot, postero-accessory spot, and postero-median spot as explained in Nava et al.[17] (Fig. 1.23).

Setae pattern of larvae of soft ticks the arrangement and number of the setae on the body surface of the larvae of soft ticks that have important taxonomic value to differentiate species. The dorsal setae are classified as anterolateral setae, central setae, and postero-lateral setae, while the ventral setae are named as postero-median setae, sternal setae, circumanal setae, and postcoxal setae following Venzal et al.[12] (Fig. 1.13). The setae on the ventral side of the basis capituli are named as posthypostomal setae (Fig. 1.13).

Spiracular plate a pair of structures situated posteriorly to coxa IV of ixodid adults and nymphs but absent in larvae, containing the spiracles or external openings of the respiratory system. The shape of spiracular plates constitutes a taxonomic character in hard ticks with three principal variants: comma-shaped, oval, and rounded (Fig. 1.17).

Trumpet-shaped sensillum structure present in tarsus I of the larva of some species of Argasidae, which extends from the capsule of Haller's organ into the lumen of the tarsus (Fig. 1.24), which has taxonomic value for soft tick larvae.

Ventral cornua ventral paired projections from the postero-lateral margins of the basis capituli, present in nymphs of some species of the genus *Amblyomma* and *Haemaphysalis* (Fig. 1.7).

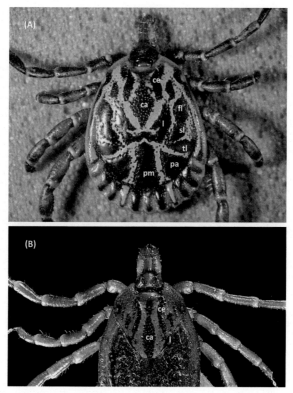

FIGURE 1.23 Scutum ornamentation of males (A) and females (B) of hard ticks (*Amblyomma*): (ca) central area, (l) limiting spots, (ce) cervical spot, (fl) first lateral spot, (sl) second lateral spot, (tl) third lateral spot, (pa) postero-accessory spot, (pm) postero-median spot.

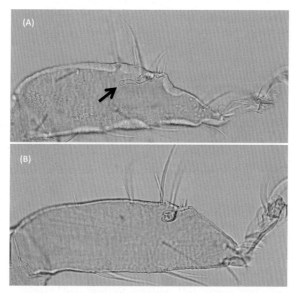

FIGURE 1.24 Trumpet-shaped sensillum is present in the tarsus I of the larvae of some species of *Argas*: (A) trumpet-shaped sensillum present, (B) trumpet-shaped sensillum absent.

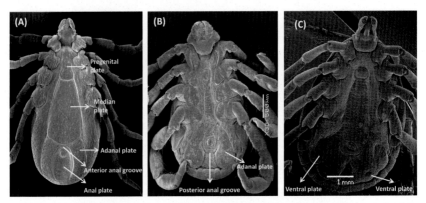

FIGURE 1.25 Ventral plates of males of the genus *Ixodes* (A) and genus *Rhipicephalus* (B). Anterior and posterior anal grove is shown. Ventral plates of the male of *Amblyomma longirostre* are also shown (C).

Ventral plates Sclerotized hard plates present on the ventral surface of males of the genus *Ixodes*. The ventral plates comprise pregenital plate, median plate, adanal plate, and anal plate (Fig. 1.25). Adanal plates are also present in males of the genus *Rhipicephalus*, and the male of *Amblyomma longirostre* has five ventral plates in the posterior field of the ventral surface (Fig. 1.25).

Ventral processes ventrally directed projections on the ventral side of basis capituli, present in some species of the genus *Amblyomma*, *Haemaphysalis*, and *Rhipicephalus* (Fig. 1.8).

Nuttalliellidae

This family is represented just by *Nuttalliella namaqua* Bedford, 1931. The larva, nymph, and male were recently described and the female was redescribed.[29] The larva has article IV of palps protruding from article III, a scutum similar to ixodid larvae, a unique anal plate with denticles, and coxae without spurs; this last characteristic is shared with the nymph and adult ticks. The capitulum is apically located and a pseudoscutum is present in the nymph and adult stages. The spiracular plates for nymph, female, and male ticks are named in Latif et al.[29] but the opening of the respiratory system is not included in a plate. Porose areas are absent in female ticks. The integument is broadly similar to argasid ticks; sexual dimorphism is evident.

BIOLOGICAL CYCLES

Ixodidae

As shown above, species of the genus *Ixodes* constitute the Prostriata group, and species of the remainder genera form the Metastriata group. Male ticks of Prostriata produce spermatids without feeding and copulation may take place off-host, while males of Metastriata produce spermatids after feeding, and usually copulate on hosts. Nevertheless, some Australasian species as *Bothriocroton hydrosauri* and *Bothriocroton concolor* produce spermatids without feeding[30] and *Amblyomma triguttatum* can even copulate off-host.[31]

Apart from the above differences, the life cycle of Ixodidae includes egg, one larva, one nymph, and adults (male and female), although a few species have only females that reproduce parthenogenetically. Larvae, nymphs, and female ticks feed for several days before molting to the next stage or lay several hundreds or thousands of eggs before dying in the case of female ticks. Males from the Metastriata group are usually intermittent feeders through their life span.

Some species (several of them of economic importance) have a **one-host tick cycle**. These ticks remain on the host for all the parasitic phase molting to nymph and adult on the same individual host and detached from it as engorged females.

Others species have a **two-host tick cycle**, characterized by larvae that feed and detached from the hosts as engorged nymphs. Nymphs molt in the environment and the resulting females attach to another host to complete the parasitic cycle.

The great majority of species of Ixodidae have a **three-host tick cycle**, where the engorged larvae detach from a host to molt in the environment, resulting nymphs seek another host to feed and detached as engorged nymphs to molt into females that complete the parasitic cycle onto another host. Schemes of these cycles are provided in Figs. 1.26 and 1.27.

Argasidae

The life cycle of Argasidae has egg, larva, usually more than one nymph, and adults (male, female) for a **multihost tick feeding cycle**, but species such as *Argas lahorensis* (formerly *Ornithodoros lahorensis*) has a two-host life cycle and *Otobius megnini* is a one-host tick, but tick feeding behavior if many species remains unknown. Immature stages feed before molting to next instar, but in species such as *Ornithodoros savignyi* and few others, the larvae molt to nymph without feeding, while in species like *Nothoaspis amazoniensis*[3] and *Ornithodoros coriaceus* first instar nymphs molt without feeding. Adult ticks feed repeatedly and copulation takes place off-host but species of *Antricola*, *Nothoaspis*, and *Otobius* have vestigial mouthparts and females laid eggs autogenically. Nymphs, adults, and larvae of some species as, *Ornithodoros erraticus*, are rapid feeders (generally takes less than an hour to complete a meal), but usually larvae of most species remain for several days on hosts. Most argasids are nidicolous and can live for years without feeding. The most common life cycle of argasids is depicted in Fig. 1.28. Most information for biology of Argasidae was obtained from Hoogstraal.[32]

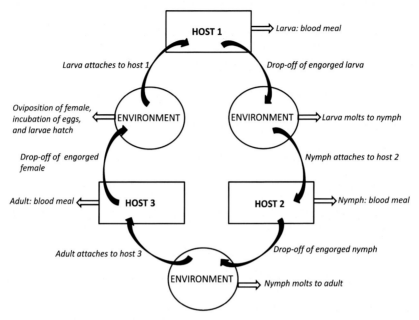

FIGURE 1.26 Three-host tick cycle in hard ticks.

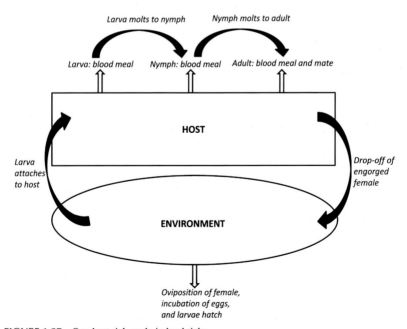

FIGURE 1.27 One-host tick cycle in hard ticks.

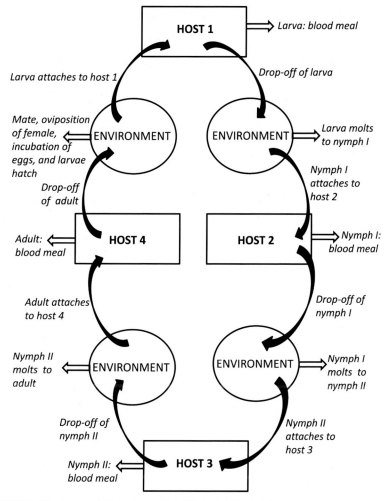

FIGURE 1.28 Representative scheme of the life cycle of soft ticks.

REFERENCES

1. Krantz GW, Walter DE. *A manual of acarology.* 3rd ed. Lubbock (TX): Texas Tech University Press; 2009.
2. Guglielmone AA, Robbins RG, Apanaskevich DA, Petney TN, Estrada-Peña A, Horak IG, et al. The Argasidae, Ixodidae and Nuttalliellidae (Acari: Ixodida) of the world: a list of valid species names. *Zootaxa* 2010;**2528**:1–28.
3. Nava S, Venzal JM, Terassini FA, Mangold AJ, Camargo LMA, Labruna MB. Description of a new argasid tick (Acari: Ixodida) from bat caves in Brazilian Amazon. *J Parasitol* 2010;**96**:1089–101.

4. Nava S, Venzal JM, Terassini FA, Mangold AJ, Camargo LMA, Casás G, et al. *Ornithodoros guaporensis* (Acari, Ixodida: Argasidae), a new tick species from the Guaporé River Basin in the Bolivian Amazon. *Zootaxa* 2013;**3666**:579−90.

5. Dantas-Torres F, Venzal JM, Bernardi LFO, Ferreira RL, Onofrio VC, Marcili A, et al. Description of a new species of bat-associated argasid tick (Acari: Argasidae) from Brazil. *J Parasitol* 2011;**98**:36−45.

6. Venzal JM, Nava S, Mangold AJ, Mastropaolo M, Casás G, Guglielmone AA. *Ornithodoros quilinensis* sp. nov. (Acari, Argasidae), a new tick species from the Chacoan region in Argentina. *Acta Parasitol* 2012;**57**:329−36.

7. Venzal JM, Nava S, González-Acuña D, Mangold AJ, Muñoz-Leal S, Lado P, et al. A new species of *Ornithodoros* (Acari: Argasidae), parasite of *Microlophus* spp. (Reptilia: Tropiduridae) from northern Chile. *Ticks Tick-borne Dis* 2013;**4**:128−32.

8. Venzal JM, González-Acuña D, Muñoz-Leal S, Mangold A, Nava S. Two new species of *Ornithodoros* (Ixodida; Argasidae) from the Southern Cone of South America. *Exp Appl Acarol* 2015;**66**:127−39.

9. Trape JF, Diatta G, Arnathau C, Bitam I, Sarih M, Belghyti D, et al. The epidemiology and geographic distribution of relapsing fever borreliosis in west and north Africa, with a review of the *Ornithodoros erraticus* complex (Acari: Ixodida). *Plos One* 2013;**8**(11) (article e78473) 19p.

10. Barros-Battesti DM, Landulfo GA, Luz HR, Marcili A, Onofrio VC, Famadas KM. *Ornithodoros faccinii* n. sp. (Acari: Ixodida: Argasidae) parasitizing the frog *Thoropa miliaris* (Amphibia: Anura: Cycloramphidae) in Brazil. *Parasit Vectors* 2015;**8** (article 268) 11p.

11. Guglielmone AA, Nava S. Names for Ixodidae (Acari: Ixodoidea): valid, synonyms, *incertae sedis, nomina dubia, nomina nuda, lapsus*, incorrect and suppressed names − with notes on confusions and misidentifications. *Zootaxa* 2014;**3767**:1−256.

12. Takada N. *A pictorial review of medical acarology in Japan*. Kyoto: Kinpodo Press; 1990.

13. Apanaskevich DA, Soarimalala V, Goodman SM. A new *Ixodes* species (Acari: Ixodidae), parasite of shrew tenrecs (Afrosoricida: Tenrecidae) in Madagascar. *J Parasitol* 2013;**99**:970−2.

14. Apanaskevich DA, Duan W, Apanaskevich MA, Filippova NA, Chen J. Redescription of *Dermacentor everestianus* Hirst (Acari: Ixodidae), a parasite of mammals in mountains of China and Nepal with synonimization of *D. abaensis* and *D. birulai* Olenev. *J Parasitol* 2014;**100**:268−74.

15. Estrada-Peña A, Nava S, Petney TN. Description of all the stages of *Ixodes inopinatus* n. sp. (Acari: Ixodidae). *Ticks Tick-borne Dis* 2014;**5**:734−43.

16. Hornok S, Kontschán J, Kováts D, Kovács R, Angyal D, Gorfol T, et al. Bat ticks revisited: *Ixodes ariadnae* sp. nov. and allopatric genotypes of *I. vespertilionis* in caves of Hungary. *Parasit Vectors* 2014;**7** (article 202) 9p.

17. Nava S, Beati L, Labruna MB, Cáceres AG, Mangold AJ, Guglielmone AA. Reassessment of the taxonomic status of *Amblyomma cajennense* (Fabricius, 1787) with the description of three new species, *Amblyomma tonelliae* n. sp., *Amblyomma interandinum* n. sp. and *Amblyomma patinoi* n. sp., and reinstatement of *Amblyomma mixtum* Koch, 1844 and *Amblyomma sculptum* Berlese, 1888 (Ixodida: Ixodidae). *Ticks and Tick-borne Diseases* 2014;**5**:252−76.

18. Nava S, Mastropaolo M, Mangold AJ, Martins TF, Venzal JM, Guglielmone AA. *Amblyomma hadanii* n. sp. (Acari: Ixodidae), a tick from northwestern Argentina previously confused with *Amblyomma coelebs* Neumann, 1899. *Syst Parasitol* 2014;**88**:261−72.

19. Apanaskevich DA, Apanaskevich MA. Description of new *Dermacentor* (Acari: Ixodidae) species from Thailand and Vietnam. *J Med Entomol* 2015;**52**:806−12.

20. Apanaskevich DA, Apanaskevich MA. Description of new *Dermacentor* (Acari: Ixodidae) species from Malaysia and Vietnam. *J Med Entomol* 2015;**52**:156−62.

21. Apanaskevich DA, Apanaskevich MA. Reinstatement of *Dermacentor bellulus* (Acari: Ixodidae) as a valid species previously confused with *D. taiwanensis* and comparisons of all parasitic stages. *J Med Entomol* 2015;**52**:573−95.

22. Apanaskevich DA, Apanaskevich MA. Description of two new species of *Dermacentor* Koch, 1844 (Acari: Ixodidae) from Oriental Asia. *Syst Parasitol* 2016;**93**:159−71.

23. Krawczak FS, Martins TF, Oliveira CS, Binder LC, Costa FB, Nunes PH, et al. *Amblyomma yucumense* n. sp. (Acari: Ixodidae), a parasite of wild mammals in Southern Brazil. *J Med Entomol* 2015;**52**:28−37.

24. Cooley RA, Kohls GM. The Argasidae of North America, Central America and Cuba. *Am Midl Nat Monogr* 1944;**1**:152.

25. Keirans JE, Clifford CM. The genus *Ixodes* in the United States: a scanning electron microscope study and key to adults. *J Med Entomol* 1978.(Suppl. 2):149.

26. Baker AS. *Mites and ticks of domestic animals, an identification guide and information source.* London: The Stationary Office; 1999.

27. Walker AR, Bouattour A, Camicas JL, Estrada-Peña A, Horak IG, Latif AA, et al. *Ticks of domestic animals in Africa: a guide to identification of species.* Houten (The Netherlands): Special Publication of the International Consortium on Ticks and Tick-borne Disease; 2003.

28. Guglielmone AA, Mangold AJ, Keirans JE. Redescription of the male and female of *Amblyomma parvum* Aragão, 1908, and description of the nymph and larva, and description of all stages of *Amblyomma pseudoparvum* sp. n. (Acari: Ixodida: Ixodidae). *Acarologia* 1990;**31**:143−59.

29. Latif AA, Putterill JF, de Klerk DG, Pienaar R, Mans BJ. *Nuttalliella namaqua* (Ixodoidea: Nuttalliellidae): first description of the male, immature stages and re-description of the female. *Plos One* 2012;**7**(7) (article e41651) 9p.

30. Oliver JH, Stone BF. Spermatid production in unfed, Metastriata ticks. *J Parasitol* 1983;**69**:420−1.

31. Guglielmone AA, Moorhouse DE. Copulation and successful insemination by unfed *Amblyomma triguttatum triguttatum* Koch. *J Parasitol* 1983;**69**:786−7.

32. Hoogstraal H. Argasid and Nuttalliellid ticks as parasites and vectors. *Adv Parasitol* 1985;**24**:135−238.

Chapter 2

Genera and Species of Ixodidae

Genus *Amblyomma*

There are 137 species of *Amblyomma* worldwide which represents 19% of Ixodidae and all of them are characterized for a three-host parasitic cycle. The great majority of *Amblyomma* are established in the lands of Gondwanian origin and the Neotropical Region contains the highest number of species, and ticks from this genus are also the most numerous in the Southern Cone of America with a total of 25 taxa. Host usage of *Amblyomma* is peculiar because they are more prone to feed on reptiles than other ixodids, and the only two species of hard ticks commonly found on amphibians are *A. dissimile* and *A. rotundatum*, and both species are established in the region. Nevertheless, most species found in Argentina, Chile, Paraguay, and Uruguay feed on Mammalia. Although the importance of Aves as hosts for the immature stages of a few species has been historically recognized, recent studies demonstrated that several more species than previously recognized depend on birds for the maintenance of larvae and nymphs. This genus has been the most extensively studied in recent years throughout the region and we expect that this is reflected in the species account below.

SPECIES OF *AMBLYOMMA* NO LONGER CONSIDERED TO BE ESTABLISHED IN THE SOUTHERN CONE OF AMERICA

A. scutatum was described by Neumann[1] from adult ticks collected from an undetermined lizard and *Iguana tuberculata*, a synonym of *I. iguana*, in Guatemala, "Cerf de Virginia" (Zoo of Hamburg) and *Bothrops lanceolatus* (Railliet collection). Neumann did not describe the nymph of *A. scutatum* but stated that this stage was found on *Noctilio albiventris* from Paraguay and on *Cathartes uruba*, a synonym of *C. aura*, *Didelphis pusilla*, a synonym of *Thylamys pusillus*, and *Dasyprocta croconota*, a synonym of *D. azarae*, from Brazil. Robinson[2] doubted that these South American nymphs belong to *A. scutatum*; nevertheless, Guglielmone et al.[3] and Nava et al.[4] accepted that the range of this species includes Brazil and Paraguay, while Guglielmone et al.[5] treat the records of nymphs in Neumann[1] as provisionally valid. Dantas Torres et al.[6] have reservations about the presence of *A. scutatum* in Brazil and this species is not listed for this country because the Brazilian records have not been confirmed. The only record of this species in the Southern Cone of America is a nymph allegedly collected from *N. albiventris* at an unknown Paraguayan locality. We consider this record doubtful and the lack of subsequent evidence of *A. scutatum* in Paraguay or others countries of the Southern Cone of America casts doubt about its presence here; therefore, we exclude *A. scutatum* from the list of species established in this region.

   A. varium was described by Koch[7] in 1844 from male specimens collected in Brazil. Later, Neumann[1] described the male of *A. varium* var. *albida* of Chilean origin but host unknown, while Argentinean records of this species were provided by Neumann[8] in 1901, who described the female of *A. varium* from an engorged female collected from an undetermined host in Corrientes. Lahille[9] recorded it from dogs and deer also from Corrientes, and Boero[10] from *Bradypus tridactylus*, probably meaning *B. variegatus*, in northwestern Chaco and northwestern Salta (Argentina). The main hosts for adults of *A. varium* are Pilosa: Bradypodidae and Megalonychidae as stated in Onofrio et al.,[11] Guglielmone et al.,[5] and several other workers. The Chilean records of *A. varium* was treated as doubtful for many workers because main hosts of *A. varium* were never established in this country, and González-Acuña and Guglielmone[12] considered Chile out of the range of this species because of mislabeling of tick specimen. The presence of this tick in Argentina is difficult to sustain in the absence of specimens for confirmation. Records from Corrientes in Neumann[8] and Lahille[9] are doubtful because main hosts of *A. varium* were never established there, a condition similar to the alleged records in Boero[10] because the localities are out of any historical range of the usual hosts of *A. varium*. Bona fide presence of *B. variegatus* in Argentina, the only main host of *A. varium* documented for the country, is supported by a record from the Province of Jujuy at the beginning of the 20th century, but it is highly probable that *B. variegatus* is currently extinct in Argentina.[13] We are skeptic about the presence of *A. varium* in Argentina. Consequently this species is not included as established in the Southern Cone of America.

REFERENCES

1. Neumann LG. Révision de la famille des ixodidés (3e mémoire). *Mém Soc Zool Fr* 1899;**12**:107–294.
2. Robinson LE. *Ticks. A monograph of the Ixodoidea. Part IV. The genus Amblyomma.* London: Cambridge University Press; 1926.

3. Guglielmone AA, Estrada-Peña A, Keirans JE, Robbins RG. Ticks (Acari: Ixodida) of the neotropical zoogeographic region. Special publication of the international consortium on ticks and tick-borne diseases-2. Houten (The Netherlands): Atalanta; 2003.

4. Nava S, Lareschi M, Rebollo C, Benítez Usher C, Beati L, Robbins RG, et al. The ticks (Acari: Ixodida: Argasidae, Ixodidae) of Paraguay. *Ann Trop Med Parasitol* 2007;**101**:255−70.

5. Guglielmone AA, Robbins RG, Apanaskevich DA, Petney TN, Estrada-Peña A, Horak IG. *The hard ticks of the world (Acari: Ixodida: Ixodidae)*. London: Springer; 2014.

6. Dantas-Torres F, Onofrio VC, Barros-Battesti DM. The ticks (Acari: Ixodida: Argasidae, Ixodidae) of Brazil. *Syst Appl Acarol* 2009;**14**:30−46.

7. Koch CL. Systematische Übersicht über die Ordnung der Zecken. *Arch Naturgesch* 1844;**10**:217−39.

8. Neumann LG. Révision de la famille des ixodidés (4ᵃ mémoire). *Mém Soc Zool Fr* 1901;**14**:249−372.

9. Lahille F. *Enumeración sistemática de los pediculidos, malófagos, pulícidos, linguatulidos y ácaros (1ᵃ parte) encontrados en la República Argentina con una nota sobre una especie de piojo*. Buenos Aires: Ministerio de Agricultura; 1920.

10. Boero JJ. *Las garrapatas de la República Argentina (Acarina: Ixodoidea)*. Buenos Aires: Universidad de Buenos Aires; 1957.

11. Onofrio VC, Barros-Battesti DM, Marques S, Faccini JLH, Labruna MB, Beati L, et al. Redescription of *Amblyomma varium* Koch, 1844 (Acari: Ixodidae) based on light and scanning electron microscopy. *Syst Parasitol* 2008;**69**:137−44.

12. González-Acuña D, Guglielmone AA. Ticks (Acari: Ixodoidea: Argasidae, Ixodidae) of Chile. *Exp Appl Acarol* 2005;**35**:147−63.

13. Barquez RM, Díaz MM, Ojeda RA. *Mamíferos de Argentina. Sistemática y distribución*. Tucumán: Sociedad Argentina para el Estudio de los Mamíferos; 2006.

Amblyomma argentinae
Neumann, 1905

Neumann, L.G. (1905) Notes sur les ixodidés. III. *Archives de Parasitologie*, 9, 225−241.

DISTRIBUTION

Argentina. The main host of *A. argentinae* is found in western Paraguay which is the natural extension of the Argentinean tick infested area and Nava et al.[1] state that *A. argentinae* is probably established there. Records of *A. argentinae* in Brazil, Chile, Cuba, Peru, Surinam, Uruguay, and Venezuela[2−4] are considered misidentifications or due to ticks introduced with tortoises imported from Argentina.

    Biogeographic distribution in the Southern Cone of America: Chaco (Neotropical Region) and Monte (South American Transition Zone) (see Fig. 1 in the Introduction).

TABLE 2.1 Hosts of Adults (A), Nymphs (N), and Larvae (L) of *A. argentinae*

AMPHIBIA		Viperidae	
ANURA		*Bothrops neuwiedi*	A
Bufonidae		*Crotalus terrificus*	AN
Rhinella sp.	A	**TESTUDINES**	
REPTILIA		**Chelidae**	
SQUAMATA		*Phrynops hilarii*	A
Boidae		**Testudinidae**	
Boa constrictor	AN	<u>*Chelonoidis chilensis*</u>	ANL
Epicrates cenchria	A		
Eunectes notaeus	A		

Main host underlined.

HOSTS

The principal host for immature and adult stages of *A. argentinae* is the tortoise *Chelonoidis chilensis*, but records on another reptile, the *Boa constrictor*, are not unusual. There are also records on other species of the classes Amphibia and Reptilia as well as an odd finding on the bird *Turdus falklandii* in Chile[5]; this last host is not included in the host list detailed in Table 2.1.

ECOLOGY

There is little information about the ecology of *A. argentinae*. The geographic distribution and seasonality pattern of this tick appear to be related to the activity of its principal host, the tortoise *C. chilensis*.[2] Analyses of data from Argentinean tick collections have shown that adults of *A. argentinae* are present in low numbers during winter months, when their principal hosts are inactive.[2] However, longitudinal studies on seasonality under field conditions are lacking.

SANITARY IMPORTANCE

The capacity of *A. argentinae* to transmit pathogens has not been investigated to date, and no records of human parasitism have been found. Heavy tick infestations of main hosts are quite common but the detrimental effect of this host—parasite relationship has not been studied.

DIAGNOSIS

Male (Fig. 2.1): large body size, total length 4.7 mm (4.5−5.0), breadth
3.9 mm (3.7−4.2). Body outline oval, scapulae rounded, cervical grooves
deep, short, comma-shaped; marginal groove absent. Eyes flat. Scutum
ornate, with small brownish spots; postero-accessory and postero-median
spots narrow, lateral spots indistinct, cervical spots present; punctuations
numerous, areolate, absent in the central and antero-median fields. Carena
absent. Basis capituli subrectangular dorsally, cornua short. Hypostome
spatulate, dental formula 3/3. Genital aperture located at level of coxa II,
U-shaped. Legs: coxae I−IV each with two distinct, short, blunt spurs; the
external spur on coxa IV larger than the internal spur; trochanters without
spurs. Spiracular plates comma-shaped.

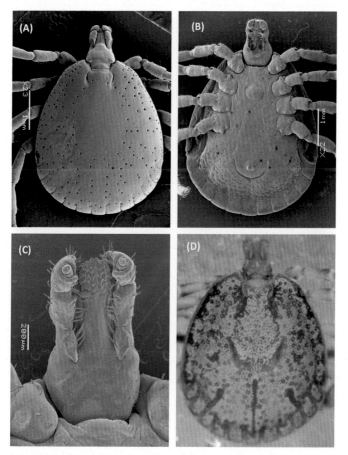

FIGURE 2.1   *A. argentinae* male: (A) dorsal view, (B) ventral view, (C) capitulum ventral
view, (D) scutum ornamentation.

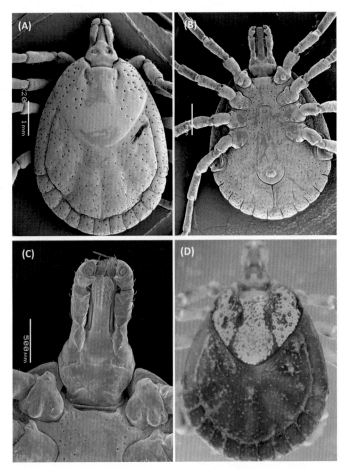

FIGURE 2.2 *A. argentinae* female: (A) dorsal view, (B) ventral view, (C) capitulum ventral view, (D) scutum ornamentation.

Female (Fig. 2.2): large body size, total length 4.7 mm (4.7−4.8), breadth 4.0 mm (4.0−4.0) in unfed specimens. Body outline oval. Scutum length 2.5 mm (2.5−2.6), breadth 2.9 mm (2.9−2.9); scapulae rounded, cervical grooves short and comma-shaped. Eyes flat. Chitinous tubercles at the postero-body margin absent. Scutum ornate, extensively pale yellowish, cervical spots elongated touching posteriorly the limiting spots; punctuations large, areolate, concentrated in the antero-lateral fields. Notum glabrous. Basis capituli dorsally subrectangular, cornua short, porose areas rounded. Hypostome spatulate, dental formula 3/3. Genital aperture located at level of coxa II, U-shaped. Legs: coxae I−IV each with two distinct, short, blunt spurs; the external spur on coxa IV larger than the internal spur; trochanters without spurs. Spiracular plates comma-shaped.

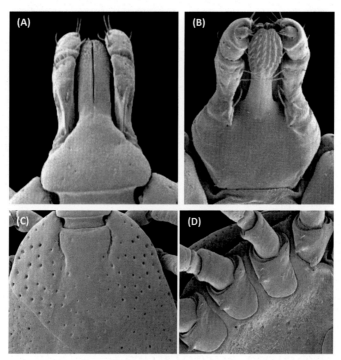

FIGURE 2.3   *A. argentinae* nymph: (A) capitulum dorsal view, (B) capitulum ventral view, (C) scutum, (D) coxae. *Figures (A) to (D) are reproductions of figure 3:* A. argentinae *nymph. Capitulum, dorsum; Capitulum, venter; Scutum; Coxae I—IV. From Martins TF, Labruna MB, Mangold AJ, Cafrune MM, Guglielmone AA, Nava S. Taxonomic key to nymphs of the genus* Amblyomma *(Acari: Ixodidae) in Argentina, with description and redescription of the nymphal stage of four* Amblyomma *species. Ticks Tick-borne Dis 2014;5:753—70, with permission of Elsevier.*

Nymph (Fig. 2.3): large body size, total length 1.7 mm (1.6—1.8), breadth 1.3 mm (1.3—1.5). Body outline oval. Chitinous tubercles at the posterior body margin absent. Scutum with punctuations evenly distributed, larger and deeper laterally. Eyes flat, located on lateral scutal angles at the level of scutal mid-length. Cervical grooves reaching mid-level of the scutum, deeper in the scutal anterior third. Basis capituli dorsally subtriangular, without cornua. Hypostome spatulate, dental formula 3/3 for most of the length, 2/2 at the base. Legs: coxa I with two pointed spurs, the external longer than the internal, coxae II—IV with a small triangular spur each.

Taxonomic notes: The name *A. testudinis* (Conil, 1877) was largely used to refer to *A. argentinae*, but the correct name for this tick species is the latter as discussed in Guglielmone et al.[2,6] Morphologically *A. argentinae* is closely related to *A. dissimile* Koch, 1844, and *A. rotundatum* Koch, 1844, but due to the lack of DNA sequences of *A. argentinae* there are no phylogenetic analyses to know whether this relationship has an evolutionary correlate or not.

DNA sequences with relevance for tick identification and phylogeny: DNA sequences of *A. argentinae* are not available.

REFERENCES

1. Nava S, Lareschi M, Rebollo C, Benítez Usher C, Beati L, Robbins RG, et al. The ticks (Acari: Ixodida: Argasidae, Ixodidae) of Paraguay. *Ann Trop Med Parasitol* 2007; **101**:255−70.
2. Guglielmone AA, Luciani CA, Mangold AJ. Aspects of the ecology of *Amblyomma argentinae* [= *Amblyomma testudinis* (Conil, 1877)] (Acari: Ixodidae). *Syst Appl Acarol Spec Publ* 2001;**8**:1−12.
3. González-Acuña D, Beldoménico PM, Venzal JM, Fabry M, Keirans JE, Guglielmone AA. Reptile trade and the risk of exotic tick introductions into southern South American countries. *Exp Appl Acarol* 2005;**35**:335−9.
4. Dantas-Torres F, Onofrio VC, Barros-Battesti DM. The ticks (Acari: Ixodida: Argasidae, Ixodidae) of Brazil. *Syst Appl Acarol* 2009;**14**:30−46.
5. González-Acuña D, Guglielmone AA. Ticks (Acari: Ixodoidea: Argasidae, Ixodidae) of Chile. *Exp Appl Acarol* 2005;**35**:147−63.
6. Guglielmone AA, Robbins RG, Apanaskevich DA, Petney TN, Estrada-Peña A, Horak IG. Comments on controversial tick (Acari: Ixodida) species names and species described or resurrected from 2003 to 2008. *Exp Appl Acarol* 2009;**48**:311−27.

Amblyomma aureolatum (Pallas, 1772)

Acarus aureolatus. Pallas, P.S. (1772) *Spicilegia Zoologica*, volume 40, fascicule 9, 87 pp.

DISTRIBUTION

Argentina, Brazil, French Guiana, Paraguay, Surinam, and Uruguay.[1,2] Records of *A. aureolatum* in other American countries (i.e., Bolivia, Venezuela) are treated as doubtful by Guglielmone et al.[3]

Biogeographic distribution in the Southern Cone of America: Chaco, Pampa, and Parana Forest (Neotropical Region) (see Fig. 1 in the introduction).

HOSTS

The principal hosts for immature and adult stages of *A. aureolatum* are passerine birds and carnivorous, respectively,[1,4−7] but the range of hosts is wide, especially for adult ticks, and includes several orders of mammals as detailed in Table 2.2.

TABLE 2.2 Hosts of Adults (A), Nymphs (N), and Larvae (L) of
A. aureolatum

MAMMALIA		Echimyidae	
ARTIODACTYLA		*Eurozygomatomys spinosus*	ANL
Bovidae		**Erethizontidae**	
Cattle	A	*Sphiggurus insidiosus*	A
Cervidae		**Sciuridae**	
Mazama gouazoubira	A	*Sciurus aestuans*	A
CARNIVORA		**AVES**	
Canidae		**PASSERIFORMES**	
Cerdocyon thous	AN	**Conopophagidae**	
Chrysocyon brachyurus	A	*Conopophaga lineata*	NL
Domestic dog	A	**Emberizidae**	
Lycalopex gymnocercus	A	*Haplospiza unicolor*	N
Felidae		*Poospiza lateralis*	N
Domestic cat	A	*Saltator similis*	NL
Herpailurus yagouaroundi	A	*Zonotrichia capensis*	N
Leopardus geoffroyi	A	**Furnariidae**	
L. wiedii	A	*Automolus leucophtalmus*	N
Puma concolor	AN	*Cranioleuca obsoleta*	NL
Mustelidae		*C. pallida*	L
Galictis cuja	A	*Furnarius rufus*	NL
G. vittata	A	*Synallaxis cinerascens*	NL
Procyonidae		*S. ruficapilla*	N
Nasua nasua	A	**Parulidae**	
Procyon cancrivorus	A	*Basileuterus leucoblepharus*	NL
DIDELPHIMORPHIA		**Thamnophilidae**	
Didelphidae		*Dysithamnus mentalis*	N
Didelphis albiventris	A	*Myrmeciza squamosa*	N
Lutreolina crassicaudata	A	*Pyriglena leucoptera*	NL

(Continued)

TABLE 2.2 (Continued)

PILOSA		Thamnophilus caerulescens	NL
Bradypodidae		T. ruficapillus	L
Bradypus sp.	AL	**Thraupidae**	
Myrmecophagidae		Tachyphonus coronatus	L
Tamandua tetradactyla	A	Trichothraupis melanops	NL
PRIMATES		**Troglodytidae**	
Atelidae		Troglodytes aedon	NL
Alouatta guariba	A	**Turdidae**	
Hominidae		Turdus albicollis	NL
Humans	A	T. amaurochalinus	N
RODENTIA		T. leucomelas	N
Caviidae		T. rufiventris	NL
Hydrochoerus hydrochaeris	A	**Tyrannidae**	
Ctenomyidae		Platyrinchus mystaceus	L
Ctenomys sp.	AL		

ECOLOGY

Apart from the range and hosts of A. aureolatum,[1] ecological information about this species is scanty for the region. Studies in southeastern Brazil show that infestations of adult ticks on dogs were recorded throughout the year without a marked seasonality trend,[8] while larvae and nymphs prevail from April to September and from October to March, respectively.[4] Guglielmone and Nava[9] speculated that the restricted range of A. aureolatum in comparison with A. ovale is probably due to humidity requirements, while Barbieri et al.[10] revealed that at 23°S (São Paulo, Brazil) A. aureolatum prevails at higher altitude but this difference disappears at 26−27°S (Santa Catarina State, Brazil) which may be due to lower temperature requirements for the cycle of A. aureolatum than for the cycle of A. ovale. Probably the effect of abiotic factors may explain range differences of these species but additional studies are needed for conclusive evidence.

SANITARY IMPORTANCE

Adults of A. aureolatum were recorded biting humans. This fact has relevance because this tick species is vector of the human pathogens Rickettsia rickettsii and Rickettsia sp. strain Atlantic rainforest.[11−14] A. aureolatum is also a common parasite of dogs in rural and periurban areas, with capacity to transmit the dog pathogen *Rangelia vitalii*.[15]

DIAGNOSIS

Male (Fig. 2.4): medium body size, total length 3.2 mm (2.7−3.7), breadth 2.2 mm (1.7−2.7). Body outline oval; scapulae small and rounded; cervical grooves sigmoid, deeper anteriorly; marginal groove absent. Eyes flat. Scutum ornate, cervical spot long, central area small, and slightly conspicuous, postero-accessory and postero-median spots narrow, lateral spots conspicuous and small; punctuations numerous, uniformly distributed. Carena present. Basis capituli dorsally subtriangular, cornua short. Hypostome spatulate, dental

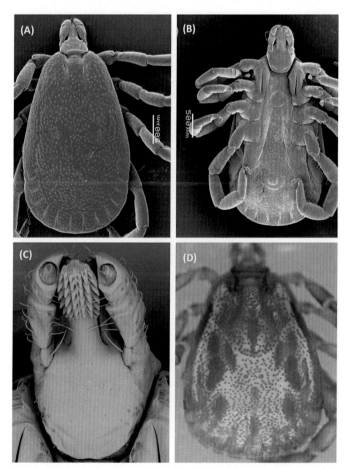

FIGURE 2.4   *A. aureolatum* male: (A) dorsal view, (B) ventral view, (C) capitulum ventral view, (D) scutum ornamentation.

formula 3/3. Genital aperture located at level of coxa II, U-shaped. Legs: coxa I with two distinct, long, triangular, sharp spurs; external spur slightly longer than internal spur; coxae II–IV with a triangular spur; trochanters without spurs. Spiracular plates comma-shaped.

Female (Fig. 2.5): medium body size, total length 3.8 mm (3.2–4.2), breadth 2.8 mm (2.3–3.3). Body outline oval. Scutum length 2.2 mm (2.1–2.3), breadth 2.1 mm (2.1–2.2); scapulae rounded; cervical grooves sigmoid, deeper anteriorly. Eyes flat. Chitinous tubercles at posterior body margin absent. Scutum ornate, extensively pale yellowish, cervical spots narrow and elongated; punctuations numerous, uniformly distributed. Notum glabrous. Basis capituli dorsally subtriangular, cornua short, porose areas oval. Hypostome spatulate, dental formula 3/3. Genital aperture located at level of

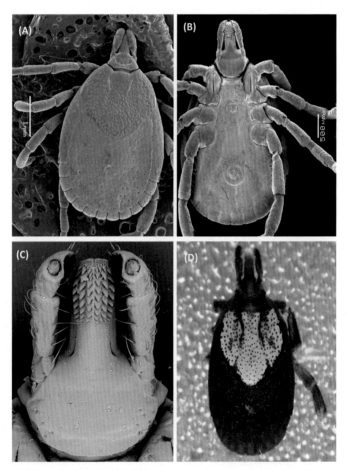

FIGURE 2.5 *A. aureolatum* female: (A) dorsal view, (B) ventral view, (C) capitulum ventral view, (D) scutum ornamentation.

coxa II, U-shaped. Legs: coxa I with two distinct, long, triangular, sharp spurs; external spur slightly longer than internal spur; coxae II–IV with a triangular spur; trochanters without spurs. Spiracular plates comma-shaped.

Nymph (Fig. 2.6): small body size, total length 1.3 mm (1.1–1.4), breadth 0.9 mm (0.8–1.0). Body outline oval. Chitinous tubercles at posterior body margin absent. Scutum with few punctuations, larger and deeper laterally. Eyes flat, located on lateral scutal angles at the level of scutal mid-length. Cervical grooves reaching the scutal mid-length, deeper at the anterior half. Basis capituli dorsally subtriangular, without cornua; pseudoauriculae present. Hypostome spatulate, dental formula 2/2. Legs: coxa I with two triangular blunt spurs, the external longer than the internal; coxae II–IV with a small triangular spur.

Taxonomic notes: the presence of carena is described for the males of *A. aureolatum*, but in some specimens this morphological feature is almost imperceptible. *A. aureolatum* and *A. ovale* Koch, 1844, are morphologically

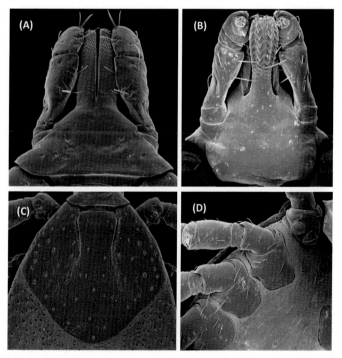

FIGURE 2.6   *A. aureolatum* nymph: (A) capitulum dorsal view, (B) capitulum ventral view, (C) scutum, (D) coxae. *Figures (A) to (D) are reproductions of figure 16: A.* aureolatum *nymph. Capitulum, dorsum; Capitulum, venter; Scutum; Coxae I–IV. From Martins TF, Labruna MB, Mangold AJ, Cafrune MM, Guglielmone AA, Nava S. Taxonomic key to nymphs of the genus* Amblyomma *(Acari: Ixodidae) in Argentina, with description and redescription of the nymphal stage of four* Amblyomma *species. Ticks Tick-borne Dis 2014;5:753–70, with permission of Elsevier.*

and phylogenetically closely related. This phylogenetic relationship was inferred based on the concatenated sequences of the small and the large ribosomal subunits 12S rDNA and 16S rDNA (Fig. 2.7). The morphological distinctiveness between *A. aureolatum* and *A. ovale* was clarified by Aragão and Fonseca.[16] The morphological diagnoses made before the work of Aragão and Fonseca[16] are not useful to differentiate these two species. *A. aureolatum* is named as *A. striatum* Koch, 1844, in the dichotomous keys of Jones et al.[17] and Guglielmone and Viñabal,[18] but *A. striatum* is a synonym of *A. aureolatum*.[19]

DNA sequences with relevance for tick identification and phylogeny: the following DNA sequences of *A. aureolatum* are available in the Gen Bank: mitochondrial genes 12S rRNA (accession number: AY277249, AY277249) and 16S rRNA (accession numbers: AF541254, KF179343, JN573301, JN800433), sequences of the nuclear genome (the fragment 5.8S ribosomal RNA gene-internal transcribed spacer 2-28S ribosomal RNA gene) (accession number: AF469611) and sequences of microsatellites (accession numbers: KF602062−KF602069).

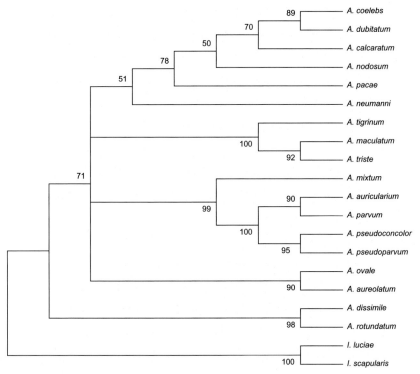

FIGURE 2.7 Maximum-likelihood analysis of the concatenated sequences of the small and the large ribosomal subunits 12S rDNA and 16S rDNA. The phylogenetic condensed tree was generated with the GTR model and a discrete Gamma-distribution.

REFERENCES

1. Guglielmone AA, Estrada-Peña A, Mangold AJ, Barros-Battesti DM, Labruna MB, Martins JR, et al. *Amblyomma aureolatum* (Pallas, 1772) and *Amblyomma ovale* Koch, 1844 (Acari: Ixodidae): DNA sequences, hosts and distribution. *Vet Parasitol* 2003;**113**: 273−88.

2. Nava S, Lareschi M, Rebollo C, Benítez Usher C, Beati L, Robbins RG, et al. The ticks (Acari: Ixodida: Argasidae, Ixodidae) of Paraguay. *Ann Trop Med Parasitol* 2007;**101**:255−70.

3. Guglielmone AA, Estrada-Peña A, Keirans JE, Robbins RG. *Ticks (Acari: Ixodida) of the neotropical zoogeographic region*. Special publication of the international consortium on ticks and tick-borne diseases-2. Houten (The Netherlands): Atalanta; 2003.

4. Arzua M, Navarro da Silva M, Famadas KM, Beati L, Barros Battesti DM. *Amblyomma aureolatum* and *Ixodes auritulus* (Acari: Ixodidae) on birds in southern Brazil, with notes on their ecology. *Exp Appl Acarol* 2003;**31**:283−6.

5. Arzua M, Onofrio VC, Barros-Battesti DM. Catalogue of the tick collection (Acari, Ixodida) of the Museu de História Natural Capao da Imbuia, Curitiba, Paraná, Brazil. *Rev Bras Zool* 2005;**22**:623−32.

6. Venzal JM, Félix ML, Olmos A, Mangold AJ, Guglielmone AA. A collection of ticks (Ixodidae) from wild birds in Uruguay. *Exp Appl Acarol* 2005;**36**:325−31.

7. Ogrzewalska M, Saraiva DG, Moraes-Filho J, Martins TF, Costa FB, Pinter A, et al. Epidemiology of Brazilian spotted fever in the Atlantic Forest, state of São Paulo, Brazil. *Parasitology* 2012;**139**:1283−300.

8. Pinter A, Dias RA, Gennari SM, Labruna MB. Study of the seasonal dynamics, life cycle, and host specificity of *Amblyomma aureolatum* (Acari: Ixodidae). *J Med Entomol* 2004;**41**:321−32.

9. Guglielmone AA, Nava S. Distribución geográfica, hospedadores y variabilidad genética de *Amblyomma ovale* y *Amblyomma aureolatum* (Acari: Ixodidae), dos vectores potenciales de rickettsias en la Argentina. In: Farjat JB, Enría D, Martino P, Rosenvitz M, Seijo A, editors. *Temas de Zoonosis VI*. Buenos Aires: Asociación Argentina de Zoonosis; 2014. pp. 183−91.

10. Barbieri JM, Rocha CMBM, Bruhn FRP, Cardoso DL, Pinter A, Labruna MB. Altitudinal assessment of *Amblyomma aureolatum* and *Amblyomma ovale* (Acari: Ixodidae), vectors of spotted fever group rickettsiosis in the State of São Paulo, Brazil. *J Med Entomol* 2015;**52**:1170−4.

11. Pinter A, Labruna MB. Isolation of *Rickettsia rickettsii* and *Rickettsia belli* in cell culture from tick *Amblyomma aureolatum* in Brazil. *Ann NY Acad Sci* 2006;**1078**:523−9.

12. Labruna MB, Ogrzewalska M, Martins TF, Pinter A, Horta MC. Comparative susceptibility of larval stages of *Amblyomma aureolatum*, *Amblyomma cajennense*, and *Rhipicephalus sanguineus* to infection by *Rickettsia rickettsii*. *J Med Entomol* 2008;**45**:1156−9.

13. Labruna MB, Ogrzewalska M, Soares JF, Martins TF, Soares HS, et al. Experimental infection of *Amblyomma aureolatum* ticks with *Rickettsia rickettsii*. *Emerg Infect Dis* 2011;**17**:829−34.

14. Barbieri ARM, Moraes-Filho JM, Nieri-Bastos FA, Souza JC, Szabó MPJ, Labruna MB. Epidemiology of *Rickettsia* sp. strain Atlantic rainforest in a spotted fever-endemic area of southern Brazil. *Ticks Tick-borne Dis* 2014;**5**:848−53.

15. Soares JF, Dall'Agnol B, Costa FB, Krawczak FS, Comerlato AT, Rossato BCD, et al. Natural infection of the wild canid, *Cerdocyon thous*, with the piroplasmid *Rangelia vitalii* in Brazil. *Vet Parasitol* 2014;**202**:156−63.

16. Aragão HB, Fonseca F. Nota de ixodologia. 9. O complexo *ovale* do género *Amblyomma*. *Mem Inst Oswaldo Cruz* 1961;**59**:131–48.
17. Jones EK, Clifford CM, Keirans JE, Kohls GM. The ticks of Venezuela (Acarina: Ixodoidea) with a key to the species of *Amblyomma* in the Western Hemisphere. *Brigham Young Univ Sci Bull Biol Ser* 1972;**17**(4):40 p.
18. Guglielmone AA, Viñabal AE. Claves morfológicas dicotómicas e información ecológica para la identificación de garrapatas del género *Amblyomma* Koch, 1844 de la Argentina. *Rev Invest Agropec* 1994;**25**(1):39–67.
19. Guglielmone AA, Nava S. Names for Ixodidae (Acari: Ixodoidea): valid, synonyms, *incertae sedis*, *nomina dubia*, *nomina nuda*, *lapsus*, incorrect and suppressed names-with notes on confusions and misidentifications. *Zootaxa* 2014;**3767**:1–256.

Amblyomma auricularium (Conil, 1878)

Ixodes auricularius. Conil, P.A. (1878) Description d'une nouvelle espèce d'ixode, *Ixodes auricularius*. *Acta de la Academia Nacional de Ciencias Exactas (Argentina)*, 3, 99–110.

DISTRIBUTION

Argentina, Belize, Bolivia, Brazil, Colombia, Costa Rica, El Salvador, Guatemala, Guyana, French Guiana, Honduras, Mexico (Neotropical), Nicaragua, Panama, Paraguay, Trinidad and Tobago, Surinam, Uruguay, and Venezuela, up to southern Nearctic in Mexico and United States.[1] Specimens of *A. auricularium* from Argentina (Salta and Chubut Provinces) deposited in the collections of INTA Rafaela (Rafaela, Argentina) and Museo de La Plata (La Plata, Argentina) were redetermined as *A. pseudoconcolor* by Guglielmone and Nava.[2] These authors stated that all records of *A. auricularium* from Argentina should be reevaluated, and this may include records from other countries too.

Biogeographic distribution in the Southern Cone of America: Chaco, Pampa, Yungas (Neotropical Region), Monte, Puna (South American Transition Zone), and Patagonia Central (Andean region) (see Fig. 1 in the introduction).

HOSTS

The principal hosts for immature and adult stages of *A. auricularium* are armadillos (Cingulata), although there also are records on other species of the classes Mammalia, Aves, and Reptilia, and Aves may play an important role as hosts for larvae and nymphs of this species[1-5] (Table 2.3).

TABLE 2.3 Hosts of Adults (A), Nymphs (N), and Larvae (L) of
A. auricularium

MAMMALIA		Hydrochoerus hydrochaeris	A
CARNIVORA		Pediolagus salinicola	AN
Canidae		Chinchillidae	
Cerdocyon thous	A	Lagostomus maximus	AN
Chrysocyon brachyurus	A	Echymyidae	
Domestic dog	AN	Thrichomys apereoides	N
Urocyon cinereoargenteus	N	Erethizontidae	
Mephitidae		Sphiggurus mexicanus	N
Conepatus semistriatus	ANL	Cricetidae	
Mustelidae		Sigmodon hispidus	N
Galictis vittata	A	AVES	
Procyonidae		PASSERIFORMES	
Nasua nasua	A	Cardinalidae	
Procyon lotor	A	Cyanacompsa brissoni	N
CINGULATA		Saltator similis	N
Dasypodidae		Emberizidae	
Cabassous centralis	A	Zonotrichia capensis	N
Chaetophractus vellerosus	A	Furnariidae	
C. villosus	AN	Furnarius leucopus	L
Dasypus hybridus	A	Megaxenops parnaguae	N
D. keppleri	A	Pipridae	
D. novemcinctus	ANL	Neopelma pallescens	N
D. sabanicola	A	Thamnophilidae	
Euphractes sexcinctus	A	Myrmochilus strigilatus	N
Tolypeutes matacus	ANL	Sakesphorus cristatus	N
T. tricinctus	A	Thamnophilus capistratus	N
Zaedyus pichyi	AN	Thraupidae	
DIDELPHIMORPHIA		Coryphospingus pileatus	L
Didelphidae		Troglodytidae	
Didelphis marsupialis	A	Troglodytes musculus	L
Monodelphis domestica	N	Turdidae	

(Continued)

TABLE 2.3 (Continued)

Philander opossum	AN	Turdus amaurochalinus	NL
PERISSODACTYLA		T. rufiventris	N
Equidae		**Tyrannidae**	
Horse	A	Cnemotriccus fuscatus	L
PILOSA		Stigmatura napensis	N
Myrmecophagidae		**REPTILIA**	
Tamandua mexicana	A	**SQUAMATA**	
T. tetradactyla	AN	**Iguanidae**	
RODENTIA		Iguana sp.	A
Caviidae			
Galea spixii	N		

ECOLOGY

A. auricularium is established in areas with divergent ecological characteristics. The distribution of this nidicolous tick species appears to be determined by that of its principal hosts, armadillos of the family Dasypodidae, and in fact it covers most of its range from Patagonia to southern Nearctic.

SANITARY IMPORTANCE

At present *A. auricularium* lacks sanitary relevance, but perhaps, this is due to lack of research of its effect on their hosts. "*Candidatus Rickettsia amblyommii*" was detected in *A. auricularium*,[6,7] but this rickettsia is currently considered of unknown pathogenicity, and there are no records of *A. auricularium* biting humans.[8,9]

DIAGNOSIS

Male (Fig. 2.8): small body size, total length 2.8 mm (2.5−3.0), breadth 1.9 mm (1.6−2.2). Body outline oval rounded, scapulae rounded, cervical grooves short, comma-shaped; marginal groove complete. Eyes flat. Scutum inornate; punctuations small and shallow, more concentrated in the anterior field. Carena absent. Basis capituli subrectangular dorsally, cornua short, article I of palps ventrally with a large blunt spur directed posteriorly. Hypostome spatulate, dental formula 3/3. Genital aperture located at level of coxa II, U-shaped. Legs: coxa I with two distinct, short, blunt spurs,

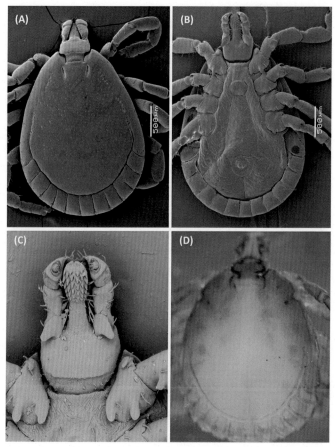

FIGURE 2.8 *A. auricularium* male: (A) dorsal view, (B) ventral view, (C) capitulum ventral view, (D) scutum ornamentation.

subequal in size; coxae II−IV with a triangular spur each; trochanters with spurs. Spiracular plates comma-shaped.

Female (Fig. 2.9): medium body size, total length 3.5 mm (3.4−3.6), breadth 2.7 mm (2.6−2.8). Body outline oval rounded. Scutum length 1.7 mm (1.5−1.8), breadth 1.7 mm (1.6−1.8); scapulae rounded, cervical grooves short, and comma-shaped. Eyes flat. Chitinous tubercles at the posterior body margin absent. Scutum inornate, punctuations small and shallow. Notum glabrous. Basis capituli subrectangular dorsally, cornua short, porose areas oval, article I of palps ventrally with a large blunt spur directed posteriorly. Hypostome spatulate, dental formula 3/3. Genital aperture located at level of coxa II, U-shaped. Legs: coxa I with two distinct, short, blunt spurs, subequal in size; coxae II−IV with a triangular spur each; trochanters with spurs. Spiracular plates comma-shaped.

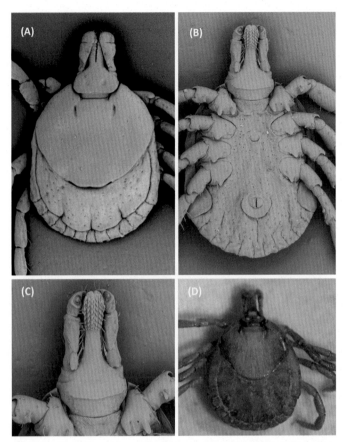

FIGURE 2.9 *A. auricularium* female: (A) dorsal view, (B) ventral view, (C) capitulum ventral view, (D) scutum ornamentation.

Nymph (Fig. 2.10): small body size, total length 1.3 mm (1.1−1.4), breadth 1.0 mm (0.8−1.1). Body outline oval. Chitinous tubercles at the posterior body margin absent. Scutum with few medium punctuations evenly distributed, deeper laterally. Eyes flat, located on lateral scutal angles at the level of scutal mid-length. Cervical grooves deep, converging anteriorly then diverging as shallow depressions at the scutal median. Basis capituli subrectangular dorsally, without cornua. Hypostome spatulate, dental formula 2/2. Legs: coxa I with two spurs, the external slightly longer than the internal; coxae II−IV with a small triangular spur each.

Taxonomic notes: *A. auricularium*, *A. parvum* Aragão, 1908, *A. pseudoconcolor* Aragão, 1908, and *A. pseudoparvum* Guglielmone, Mangold and Keirans, 1990, form a phylogenetically closely related group.[10] This phylogenetic relationship is supported by nuclear (18S rRNA) and mitochondrial genetic data (12S rRNA and 16S rRNA) (Figs. 2.7 and 2.11). Also, these four species are

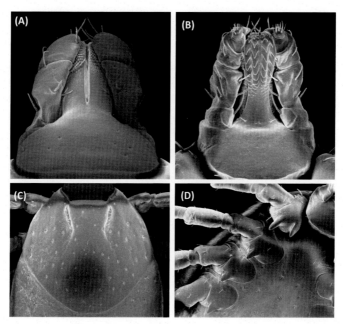

FIGURE 2.10 *A. auricularium* nymph: (A) capitulum dorsal view, (B) capitulum ventral view, (C) scutum, (D) coxae. *Figures (A) to (D) are reproductions of figure 20:* A. auricularium *nymph. Capitulum, dorsum; Capitulum, venter; Scutum; Coxae I–IV. From Martins TF, Labruna MB, Mangold AJ, Cafrune MM, Guglielmone AA, Nava S. Taxonomic key to nymphs of the genus* Amblyomma *(Acari: Ixodidae) in Argentina, with description and redescription of the nymphal stage of four* Amblyomma *species.* Ticks Tick-borne Dis *2014;5:753–70, with permission of Elsevier.*

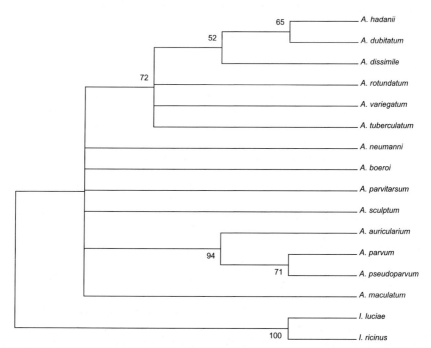

FIGURE 2.11 Maximum-likelihood condensed tree constructed from 18S rDNA sequences. The phylogenetic tree was generated with Kimura 2-parameter (K2) model by using Gamma-distribution with Invariant sites (G + I).

characterized by the presence of a unique character among the Neotropical species of *Amblyomma*, the presence of spurs on trochanters in both male and female ticks. Records of *A. auricularium* from Guyana and Surinam are under the name *A. concolor* Neumann, 1899, in Rawlins et al.,[11] but this name is a synonym of *A. auricularium* as stated in Keirans and Hillyard.[12]

DNA sequences with relevance for tick identification and phylogeny: DNA sequences of *A. auricularium* are available in the Gen Bank: mitochondrial genes 12S rRNA (accession numbers AY342292, JF523334) and 16S rRNA (accession numbers KC202817, FJ627951), cytochrome oxidase I (accession numbers KF200126, KF200137), and sequences of the nuclear marker 18S rRNA (accession number FJ464426).

REFERENCES

1. Guglielmone AA, Estrada-Peña A, Luciani CA, Mangold AJ, Keirans JE. Hosts and distribution of *Amblyomma auricularium* (Conil 1878) and *Amblyomma pseudoconcolor* Aragão, 1908 (Acari: Ixodidae). *Exp Appl Acarol* 2003;**29**:131−9.
2. Guglielmone AA, Nava S. Las garrapatas argentinas del género *Amblyomma* (Acari: Ixodidae): distribución y hospedadores. *Rev Invest Agropec* 2006;**35**(3):135−55.
3. Fairchild GB, Kohls GM, Tipton VJ. The ticks of Panama (Acarina: Ixodoidea). In: Wenzel WR, Tipton VJ, editors. *Ectoparasites of Panama*. Chicago (IL): Field Museum of Natural History; 1966. pp. 167−219.
4. Jones EK, Clifford CM, Keirans JE, Kohls GM. The ticks of Venezuela (Acarina: Ixodoidea) with a key to the species of *Amblyomma* in the Western Hemisphere. *Brigham Young Univ Sci Bul Biol Ser* 1972;**17**(4):40 p.
5. Lugarini C, Martins TF, Ogrzewalska M, Vasconcelos NCT, Ellis VA, Oliveira JB, et al. Rickettsial agents in avian ixodid ticks in northeast Brazil. *Ticks Tick-borne Dis* 2015;**6**:364−75.
6. Saraiva DG, Nieri-Bastos F, Horta MC, Soraes HS, Nicola PA, Pereira LCM, et al. *Rickettsia amblyommii* infecting *Amblyomma auricularium* ticks in Pernambuco, northeastern Brazil: isolation, transovarial transmission, and transstadial perpetuation. *Vector-Borne Zoon Dis* 2013;**13**:615−18.
7. Soares HS, Barbieri ARM, Martins TF, Minervino AHH, de Lima JTR, Marcili A, et al. Ticks and rickettsial infection in the wildlife of two regions of the Brazilian Amazon. *Exp Appl Acarol* 2014;**65**:125−40.
8. Guglielmone AA, Beati L, Barros-Battesti DM, Labruna MB, Nava S, Venzal JM, et al. Ticks (Ixodidae) on humans in South America. *Exp Appl Acarol* 2006;**40**:83−100.
9. Guglielmone AA, Robbins RG, Apanaskevich DA, Petney TN, Estrada-Peña A, Horak IG. *The hard ticks of the world (Acari: Ixodida: Ixodidae)*. London: Springer; 2014.
10. Nava S, Szabó MPJ, Mangold AJ, Guglielmone AA. Distribution, hosts, 16S rDNA sequences and phylogenetic position of the Neotropical tick *Amblyomma parvum* (Acari: Ixodidae). *Ann Trop Med Parasitol* 2008;**102**:409−25.
11. Rawlins SC, Mahadeo S, Martínez R. A list of ticks affecting man and animals in the Caribbean. *CARAPHIN News* 1993;**6**:8−9.
12. Keirans JE, Hillyard PD. A catalogue of the type specimens of Ixodida (Acari: Argasidae, Ixodidae) deposited in The Natural History Museum, London. *Occas Pap Syst Entomol* 2001;**13**:74.

Amblyomma boeroi
Nava, Mangold, Mastropaolo, Venzal, Oscherov, and Guglielmone, 2009

Nava, S., Mangold, A.J., Mastropaolo, M., Venzal, J.M., Oscherov, E.B. & Guglielmone, A.A. (2009) *Amblyomma boeroi* n. sp. (Acari: Ixodidae), a parasite of the Chacoan peccary *Catagonus wagneri* (Rusconi) (Artiodactyla: Tayassuidae) in Argentina. *Systematic Parasitology*, 73, 161−174.

DISTRIBUTION

Argentina, Paraguay.[1,2]

    Biogeographic distribution in the Southern Cone of America: Chaco (Neotropical Region) (see Fig. 1 in the introduction).

HOSTS

To date, the only host for immature and adult stages of *A. boeroi* is *Catagonus wagneri* (Artiodactyla: Tayassuidae).

ECOLOGY

Besides the few data on host association and geographical distribution, there is no information on the ecology of *A. boeroi*.

SANITARY IMPORTANCE

The capacity of *A. boeroi* to transmit pathogens has not been investigated to date, and no records on humans and domestic animals have been found.

DIAGNOSIS

Male (Fig. 2.12): large body size, total length 4.9 mm (4.7−5.1), breadth 3.3 mm (3.1−3.5). Body outline pyriform, scapulae pointed, cervical grooves deep and almost parallel, diverging as shallow depressions posteriorly. Eyes orbited. Scutum ornate, light gray to very pale ivory colored, with single bilateral white stripe converging on the middle level of scutum *then diverging posteriorly*. Carena absent. Basis capituli rectangular dorsally, cornua long. Hypostome spatulate, dental formula 2/2. Genital aperture located at level of coxa II, U-shaped. Legs: coxa I with two unequal spurs, the external spur longer than the internal spur and reaching anterior margin

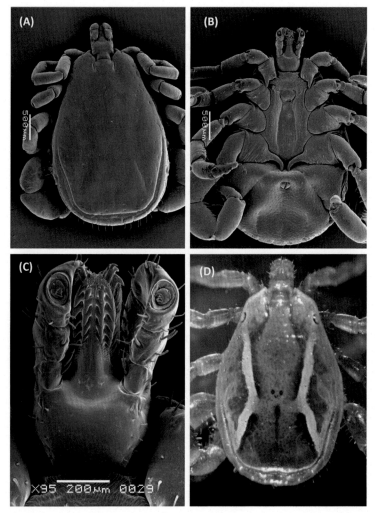

FIGURE 2.12 *A. boeroi* male: (A) dorsal view, (B) ventral view, (C) capitulum ventral view, (D) scutum ornamentation.

of coxa II; coxae II−III each with small triangular spur; coxa IV considerably larger than coxae I−III, with a long, sickle-shaped, medially directed spur arising from its internal margin; trochanters without spurs. Spiracular plates comma-shaped.

Female (Fig. 2.13): large body size, total length 4.9 mm (4.7−5.1), breadth 3.3 mm (3.2−3.5). Body outline pyriform. Scutum length 1.6 mm (1.5−1.7), breadth 1.6 mm (1.5−1.8); scapulae pointed, cervical grooves deep, converging anteriorly then diverging as shallow depressions. Eyes

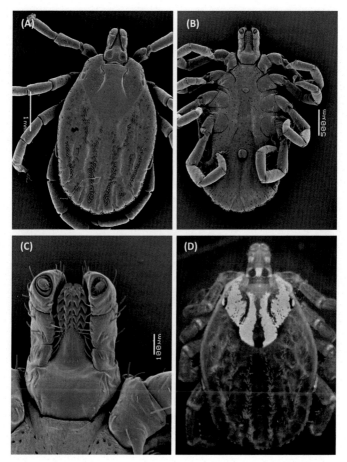

FIGURE 2.13 *A. boeroi* female: (A) dorsal view, (B) ventral view, (C) capitulum ventral view, (D) scutum ornamentation.

orbited. Chitinous tubercles at the posterior body margin absent. Scutum ornate, central area reaching the posterior scutal margin and cervical spots externally concave. Notum with setae short, coarse, ivory in color, slightly curved. Basis capituli rectangular dorsally, cornua short, porose areas oval. Hypostome spatulate, dental formula 2/2. Genital aperture located at level of coxa II, U-shaped. Legs: coxa I with two spurs, the external longer than the internal; coxae II−IV with a small triangular spur each; trochanters without spurs. Spiracular plates comma-shaped.

Nymph (Fig. 2.14): large body size, total length 1.8 mm (1.6−2.0), breadth 1.2 mm (1.1−1.3). Body outline oval. Chitinous tubercles at the posterior body margin absent. Scutum with few medium punctuations, deeper

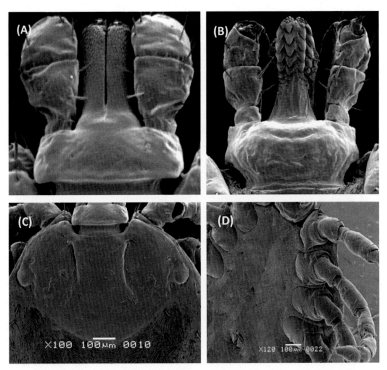

FIGURE 2.14 *A. boeroi* nymph: (A) capitulum dorsal view, (B) capitulum ventral view, (C) scutum, (D) coxae. *Figures (A) to (D) are reproductions of figure 13:* A. boeroi *nymph. Capitulum, dorsum; Capitulum, venter; Scutum; Coxae I–IV. From Martins TF, Labruna MB, Mangold AJ, Cafrune MM, Guglielmone AA, Nava S. Taxonomic key to nymphs of the genus* Amblyomma *(Acari: Ixodidae) in Argentina, with description and redescription of the nymphal stage of four* Amblyomma *species.* Ticks Tick-borne Dis *2014;5:753–70, with permission of Elsevier.*

laterally. Eyes orbited. Cervical grooves deep, parallel at mid-length and then slightly divergent. Basis capituli rectangular dorsally, with small cornua. Hypostome spatulate, dental formula 2/2. Legs: coxa I with a triangular, large, external spur and a minute, rounded, internal spur; coxae II–IV with a small triangular spur each.

Taxonomic notes: the phylogenetic analyses constructed with 18S rDNA sequences (Fig. 2.11) and 16S rDNA sequences (Fig. 2.15) indicate that *A. boeroi* is not phylogenetically related to any of the Neotropical species of *Amblyomma*.

DNA sequences with relevance for tick identification and phylogeny: DNA sequences of *A. boeroi* genes are available in the Gen Bank as follows: 16S rDNA (accession numbers FJ464416–FJ464419, JN828797) and 18S rDNA (accession number FJ464420).

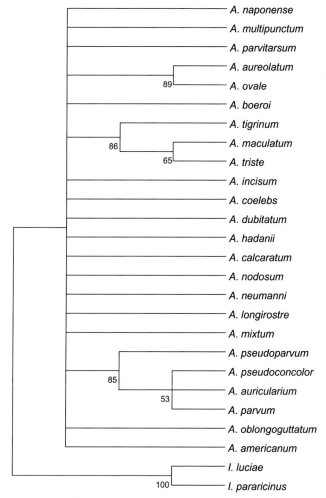

FIGURE 2.15   Maximum-likelihood condensed tree constructed from 16S rDNA sequences. The phylogenetic tree was generated with GTR model and a discrete Gamma-distribution.

REFERENCES

1. Mastropaolo M, Nava S, Schmidt E, Guglielmone AA, Mangold AJ. First report of *Amblyomma boeroi* Nava, Mangold, Mastropaolo, Venzal, Oscherov and Guglielmone (Acari: Ixodidae) from the Chacoan peccary, *Catagonus wagneri* (Artiodactyla: Tayassuidae), in Paraguay. *Syst Appl Acarol* 2012;**17**:7–9.
2. Nava S, Mangold AJ, Mastropaolo M, Venzal JM, Oscherov EB, Guglielmone AA. *Amblyomma boeroi* n. sp. (Acari: Ixodidae), a parasite of the Chacoan peccary *Catagonus wagneri* (Rusconi) (Artiodactyla: Tayassuidae) in Argentina. *Syst Parasitol* 2009; **73**:161–74.

Amblyomma brasiliense
Aragão, 1908

Aragão, H.B. (1908) Algunas novas especies de carrapatos brazileiros. *Brazil Medico*, 22, 111–115.

DISTRIBUTION

Argentina, Brazil, and Paraguay.[1–4] Records of *A. brasiliense* from northwestern Argentina in the Yungas Biogeographic Province[5–9] were not confirmed after many years of extensive studies in this province. Welschen et al.[10] treat those records as uncertain and call the attention of weak morphological support for *A. brasiliense* in Dios and Knoppoff.[5,6] We decided to exclude Yungas in the distribution of this species. A larval record of *A. brasiliense* from Bolivia in 1972 by Squire[11] is also excluded because this tick stage was described recently.[12]

Biogeographic distribution in the Southern Cone of America: Chaco and Parana Forest (Neotropical Region) (see Fig. 1 in the introduction).

HOSTS

Both immature and adult stages of *A. brasiliense* appear to be ticks with low host specificity, although some authors assert that species of Tayassuidae are the main hosts for adult ticks.[13] *A. brasiliense* is principally associated with mammals[13,14] with just one record on birds as detailed in Table 2.4.

ECOLOGY

A. brasiliense has a 1-year life cycle under field conditions,[13] but not definitive seasonal peaks were determined for nymphs and adult ticks; larvae were found in autumn.[13] This tick species is principally associated to tropical areas with hot and humid conditions.

SANITARY IMPORTANCE

All parasitic stages of *A. brasiliense* are aggressive to humans.[4,13,15,16] Trials under laboratory conditions have suggested that *A. brasiliense* is competent to transmit the highly pathogenic *Rickettsia rickettsii*,[17] but Sabatini et al.[18] and Szabó et al.[19] did not find natural infections with *Rickettsia* in 50 ticks analyzed.

TABLE 2.4 Hosts of Adults (A), Nymphs (N), and Larvae (L) of
A. brasiliense

MAMMALIA		PILOSA	
ARTIODACTYLA		Myrmecophagidae	
Cervidae		*Myrmecophaga tridactyla*	AN
Ozotoceros bezoarticus	AN	*Tamandua tetradactyla*	N
Tayassuidae		PERISSODACTYLA	
Pecari tajacu	A	Tapiridae	
Tayassu pecari	AN	*Tapirus terrestris*	AN
CARNIVORA		PRIMATES	
Canidae		Hominidae	
Cerdocyon thous	N	Humans	ANL
Chrisocyon brachyurus	N	RODENTIA	
Felidae		Caviidae	
Panthera onca	N	*Hydrochoerus hydrochaeris*	AN
Puma concolor	N	Cuniculidae	
Procyonidae		*Cuniculus paca*	N
N. nasua	N	Dasyproctidae	
DIDELPHIMORPHIA		*Dasyprocta azarae*	N
Didelphidae		AVES	
D. aurita	N	GALLIFORMES	
		Cracidae	
		Penelope obscura	N

DIAGNOSIS

Male (Fig. 2.16): medium body size, total length 3.3 mm (3.1−3.5), breadth
2.5 mm (2.3−2.8). Body outline oval, scapulae rounded, cervical grooves
short, comma-shaped; marginal groove incomplete. Eyes flat. Scutum ornate,
cervical spots elongated posteriorly, small limiting spots, lateral spots
fused but distinct, postero-accessory spot small, postero-median spot long;
punctuations numerous and shallow, larger in the antero-lateral field. Carena
present. Basis capituli subrectangular dorsally, cornua long. Hypostome
spatulate, dental formula 3/3. Genital aperture located at level of coxa II,
U-shaped. Legs: coxa I with two distinct, long, triangular, sharp spurs,

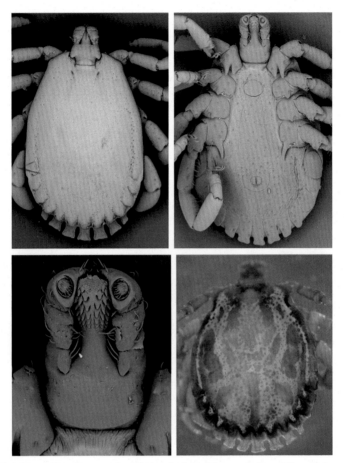

FIGURE 2.16   *A. brasiliense* male: (A) dorsal view, (B) ventral view, (C) capitulum ventral view, (D) scutum ornamentation.

the external spur longer than the internal; coxae II—III each with two triangular, short spurs; coxa IV with two spurs, the internal spur long and triangular and the external spur very short; trochanters without spurs. Spiracular plates comma-shaped.

Female (Fig. 2.17): medium body size, total length 3.7 mm (3.6—3.8), breadth 2.7 mm (2.5—3.0). Body outline oval. Scutum length 1.7 mm (1.6—1.8), breadth 2.0 mm (1.9—2.0); scapulae rounded, cervical grooves short. Eyes flat. Chitinous tubercles at the posterior body margin present. Scutum ornate, cervical spots short, externally concave; punctuations numerous and shallow, larger in the antero-lateral field. Notum glabrous. Basis capituli subrectangular dorsally, cornua short, porose areas rounded. Hypostome spatulate, dental formula 4/4. Genital aperture located at level of coxa II, U-shaped.

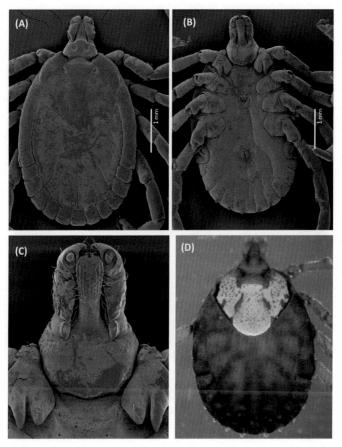

FIGURE 2.17   *A. brasiliense* female: (A) dorsal view, (B) ventral view, (C) capitulum ventral view, (D) scutum ornamentation.

Legs: coxa I with two distinct, triangular, sharp spurs, the external spur longer than the internal; coxae II−IV each with two triangular, short spurs, internal spurs very small; trochanters without spurs. Spiracular plates comma-shaped.

Nymph (Fig. 2.18): large body size, total length 1.7 mm (1.6−1.8), breadth 1.3 mm (1.2−1.4). Body outline oval. Chitinous tubercles at the posterior body margin present. Few medium and shallow punctuations evenly distributed, deeper laterally. Eyes slightly bulging, located on lateral scutal angles at the level of scutal mid-length. Cervical grooves broad and long, covering the first two-thirds of the scutum, deeper at the anterior third. Basis capituli rectangular dorsally, with cornua. Hypostome spatulate, dental formula 2/2. Legs: coxa I with two spurs, the internal broader than the external, this external spurs slender about twice as long as the internal spur; coxae II−IV with a pointed triangular spur each.

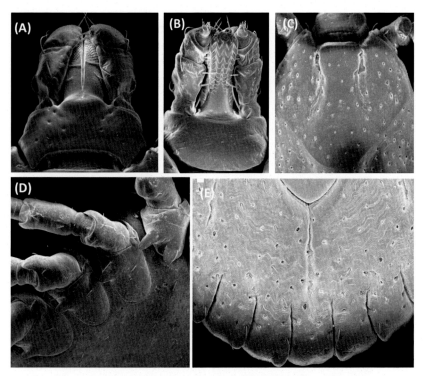

FIGURE 2.18   *A. brasiliense* nymph: (A) capitulum dorsal view, (B) capitulum ventral view, (C) scutum, (D) coxae, (E) chitinous tubercles at the posterior body margin. *Figures (A) to (E) are reproductions of figure 9:* A. brasiliense *nymph. Capitulum, dorsum; Capitulum, venter; Scutum; Coxae I–IV; Ventral festoons. From Martins TF, Labruna MB, Mangold AJ, Cafrune MM, Guglielmone AA, Nava S. Taxonomic key to nymphs of the genus* Amblyomma *(Acari: Ixodidae) in Argentina, with description and redescription of the nymphal stage of four* Amblyomma *species.* Ticks Tick-borne Dis *2014;5:753–70, with permission of Elsevier.*

DNA sequences with relevance for tick identification and phylogeny: the following sequences of the mitochondrial 16S rDNA gene of *A. brasiliense* are available in the Gen Bank: accession numbers KM159940-41; FJ424329.

REFERENCES

1. Aragão HB. Ixodidas brasileiros e de algunos paízes limitrophes. *Mem Inst Oswaldo Cruz* 1936;**31**:759–843.
2. Guglielmone AA, Nava S. Las garrapatas argentinas del género *Amblyomma* (Acari: Ixodidae): distribución y hospedadores. *Rev Invest Agropec* 2006;**35**(3):135–55.
3. Nava S, Lareschi M, Rebollo C, Benítez Usher C, Beati L, Robbins RG, et al. The ticks (Acari: Ixodida: Argasidae, Ixodidae) of Paraguay. *Ann Trop Med Parasitol* 2007;**101**:255–70.
4. Lamattina D, Tarragona EL, Costa SA, Guglielmone AA, Nava S. Ticks (Acari: Ixodidae) of northern Misiones Province, Argentina. *Syst Appl Acarol* 2014;**19**:393–8.
5. Dios RL, Knopoff R. Sobre Ixodoidea de la República Argentina. *Rev Soc Argent Biol* 1930;**6**:593–627.

6. Dios RL, Knopoff R. Sobre Ixodoidea de la República Argentina. *Rev Inst Bacteriol Dep Nac Hyg* 1934;**6**:359−412.

7. Boero JJ. Los Ixodideos de la República Argentina. *Rev Med Vet (Buenos Aires)* 1945;**26**:1−10.

8. Boero JJ. Los Ixodoideos de la República Argentina y sus huéspedes. *Rev Fac Agr Vet (Buenos Aires)* 1955;**18**:505−14.

9. Boero JJ. Acarina-Ixodoidea. *Primeras Jornadas Entomoepidemiológicas Argentinas.* Buenos Aires: Dirección General de Sanidad de la Secretraría de Guerra; 1959, pp. 595−6.

10. Welschen NM, Tarragona EL, Nava S, Guglielmone AA. Confirmación de la presencia de *Amblyomma brasiliense* Aragão, 1908 (Acari: Ixodidae) en la Argentina. *Rev FAVE Cienc Vet* 2012;**11**:65−9.

11. Squire FA. Entomological problems in Bolivia. *Pest Artic News Summ* 1972;**18**:249−68.

12. Sanches GS, Bechara GH, Camargo-Mathias MI. Morphological description of *Amblyomma brasiliense* (Acari: Ixodidae) larva and nymphs. *Rev Bras Parasitol Vet* 2012;**18**:15−21.

13. Szabó MPJ, Labruna MB, García MV, Pinter A, Castagnolli KC, Pacheco RP, et al. Ecological aspects of the free-living ticks (Acari: Ixodidae) on animal trails within Atlantic rainforest in South-eastern Brazil. *Ann Trop Med Parasitol* 2009;**103**:57−72.

14. Martins TF, Teixeira RHF, Labruna MB. Ocorrência de carrapatos em animais silvestres recebidos e atendidos pelo Parque Zoológico Municipal Quinzinho de Barros, Sorocaba, São Paulo, Brasil. *Braz J Vet Res Anim Sci* 2015;**52**:319−24.

15. Guglielmone AA, Beati L, Barros-Battesti DM, Labruna MB, Nava S, Venzal JM, et al. Ticks (Ixodidae) on humans in South America. *Exp Appl Acarol* 2006;**40**:83−100.

16. Lamattina D, Nava S. Ticks (Acari: Ixodidae) infesting humans in northern Misiones Province, Argentina. *Medicina (Buenos Aires)* 2016;**76**:89−92.

17. Dias E, Martins AV. Spotted fever in Brazil. *Am J Trop Med* 1939;**19**:103−8.

18. Sabatini GS, Pinter A, Nieri-Bastos FA, Marcilli A, Labruna MB. Survey of ticks (Acari: Ixodidae) and their *Rickettsia* in an Atlantic rain forest reserve in the State of São Paulo, Brazil. *J Med Entomol* 2010;**47**:913−16.

19. Szabó MPJ, Nieri-Bastos FA, Spolidorio MG, Martins TF, Barbieri AM, Labruna MB. *In vitro* isolation from *Amblyomma ovale* (Acari: Ixodidae) and ecological aspects of the Atlantic rainforest *Rickettsia*, the causative agent of a novel spotted fever rickettsiosis in Brazil. *Parasitology* 2013;**140**:719−28.

Amblyomma calcaratum Neumann, 1899

Neumann, L.G. (1899) Révision de la famille des ixodidés (3e mémoire). *Mémoires de la Société Zoologique de France*, 12, 107−294.

DISTRIBUTION

Argentina, Belize, Bolivia, Brazil, Colombia, Costa Rica, Ecuador, French Guiana, Mexico, Panama, Paraguay, Peru, Surinam, Trinidad and Tobago, and Venezuela.[1−3]

Biogeographic distribution in the Southern Cone of America: Chaco and Parana Forest (Neotropical Region) (see Fig. 1 in the introduction).

HOSTS

Anteaters (Pilosa: Myrmecophagidae) and passerine birds are the principal hosts for adults and immature stages of *A. calcaratum*, respectively,[4–8] but the range of hosts for adult ticks also includes several orders of mammals (Table 2.5), while immature stages were found on two orders of mammals and two orders of birds.

TABLE 2.5 Hosts of Adults (A), Nymphs (N), and Larvae (L) of
A. calcaratum

MAMMALIA			Furnariidae	
ARTIODACTYLA			*Anabazenops fuscus*	NL
Cervidae			*Automolus leucophtalmus*	N
Mazama americana	A		*Syndactyla rufosuperciliata*	N
CARNIVORA			**Parulidae**	
Canidae			*Geothlypis aequinoctialis*	N
Domestic dog	A		**Pipridae**	
Procyonidae			*Chiroxiphia caudata*	N
Procyon cancrivorus	A		*Dixiphia pipra*	N
CHIROPTERA			*Manacus manacus*	N
Phyllostomidae			*Pipra fasciicauda*	N
Macrophyllus macrophyllus	A		**Thamnophilidae**	
Platyrrhinchus umbratus	A		*Cymbilaimus lineatus*	N
PERISSODACTYLA			*Dysithamnus mentalis*	N
Tapiridae			*Gymnopithys leucaspis*	N
Tapirus terrestris	ANL		*Hylophylax naevius*	N
PILOSA			*Pyriglena leucoptera*	N
Megalonychidae			*Thamnophilus aethiops*	N
Choloepus hoffmanni	A		*T. caerulescens*	NL
Myrmecophagidae			*T. pelzeni*	N
Myrmecophaga tridactyla	AN		*Willisornis poecilinotus*	N

(*Continued*)

TABLE 2.5 (Continued)

T. mexicana	A	**Thraupidae**	
T. tetradactyla	AN	*Pyrrhocoma ruficeps*	N
PRIMATES		*Ramphocelus carbo*	N
Hominidae		*Tachyphonus coronatus*	N
Human	A	*Tangara cayana*	N
AVES		*Trhicothraupis melanops*	N
CORACIIFORMES		**Troglodytidae**	
Momotidae		*Cantorchilus modestus*	L
Baryphthengus ruficapillus	N	**Turdidae**	
PASSERIFORMES		*Turdus albicollis*	N
Cardinalidae		*T. amaurochalinus*	N
Cyanacompsa cyanoides	N	*T. melanops*	L
Saltator similis	N	*T. leucomelas*	N
Conopophagidae		*T. subalaris*	N
Conopophaga lineata	NL	**Tyrannidae**	
C. melanops	N	*Cnemotriccus fuscatus*	N
Dendrocolaptidae		*Hemitriccus margaritaceiventer*	N
Dendrocolaptes platyrostris	N	*Lathrotriccus euleri*	N
Glyphorynchus spirurus	N	*Mionectes oleagineus*	N
Hylexetastes brigidai	NL	*Myiarchus tyranullus*	N
Xiphocolaptes albicollis	N	*Platyrinchus mystaceus*	L
Xiphorhynchus ocellatus	N	*Poecilotriccus latirostris*	N
Emberizidae		**Vireonidae**	
Arremon flavirostris	N	*Cyclarhis gujanensis*	N
A. taciturnus	N		
Coryphospingus cucullatus	N		
Oryzoborus funereus	N		

Main hosts for adult ticks underlined.

ECOLOGY

Ecological information about *A. calcaratum* refers just to distribution and host association.

SANITARY IMPORTANCE

There is just one record of *A. calcaratum* feeding on humans,[9] but this species is not involved as a potential vector of human and animal pathogens to date.

DIAGNOSIS

Male (Fig. 2.19): medium body size, total length 3.9 mm (3.8−4.0), breadth 2.8 mm (2.7−3.0). Body outline oval, scapulae rounded; cervical grooves

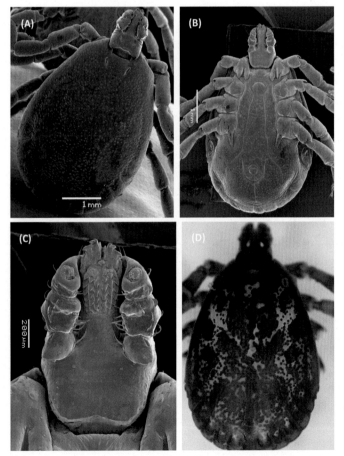

FIGURE 2.19 *A. calcaratum* male: (A) dorsal view, (B) ventral view, (C) capitulum ventral view, (D) scutum ornamentation.

deep and short, comma-shaped; marginal groove absent. Eyes flat. Scutum ornate, with irregular pale spots Y-shaped in the antero-lateral fields, postero-accessory and postero-median spots narrow, lateral spots conspicuous and small; punctuations numerous, uniformly distributed. Carena absent. Basis capituli dorsally rectangular, cornua long, article II of palps with a postero-dorsal projection. Hypostome spatulate, dental formula 3/3. Genital aperture located at level of coxa II, U-shaped. Legs: coxa I with two distinct, long, triangular blunt spurs, subequal in size; coxae II—III with one triangular short spur each; coxa IV with one long, triangular sharp spur; trochanters without spurs. Spiracular plates comma-shaped.

Female (Fig. 2.20): large body size, total length 4.8 mm (4.6—5.0), breadth 4.0 mm (4.0—4.0). Body outline oval. Scutum length 3.1 mm (2.8—3.4), breadth 2.9 mm (2.7—3.0); scapulae rounded, cervical grooves

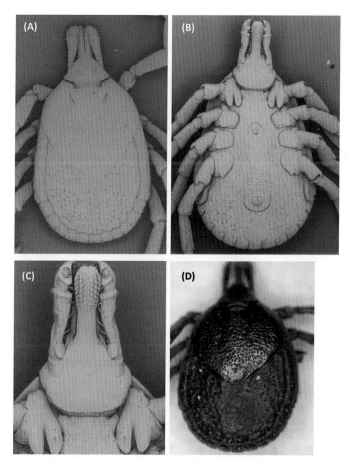

FIGURE 2.20   *A. calcaratum* female: (A) dorsal view, (B) ventral view, (C) capitulum ventral view, (D) scutum ornamentation.

deep and short, comma-shaped. Eyes flat. Chitinous tubercles at posterior body margin absent. Scutum ornate, with an irregular pale spot in the posterior field; punctuations numerous, uniformly distributed. Notum with small setae, barely perceptible. Basis capituli dorsally subrectangular, cornua short, porose areas oval. Hypostome spatulate, dental formula 3/3. Genital aperture located at level of coxa II, U-shaped. Legs: coxa I with two distinct, triangular blunt spurs, equal in size; coxae II−IV with one triangular short spur each; spur on coxa IV larger than spurs of coxae II−III; trochanters without spurs. Spiracular plates comma-shaped.

Nymph (Fig. 2.21): medium body size, total length 1.4 mm (1.3−1.5), breadth 1.2 mm (1.1−1.3). Body outline oval. Chitinous tubercles at posterior body margin absent. Scutum with surface rugose; deep punctuations, more numerous on lateral and posterior fields. Eyes slightly bulging, located on lateral scutal angles at the level of scutal mid-length. Cervical grooves

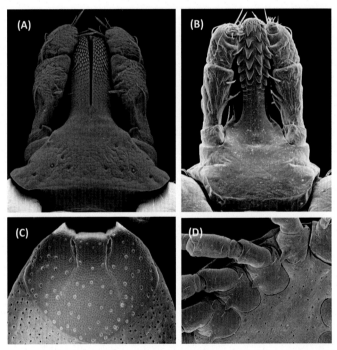

FIGURE 2.21 *A. calcaratum* nymph: (A) capitulum dorsal view, (B) capitulum ventral view, (C) scutum, (D) coxae. *Figures (A) to (D) are reproductions of figure 18:* A. calcaratum *nymph. Capitulum, dorsum; Capitulum, venter; Scutum; Coxae I−IV. From Martins TF, Labruna MB, Mangold AJ, Cafrune MM, Guglielmone AA, Nava S. Taxonomic key to nymphs of the genus* Amblyomma *(Acari: Ixodidae) in Argentina, with description and redescription of the nymphal stage of four* Amblyomma *species. Ticks Tick-borne Dis 2014;5:753−70, with permission of Elsevier.*

long, reaching the scutal posterior third, deeper anteriorly. Basis capituli dorsally subtriangular, without cornua. Hypostome spatulate, dental formula 2/2. Legs: coxa I with two triangular spurs, the external longer and pointed; coxae II−IV with one small triangular spur each.

Taxonomic notes: *A. calcaratum* is morphologically closely related to *A. nodosum* Neumann, 1899. These two tick species have been confused with each other due to their morphological similarities.[10]

DNA sequences with relevance for tick identification and phylogeny: the following DNA sequences of *A. calcaratum* are available in the Gen Bank: mitochondrial genes 12S rRNA (accession numbers JX192880−882, AY225322) and 16S rRNA (accession numbers FJ424400, KF702342, JN573302), cytochrome oxidase I gene (accession numbers KF200144−145; KF200078), and sequences of the nuclear marker 28S rRNA gene (accession number AY225324).

REFERENCES

1. Cáceres AG, Beati L, Keirans JE. First evidence of the occurrence of *Amblyomma calcaratum* Neumann, 1899 in Peru. *Rev Peru Biol* 2002;**9**:116−17.
2. Guglielmone AA, Estrada-Peña A, Keirans JE, Robbins, RG. *Ticks (Acari: Ixodida) of the neotropical zoogeographic region*. Special publication of the integrated consortium on ticks and tick-borne diseases-2. Houten (The Netherlands): Atalanta; 2003.
3. Guzmán Cornejo C, Pérez TM, Nava S, Guglielmone AA. First records of the ticks *Amblyomma calcaratum* and *A. pacae* (Acari: Ixodidae) parasitizing mammals of Mexico. *Rev Mex Biodiversidad* 2006;**77**:123−7.
4. Jones EK, Clifford CM, Keirans JE, Kohls GM. The ticks of Venezuela (Acarina: Ixodoidea) with a key to the species of *Amblyomma* in the Western Hemisphere. *Brigham Young Univ Sci Bull Biol Ser* 1970;**17**(4):40 p.
5. Labruna MB, Sanfilippo LE, Demetrio C, Menezes AC, Pinter A, Guglielmone AA, et al. Ticks collected on birds in the state of São Paulo, Brazil. *Exp Appl Acarol* 2007;**43**:147−60.
6. Ogrzewalska M, Pacheco R, Uezu A, Richtzenhain LJ, Ferreira F, Labruna MB. Ticks (Acari: Ixodidae) infesting birds in an Atlantic rain forest region of Brazil. *J Med Entomol* 2009;**46**:1225−9.
7. Guglielmone AA, Robbins RG, Apanaskevich DA, Petney TN, Estrada-Peña A, Horak IG. *The hard ticks of the world (Acari: Ixodida: Ixodidae)*. London: Springer; 2014.
8. Martins TF, Fecchio A, Labruna MB. Ticks of the genus *Amblyomma* (Acari: Ixodidae) on wild birds in Brazilian Amazon. *Syst Appl Acarol* 2014;**19**:385−92.
9. Smith MW. A survey of the distribution of the ixodid ticks *Boophilus microplus* (Canestrini, 1888) and *Amblyomma cajennense* (Fabricius, 1787) in Trinidad & Tobago and the possible influence of the survey results on planned livestock development. *Trop Agric* 1974;**51**:559−67.
10. Fairchild GB, Kohls GM, Tipton VJ. The ticks of Panama (Acarina: Ixodoidea). In: Wenzel WR, Tipton VJ, editors. *Ectoparasites of Panama*. Chicago (IL): Field Museum of Natural History; 1966. pp. 167−219.

Amblyomma coelebs
Neumann, 1899

Neumann, L.G. (1899) Révision de la famille des ixodidés (3e mémoire). *Mémoires de la Société Zoologique de France*, 12, 107−294.

DISTRIBUTION

Argentina, Belize, Bolivia, Brazil, Colombia, Costa Rica, French Guiana, Guatemala, Guyana, Honduras, Mexico, Nicaragua, Panama, Paraguay, Peru, Surinam and Venezuela.[1] *A. coelebs* is also established in the Nearctic Region.[1]

Biogeographic distribution in the Southern Cone of America: Chaco and Parana Forest (Neotropical Region) (see Fig. 1 in the introduction).

HOSTS

Although tapirs are considered usual hosts for adults of *A. coelebs*,[2,3] adults and immature stages of this tick present low host specificity.[4] Adults were recorded on different mammal species, and larvae and nymphs were also found on aves[4−16] (Table 2.6).

ECOLOGY

With the exception of the reports on distribution and host association, there is no information on ecological aspects of *A. coelebs*.

SANITARY IMPORTANCE

All parasitic stages of *A. coelebs* have been recorded parasitizing humans[12,17] and also domestic animals (horses). Specimens of this tick were found infected with *"Candidatus Rickettsia amblyommii,"*[18] a bacteria with unknown pathogenicity.

DIAGNOSIS

Male (Fig. 2.22): large body size, total length 5.4 mm (5.2−5.5), breadth 4.2 mm (4.0−4.4). Body outline oval, scapulae rounded; cervical grooves deep and short, comma-shaped; marginal groove complete. Eyes flat. Scutum ornate, with two bright red-orange patches in the scapular area, patches converge posteriorly forming the outline of a pseudoscutum; postero-accessory and postero-median spots conspicuous; punctuations numerous, uniformly distributed. Carena present. Basis capituli dorsally quadrangular, cornua short. Hypostome spatulate, dental formula 3/3. Genital aperture located at

TABLE 2.6 Hosts of Adults (A), Nymphs (N), and Larvae (L) of *A. coelebs*

MAMMALIA		Cricetidae	
ARTIODACTYLA		*Naecomys spinosus*	N
Tayassuidae		*Rhipidomys macrurus*	N
Tayassu pecari	N	**Cuniculidae**	
CARNIVORA		*Cuniculus paca*	N
Canidae		**Dasyproctidae**	
Cerdocyon thous	N	*Dasyprocta azarae*	N
Felidae		**Echimyidae**	
Herpailurus yagouaroundi	N	*Thrichomys fosteri*	N
Panthera onca	AN	**AVES**	
Puma concolor	N	**COLUMBIFORMES**	
Procyonidae		**Columbidae**	
Nasua nasua	N	*Leptotila verrauxi*	L
CINGULATA		**CORACIIFORMES**	
Dasypodidae		**Momotidae**	
Dasypus novemcinctus	N	*Baryphthengus ruficapillus*	N
DIDELPHIMORPHIA		*Malacoptila striata*	N
Didelphidae		**PASSERIFORMES**	
Didelphis albiventris	N	**Conopophagidae**	
D. marsupialis	N	*Conopophaga lineata*	N
Lutreolina crassicaudata	N	**Pipridae**	
PERISSODACTYLA		*Ceratopira mentalis*	N
Equidae		*Chiroxiphia linearis*	N
Horse	A	**Thamnophilidae**	
Tapiridae		*Phlegopsis nigromaculata*	L
Tapirus bairdii	A	*Pyriglena leuconota*	L
T. terrestris	A	*Thamnophilus pelzenii*	N
PRIMATES		**Thraupidae**	
Hominidae		*Mitrospingus casinii*	N

(Continued)

TABLE 2.6 (Continued)

Human	ANL	Turdidae	
RODENTIA		*Turdus amaurochalinus*	N
Caviidae		**PICIFORMES**	
Hydrochoerus hydrochaeris	AN	**Bucconidae**	
		Malacoptila panamensis	N
		M. striata	N

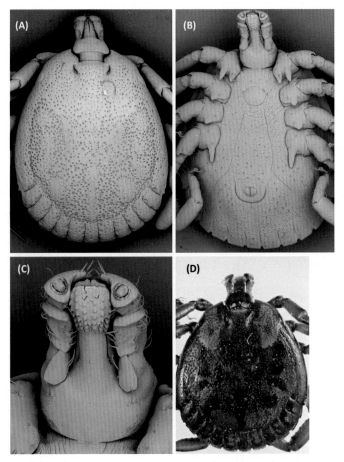

FIGURE 2.22 *A. coelebs* male: (A) dorsal view, (B) ventral view, (C) capitulum ventral view, (D) scutum ornamentation.

level of coxa II, U-shaped. Legs: coxa I with two distinct, triangular blunt spurs, the external spur slightly longer than the internal spur; coxae II–III with a triangular short spur each; coxa IV with a triangular sharp spur, longer than spurs on coxae II–III; trochanters without spurs. Spiracular plates comma-shaped.

Female (Fig. 2.23): large body size, total length 5.0 mm (4.0–6.0), breadth 4.0 mm (3.1–4.8). Body outline oval. Scutum length 2.7 mm (2.5–2.9), breadth 3.0 mm (2.8–3.2); scapulae rounded, cervical grooves deep anteriorly, shallow posteriorly, sigmoid in shape. Eyes flat. Chitinous tubercles at posterior body margin absent. Scutum ornate, with a yellow enameled posterior spot without central stripe; cervical spots elongated, merging posteriorly with limiting spots; punctuations numerous, uniformly distributed. Notum glabrous. Basis capituli dorsally subrectangular, cornua

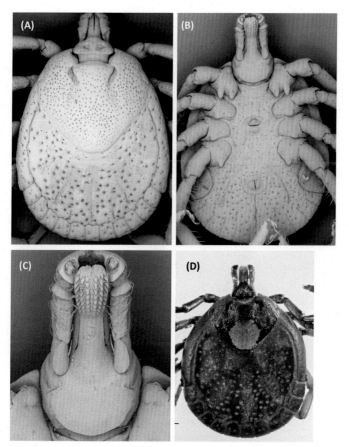

FIGURE 2.23   *A. coelebs* female: (A) dorsal view, (B) ventral view, (C) capitulum ventral view, (D) scutum ornamentation.

short, porose areas oval. Hypostome spatulate, dental formula 3/3. Genital aperture located at level of coxa II, U-shaped. Legs: coxa I with two distinct, triangular blunt spurs, the external spur slightly longer than the internal spur; coxae II–III with a triangular short spur each; coxa IV with a triangular short spur, longer than spurs on coxae II–III; trochanters without spurs. Spiracular plates comma-shaped.

Nymph (Fig. 2.24): medium body size, total length 1.6 mm (1.5–1.7), breadth 1.3 mm (1.2–1.4). Body outline oval. Chitinous tubercles at posterior body margin absent. Scutum with numerous large and deep punctuations evenly distributed. Eyes slightly bulging, located on lateral scutal angles at the level of scutal mid-length. Cervical grooves deep and short, ending as small shallow depressions at the level of the posterior margin of the eyes. Basis capituli dorsally broadly hexagonal, without cornua. Hypostome

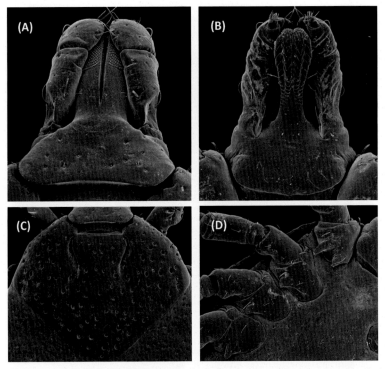

FIGURE 2.24 *A. coelebs* nymph: (A) capitulum dorsal view, (B) capitulum ventral view, (C) scutum, (D) coxae. *Figures (A) to (D) are reproductions of figure 24: A.* coelebs *nymph. Capitulum, dorsum; Capitulum, venter; Scutum; Coxae I–IV. From Martins TF, Labruna MB, Mangold AJ, Cafrune MM, Guglielmone AA, Nava S. Taxonomic key to nymphs of the genus* Amblyomma *(Acari: Ixodidae) in Argentina, with description and redescription of the nymphal stage of four* Amblyomma *species. Ticks Tick-borne Dis 2014;5:753–70, with permission of Elsevier.*

spatulate, dental formula 2/2. Legs: coxa I with two pointed spurs, the external spur slightly longer than the internal spur; coxae II—IV with a small triangular spur each.

Taxonomic notes: the presence of carena is described for the males of *A. coelebs*, but in some specimens this morphological feature is almost imperceptible. *A. coelebs* is phylogenetically related to *A. dubitatum* (Fig. 2.7). Specimens of *A. hadanii* Nava, Mastropaolo, Mangold, Martins, Venzal, and Guglielmone, 2014, have been misidentified as *A. coelebs*.[19] In this sense, all records of *A. coelebs* from the Yungas Biogeographic Province in northwestern Argentina[20,21] correspond in fact to *A. hadanii*.[19]

DNA sequences with relevance for tick identification and phylogeny: the following DNA sequences of *A. coelebs* are available in the Gen Bank: mitochondrial genes 12S rRNA (accession number AY342259) and 16S rRNA (accession numbers GQ891948, FJ424408, KM519937), and sequences of the nuclear genome (the fragment 5.8S ribosomal RNA gene-internal transcribed spacer 2-28S ribosomal RNA gene) (accession number AY887110).

REFERENCES

1. Guglielmone AA, Estrada-Peña A, Keirans JE, Robbins RG. Ticks (Acari: Ixodida) of the neotropical zoogeographic region. Special publication of the integrated consortium on ticks and tick-borne diseases-2. Houten (The Netherlands): Atalanta; 2003.
2. Labruna MB, Guglielmone AA. Ticks of new world tapirs. *Tapir Conserv* 2009;**18**:21—8.
3. Guglielmone AA, Robbins RG, Apanaskevich DA, Petney TN, Estrada-Peña A, Horak IG. *The hard ticks of the world (Acari: Ixodida: Ixodidae)*. London: Springer; 2014.
4. Nava S, Guglielmone AA. A meta-analysis of host specificity in Neotropical hard ticks (Acari: Ixodidae). *Bull Entomol Res* 2013;**103**:216—24.
5. Floch H, Fauran P. Ixodidés de la Guyane et des Antilles Françaises. *Arch Inst Pasteur Guyane Inini* 1958;**446**:94.
6. Fairchild GB, Kohls GM, Tipton VJ. The ticks of Panama (Acarina: Ixodoidea). In: Wenzel WR, Tipton VJ, editors. *Ectoparasites of Panama*. Chicago (IL): Field Museum of Natural History; 1966. pp. 167—219.
7. Jones EK, Clifford CM, Keirans JE, Kohls GM. The ticks of Venezuela (Acarina: Ixodoidea) with a key to the species of *Amblyomma* in the Western Hemisphere. *Brigham Young Univ Sci Bull Biol Ser* 1972;**17**(4):1—40.
8. Labruna MB, Camargo LMA, Terrassini FA, Ferreira F, Schumaker TST, Camargo EP. Ticks (Acari: Ixodidae) from the state of Rondônia, western Amazon, Brazil. *Syst Appl Acarol* 2005;**10**:17—32.
9. Labruna MB, Jorge RSP, Sana DA, Jácomo ATA, Kashivakura CK, Furtado MM, et al. Ticks (Acari: Ixodida) on wild carnivores in Brazil. *Exp Appl Acarol* 2005;**36**:149—63.
10. Martins TF, Furtado MM, Jácomo AA, Silveira L, Sollmann R, Torres NM, et al. Ticks on free-living wild mammals in Emas National Park, Goiás State, central Brazil. *Syst Appl Acarol* 2011;**16**:201—6.
11. Saraiva DG, Fournier GFSR, Martins TF, Leal KPG, Vieira FN, Camara EMVC, et al. Ticks (Acari: Ixodidae) associated with terrestrial mammals in the state of Minas Gerais, southeastern Brazil. *Exp Appl Acarol* 2012;**58**:159—66.

12. Lamattina D, Tarragona EL, Costa SA, Guglielmone AA, Nava S. Ticks (Acari: Ixodidae) of northern Misiones Province, Argentina. *Syst Appl Acarol* 2014;**19**:393−8.

13. Ogrzewalska M, Pacheco R, Uezu A, Richtzenhain LJ, Ferreira F, Labruna MB. Ticks (Acari: Ixodidae) infesting birds in an Atlantic rain forest region of Brazil. *J Med Entomol* 2009;**46**:1225−9.

14. Ogrzewalska M, Literak I, Capek M, Sychra O, Álvarez Calderón V, Rodríguez BC, et al. Bacteria of the genus *Rickettsia* in ticks (Acari: Ixodidae) collected from birds in Costa Rica. *Ticks Tick-borne Dis* 2015;**6**:478−82.

15. Sponchiado J, Melo GL, Martins TF, Krawczak FS, Labruna MB, Cáceres NC. Association patterns of ticks (Acari: Ixodida: Ixodidae, Argasidae) of small mammals in Cerrado fragments, western Brazil. *Exp Appl Acarol* 2015;**65**:389−91.

16. Witter R, Martins TF, Campos AK, Melo ALT, Correa SHR, Morgado TO, et al. Rickettsial infection in ticks (Acari: Ixodidae) of wild animals in Midwestern Brazil. *Ticks Tick-borne Dis* 2016;**7**:415−23.

17. Guglielmone AA, Beati L, Barros-Battesti DM, Labruna MB, Nava S, Venzal JM, et al. Ticks (Ixodidae) on humans in South America. *Exp Appl Acarol* 2006;**40**:83−100.

18. Labruna MB, Whitworth T, Bouyer DH, McBride J, Camargo LMA, Camargo EP, et al. *Rickettsia belli* and *Rickettsia amblyommii* in *Amblyomma* ticks from the state of Rondônia, Western Amazon, Brazil. *J Med Entomol* 2004;**41**:1073−81.

19. Nava S, Mastropaolo M, Mangold AJ, Martins TF, Venzal JM, Guglielmone AA. *Amblyomma hadanii* n. sp. (Acari: Ixodidae), a tick from northwestern Argentina previously confused with *Amblyomma coelebs* Neumann, 1899. *Syst Parasitol* 2014;**88**:261−72.

20. Beldoménico PM, Baldi JC, Antoniazzi LR, Orduna GM, Mastropaolo M, Macedo AC, et al. Ixodid ticks (Acari: Ixodidae) present at Parque Nacional El Rey, Argentina. *Neotr Entomol* 2003;**32**:273−7.

21. Nava S, Lareschi M, Mangold AJ, Guglielmone AA. Registros de garrapatas de importancia médico-veterinaria detectadas ocasionalmente en Argentina. *Rev FAVE* 2004;**3**:61−5.

Amblyomma dissimile
Koch, 1884

Koch, C.L. (1844) Systematische Übersicht über die Ordnung der Zecken. *Archiv für Naturgeschichte*, 10, 217−239.

DISTRIBUTION

Antigua and Barbuda, Argentina, Bahamas, Barbados, Belize, Brazil, Colombia, Costa Rica, Cuba, Ecuador, El Salvador, Grenada, Guadalupe, Guatemala, Guyana, French Guiana, Haiti, Honduras, Jamaica, Mexico, Nicaragua, Panama, Paraguay, Peru, Puerto Rico, Dominican Republic, St. Lucia, Surinam, Trinidad and Tobago, and Venezuela.[1,2] *A. dissimile* is also found in the Nearctic Region.[1] Pietzsch et al.[3] have mentioned the introduction of *A. dissimile* into the Palearctic Region, but there is no evidence of its establishment in this region.[4] The records of *A. dissimile* from the

Philippines and India[5,6] are considered a result of incorrect labeling or accidental introduction that did not result in populations of this species established in the Oriental Zoogeographic Region.[4]

Biogeographic distribution in the Southern Cone of America: Chaco (Neotropical Region) (see Fig. 1 in the introduction).

HOSTS

Amphibians and reptiles are the principal hosts for all parasitic stages of *A. dissimile*, but most records on free-ranging hosts correspond to toads from the family Bufonidae, and the reptiles *Boa constrictor* and *Iguana iguana* (Table 2.7).[4] There are also records of *A. dissimile* on mammals belonging to different orders and a record on an avian species (Table 2.7).[4,7]

ECOLOGY

Bodkin[8] has mentioned that *A. dissimile* is a parthenogenetic tick, but this statement comes from an apparent confusion with *A. rotundatum*.[4] *A. dissimile* has a three-host life cycle, but a facultative two-host life cycle has also been observed in this tick.[9] The wide geographic distribution of *A. dissimile* has been explained in function of the long period that all stages of this tick spends attached to their hosts, which allows a relative independence of the environment conditions,[10] but this hypothesis has not been confirmed.

SANITARY IMPORTANCE

High infestations with *A. dissimile* in snakes and toads under laboratory conditions have resulted in lethal or deleterious effects to these hosts.[11,12] Lampo and Bayliss[13] suggested that high infestations of *A. dissimile* may regulate the densities of toads in native habitats. *A. dissimile* has shown capacity to transmit *Hepatozoon fusifex* to snakes,[14] a parasite that drastically changes the morphology of infected red cells; this tick species is also able to maintain transtadially infection with the agent of hydropericarditis (hearthwater disease) of ruminants, *Ehrlichia ruminantium* (mentioned as *Cowdria ruminantium*).[15] *Rickettsia* sp. strain Colombianensi (pathogenicity undetermined) was detected in biological material obtained from *A. dissimile* ticks.[16] There are odd records of adults and immature stages of *A. dissimile* on humans and cattle, and just adult ticks on sheep.

DIAGNOSIS

Male (Fig. 2.25): medium body size, total length 3.6 mm (2.8−4.5), breadth 3.2 mm (2.4−3.8). Body outline rounded, scapulae rounded, cervical grooves deep, short, sigmoids; marginal groove absent. Eyes flat. Scutum ornate, with postero-accessory and postero-median spots narrow, lateral spots small

TABLE 2.7 Hosts of Adults (A), Nymphs (N), and Larvae (L) of *A. dissimile*

AMPHIBIA		*Leiocephalus carinatus*	A
ANURA		**Viperidae**	
Bufonidae		*Bothrops asper*	A
Peltophryne fustiger	ANL	*B. jararaca*	A
Rhinella bergi	A	*B. lanceolatus*	A
R. marina	ANL	*Crotalus adamanteus*	AN
R. schneideri	A	*C. durissus*	AN
REPTILIA		*Lachesis muta*	A
CROCODYLIA		*Sistrurus miliaris*	A
Alligatoridae		*Tropidolaemus wagleri*	A
Caiman crocodilus	A	**TESTUDINES**	
C. yacare	A	**Chelidae**	
Crocodylidae		*Acanthochelys macrocephala*	AN
Crocodylus moreletii	A	*Phrynops geoffroanus*	A
SQUAMATA		**Emydidae**	
Boidae		*Graptemys geographica*	A
Boa constrictor	ANL	*Mauremys leprosa*	A
Epicrates cenchria	AN	*Rhinoclemmys areolata*	A
E. striatus	ANL	*R. pulcherrima*	A
Eunectes murinus	AN	*R. rubida*	A
E. notaeus	AN	*Terrapene carolina*	ANL
Colubridae		*Trachemys scripta*	ANL
Chironius laurenti	A	**Kinosternidae**	
Drymarchon corais	AN	*Kinosternon leucostomum*	A
Elaphe guttata	A	*K. scorpiodes*	AN
E. obsoleta	AN	**Testudinidae**	
Erythrlamprus melanotus	N	*Chelonoidis carbonaria*	A
Hydrodynastes gigas	AN	*C. denticulata*	AN
Lamproptelis getula	A	**MAMMALIA**	
Leptodeira annulata	A	**ARTIODACTYLA**	

(Continued)

TABLE 2.7 (Continued)

Mastigodryas bifossatus	A	**Bovidae**	
Mussurana bicolor	AN	Cattle	ANL
Oxybelis aeneus	ANL	Sheep	A
O. fulgidus	ANL	**CARNIVORA**	
Philodryas baroni	N	**Procyonidae**	
Pituophis melanoleucus	A	*Nasua narica*	A
Pseudoboa neuwiedi	A	**DIDELPHIMORPHIA**	
Pseutes poecilonotus	A	**Didelphidae**	
Spilotes pullatus	A	*Didelphis albiventris*	N
Waglerophis merremi	A	*Monodelphis domestica*	N
Xenodon sp.	A	**PRIMATES**	
Elapidae		**Hominidae**	
Micrurus sp.	A	Human	AN
Iguanidae		**RODENTIA**	
Ctenosaura acanthura	A	**Caviidae**	
C. bakeri	A	*Cavia aperea*	N
C. pectinata	A	*Hydrochoerus hydrochaeris*	AN
Cyclura carinata	A	**Cricetidae**	
C. cornuta	A	*Oryzomys subflavus*	L
C. cychlura	ANL	*Peromyscus gossypinus*	N
Iguana iguana	ANL	**Dasyproctidae**	
Phrynosomatidae		*Dasyprocta punctata*	A
Sceloporus malachiticus	N	**Echimyidae**	
S. undulatus	L	*Proechimys semispinosus*	ANL
Teiidae		*Thrichomys apereoides*	N
Ameiva ameiva	A	**AVES**	
Tupinambis teguixin	ANL	**CICONIIFORMES**	
Tropiduridae		**Ardeidae**	
Tropidurus hispidus	N	*Cochleaurius cochleaurius*	A

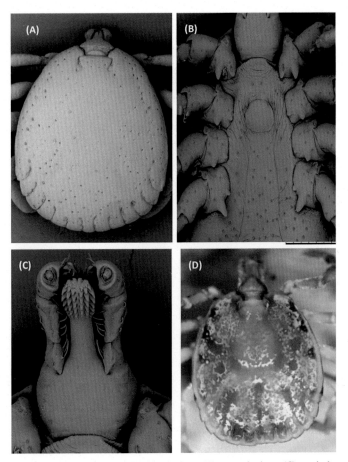

FIGURE 2.25   *A. dissimile* male: (A) dorsal view, (B) ventral view, (C) capitulum ventral view, (D) scutum ornamentation.

but distinct, cervical spots present, posterior branches of limiting spots fused posteriorly; punctuations numerous, areolate, interspersed with fine punctuations, absent in the antero-median field. Carena absent. Basis capituli rectangular dorsally, cornua short. Hypostome spatulate, dental formula 3/3. Genital aperture located at level of coxa II, U-shaped. Legs: coxae I—IV each with two distinct, short blunt spurs; external spurs on coxae I—IV longer than internal spurs; trochanters without spurs. Spiracular plates comma-shaped.

Female (Fig. 2.26): large body size, total length 4.1 mm (3.6—4.5), breadth 3.4 mm (3.0—3.8) in unfed specimens. Body outline oval. Scutum length 2.5 mm (2.4—2.5), breadth 2.7 mm (2.5—2.8); scapulae rounded, cervical grooves long and deep, sigmoid. Eyes flat. Chitinous tubercles at the

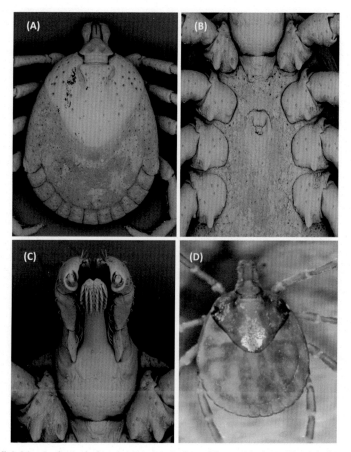

FIGURE 2.26   *A. dissimile* female: (A) dorsal view, (B) ventral view, (C) capitulum ventral view, (D) scutum ornamentation.

postero-body margin absent. Scutum ornate, with an irregular pale spot in each antero-lateral field, a large pale spot in the posterior field; punctuations few and large, concentrated in the antero-lateral fields. Notum glabrous. Basis capituli dorsally subrectangular, cornua absent, porose areas oval. Hypostome spatulate, dental formula 3/3. Genital aperture located at level of coxa II, U-shaped. Legs: coxae I–IV each with two distinct, short blunt spurs; external spurs on coxae I–IV longer than internal spurs; trochanters without spurs. Spiracular plates comma-shaped.

Nymph (Fig. 2.27): medium body size, total length 1.6 mm (1.5–1.8), breadth 1.3 mm (1.1–1.4). Body outline oval. Chitinous tubercles at the posterior body margin absent. Scutum with few punctuations, larger and deeper in the lateral fields than elsewhere. Eyes flat, located on lateral scutal angles

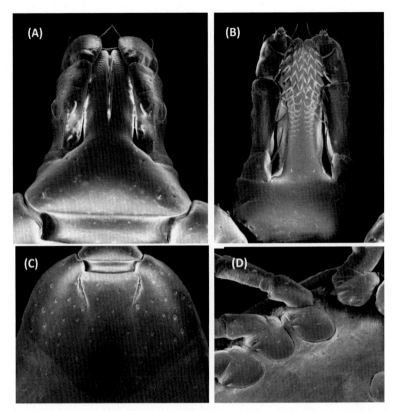

FIGURE 2.27 *A. dissimile* nymph: (A) capitulum dorsal view, (B) capitulum ventral view, (C) scutum, (D) coxae. *Figures (A) to (D) are reproductions of figure 15: A.* dissimile *nymph. Capitulum, dorsum; Capitulum, venter; Scutum; Coxae I—IV. From Martins TF, Labruna MB, Mangold AJ, Cafrune MM, Guglielmone AA, Nava S. Taxonomic key to nymphs of the genus* Amblyomma *(Acari: Ixodidae) in Argentina, with description and redescription of the nymphal stage of four* Amblyomma *species. Ticks Tick-borne Dis 2014;5:753—70, with permission of Elsevier.*

at the level of scutal mid-length. Cervical grooves reaching the scutal mid-length, deep anteriorly. Basis capituli dorsally subtriangular, without cornua. Hypostome spatulate, dental formula 3/3 for most of its length, 2/2 at the base. Legs: coxa I with two spurs, the external longer than the internal; coxae II—IV with a small triangular spur.

Taxonomic notes: although *A. dissimile* and *A. rotundatum* Koch, 1844, are morphologically very similar, and they may be found in sympatry, their genetic differences in both mitochondrial and nuclear DNA sequences are marked (Figs. 2.7 and 2.11). Principal difficulties to morphologically differentiate *A. dissimile* and *A. rotundatum* is related to the diagnosis of females. Lampo et al.[17] have stated that differences in the spurring of coxae I—IV and scutal punctuations are useful to distinguish them, while the authors of the present study stress the relevance of scutum ornamentation to differentiate females of *A. dissimile* and *A. rotundatum* and this is clearly stated in the key presented in Chapter 4, "Morphological Keys for Genera and Species of Ixodidae and Argasidae." See also the Taxonomic notes of *A. argentinae* Neumann, 1905.

DNA sequences with relevance for tick identification and phylogeny: DNA sequences of *A. dissimile* are available in the Gen Bank as follows: mitochondrial genes 12S rRNA (accession number AY342249) and 16S rRNA (accession number KJ569692), cytochrome oxidase I (accession numbers KF200170, KF200168, KF200117, KF200116, KF200114), cytochrome oxidase II (accession numbers KT630578; FJ917644−FJ917653) and sequences of the nuclear marker 18S rRNA (accession number KJ569695).

REFERENCES

1. Guglielmone AA, Estrada-Peña A, Keirans JE, Robbins, R.G. Ticks (Acari: Ixodida) of the neotropical zoogeographic region. Special publication of the integrated consortium on ticks and tick-borne diseases-2. Houten (The Netherlands): Atalanta; 2003.
2. Durden LA, Knapp CR. Ticks parasitizing reptiles in the Bahamas. *Med Vet Entomol* 2005;**19**:326−8.
3. Pietzsch M, Quest R, Hillyard PD, Medlock JM, Leach S. Importation of exotic ticks into the United Kingdom via the international trade in reptiles. *Exp Appl Acarol* 2006;**38**:59−65.
4. Guglielmone AA, Nava S. Hosts of *Amblyomma dissimile* Koch, 1844 and *Amblyomma rotundatum* Koch, 1844. *Zootaxa* 2010;**2541**:27−49.
5. Neumann LG. Révision de la familie des ixodidés (3e mémorie). *Mém Soc Zool Fr* 1899;**12**:107−294.
6. Keirans JE. George Henry Falkiner Nuttall and the Nuttall tick catalogue. *US Dep Agric Agric Res Serv Miscel Publ* 1985;**1438**:1785.
7. Witter R, Martins TF, Campos AK, Melo ALT, Correa SHR, Morgado TO, et al. Rickettsial infection in ticks (Acari: Ixodidae) of wild animals in Midwestern Brazil. *Ticks Tick-borne Dis* 2016;**7**:415−23.
8. Bodkin G. The biology of *Amblyomma dissimile* Koch, with an account of its power to reproducing parthenogenetically. *Parasitology* 1918;**1**:10−17.
9. Dunn LH. Studies on the iguana tick, *Amblyomma dissimile*, in Panama. *J Parasitol* 1918;**5**:1−10.
10. Lampo M, Rangel Y, Mata A. Population genetic structure of a three-host tick, *Amblyomma dissimile*, in eastern Venezuela. *J Parasitol* 1998;**84**:1137−42.
11. Fairchild GB. An annotated list of the blood sucking insects, ticks and mites known form Panama. *Am J Trop Med Hyg* 1943;**23**:569−91.
12. Jakowska S. Lesions produced by ticks, *Amblyomma dissimile*, in *Bufo marinus* toads from the Dominican Republic. *Am Zool* 1943;**12**:731.
13. Lampo M, Bayliss P. The impact of ticks on *Bufo marinus* from native habitats. *Parasitology* 1996;**113**:199−206.

14. Ball GH, Chao J, Telford SR. *Hepatozoon fusifex* sp. n., a hemogregarine from *Boa constrictor* producing marked morphological changes in infected erythrocytes. *J Parasitol* 1969;**55**:800−13.

15. Jongejan F. Experimental transmission of *Cowdria ruminantium* (Rickettsiales) by the American reptile tick *Amblyomma dissimile* Koch, 1944. *Exp Appl Acarol* 1992;**15**:117−21.

16. Miranda J, Portillo A, Oteo J, Mattar S. *Rickettsia* sp. Strain Colombianensi (Rickettsiales: Rickettsiaceae): a new proposed *Rickettsia* detected in *Amblyomma dissimile* (Acari: Ixodidae) from Iguanas and free-living larvae ticks from vegetation. *J Med Entomol* 2012;**49**:960−5.

17. Lampo M, Rangel Y, Mata A. Genetic markers for the identification of two tick species, *Amblyomma dissimile* and *Amblyomma rotundatum*. *J Parasitol* 1997;**83**:382−6.

Amblyomma dubitatum
Neumann, 1899

Neumann, L.G. (1899) Révision de la famille des ixodidés (3e mémoire). *Mémoires de la Société Zoologique de France*, 12, 107−294.

DISTRIBUTION

Argentina, Brazil, Paraguay, and Uruguay.[1] The records of *A. dubitatum* from Venezuela and Bolivia are considered doubtful.[1]

Biogeographic distribution in the Southern Cone of America: Chaco, Pampa, and Parana Forest (Neotropical Region) (see Fig. 1 in the introduction). The findings of *A. dubitatum* from Yungas Biogeographic Province (Tucumán, Argentina) are not included because these records correspond to ticks collected on hosts kept under semi-captivity conditions.[2]

HOSTS

The capybara *Hydrochoerus hydrochaeris* is the most usual host for all parasitic stages of *A. dubitatum*[1], but both immature and adult stages of this tick were collected on several mammal species belonging to different orders[1,3−7] (Table 2.8). Larvae and nymphs of *A. dubitatum* were also found on a few species of birds[1,8] (Table 2.8).

ECOLOGY

A. dubitatum has a 1-year life cycle, where larvae have the peak of abundance from May to July, nymphs from July to October, and females from November to March.[4] However, all stages can be found active along the entire year, and

TABLE 2.8 Hosts of Adults (A), Nymphs (N), and Larvae (L) of *A. dubitatum*

MAMMALIA			
ARTIODACTYLA		**PRIMATES**	
Bovidae		**Atelidae**	
Cattle	N	*Alouatta caraya*	N
Domestic buffalo	N	**Hominidae**	
Cervidae		Human	ANL
Axis axis	L	**RODENTIA**	
Mazama gouazoubira	AN	**Caviidae**	
Suidae		*Cavia aperea*	NL
Feral pig	ANL	*Hydrochoerus hydrochaeris*	ANL
Tayassuidae		**Chinchillidae**	
Tayassu pecari	A	*Lagostomus maximus*	N
CARNIVORA		**Cricetidae**	
Canidae		*Akodon azarae*	NL
Cerdocyon thous	A	*Cerradomys maracajuensis*	N
Chrysocyon brachyurus	N	*Hylaeamys megacephalus*	N
Domestic dog	A	*Lundomys molitor*	NL
CHIROPTERA		*Necromys lasiurus*	N
Phyllostomidae		*Nectomys rattus*	NL
Glossophaga soricina	A	*Oligoryzomys flavescens*	L
DIDELPHIMORPHIA		*Rhipidomys* sp.	N
Didelphidae		*Scapteromys aquaticus*	L
Didelphis aurita	N	*S. tumidus*	NL
D. albiventris	N	**Echimyidae**	
Lutreolina crassicaudata	N	*Thrichomys fosteri*	N
Monodelphis dimidiata	NL	**Myocastoridae**	
LAGOMORPHA		*Myocastor coypus*	N
Leporidae		**AVES**	
Lepus europaeus	NL	**PASSERIFORMES**	
PERISSODACTYLA		**Emberizidae**	
Equidae		*Zonotrichia capensis*	L

(Continued)

TABLE 2.8 (Continued)

Horse	A	Furnariidae	
Tapiridae		Furnarius rufus	L
T. terrestris	AN	Turdidae	
PILOSA		Turdus rufiventris	L
Myrmecophagidae		STRUTHIONIFORMES	
Myrmecophaga tridactyla	ANL	Rheidae	
Tamandua tetradactyla	NL	Rhea americana	N

Main hosts for adult ticks and immatures stages are underlined.

more than one cohort can coexist within the same population.[4] This tick species prevails in riparian environments and areas prone to flooding.

SANITARY IMPORTANCE

All parasitic stages of *A. dubitatum* were recorded parasitizing humans.[1,9] Larvae and nymphs are particularly very aggressive to humans. Specimens of *A. dubitatum* were found infected with the human pathogens *Rickettsia parkeri*[10] and *Rickettsia* sp. strain Atlantic rainforest,[11] but the role of *A. dubitatum* in the transmission of pathogenic rickettsia to humans is unknown. *A. dubitatum* ticks have been also found infected with *Rickettsia* sp. strain Cooperi, *Rickettsia* sp. strain Pampulha, and *Rickettsia bellii*,[12,13] all of them of unknown pathogenicity. Nymphs of *A. dubitatum* can be occasionally found on cattle.

DIAGNOSIS

Male (Fig. 2.28): medium body size, total length 3.8 mm (3.5−4.0), breadth 2.8 mm (2.6−3.0). Body outline oval, narrower in the anterior part; scapulae rounded; cervical grooves deep and short, comma-shaped; marginal groove complete. Eyes flat. Scutum ornate, with a pale strip extending from the level of eyes to the level of the external festoon, postero-median spot and postero-accessory spots narrow, lateral spots small and distinct, cervical spots narrow and divergent posteriorly; punctuations numerous, uniformly distributed. Carena present. Basis capituli dorsally rectangular, cornua short. Hypostome spatulate, dental formula 3/3. Genital aperture located at level of coxa II, U-shaped. Legs: coxa I with two distinct, triangular short subequal

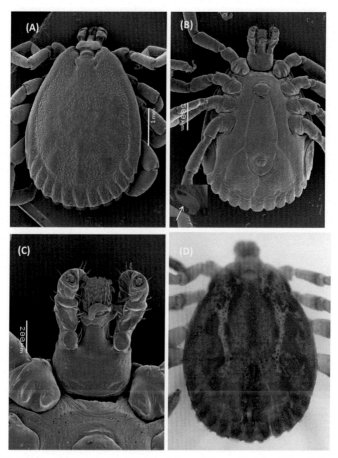

FIGURE 2.28 *A. dubitatum* male: (A) dorsal view, (B) ventral view, (C) capitulum ventral view, (D) scutum ornamentation. *White arrow* in (B) shows the shape of the spiracular plate.

spurs; coxae II–III each with a triangular, short, blunt spur; coxa IV with a triangular sharp spur, longer than spurs on coxae II–III; trochanters without spurs. Spiracular plates oval.

Female (Fig. 2.29): large body size, total length 4.2 mm (4.0–4.3), breadth 3.1 mm (3.0–3.3). Body outline oval. Scutum length 2.3 mm (2.1–2.5), breadth 2.5 mm (2.3–2.6); scapulae rounded, cervical grooves deep anteriorly, shallow posteriorly, sigmoid in shape. Eyes flat. Chitinous tubercles at posterior body margin absent. Scutum ornate, with a broad and longitudinal central area, extending to the posterior angle of the scutum; cervical spots elongated, merging posteriorly with limiting spots; punctuations numerous, uniformly distributed. Notum glabrous. Basis capituli dorsally subrectangular, cornua absent, porose areas oval. Hypostome spatulate, dental formula 3/3; article I of the palp with a flat, retrograde spur. Genital

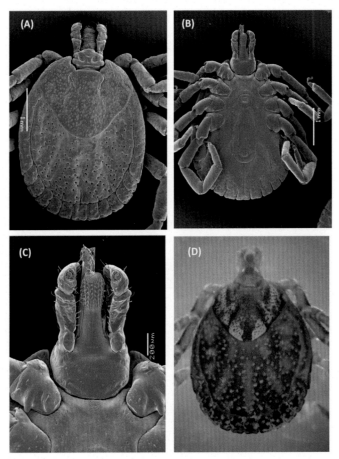

FIGURE 2.29 *A. dubitatum* female: (A) dorsal view, (B) ventral view, (C) capitulum ventral view, (D) scutum ornamentation.

aperture located at level of coxa II, U-shaped. Legs: coxa I with two distinct, triangular, short subequal spurs; coxae II–IV each with a triangular, short blunt spur; trochanters without spurs. Spiracular plates oval.

Nymph (Fig. 2.30): large body size, total length 1.7 mm (1.6–1.8), breadth 1.4 mm (1.3–1.5). Body outline oval. Chitinous tubercles at posterior body margin absent. Scutum with punctuations evenly distributed, larger laterally, smaller centrally. Eyes slightly bulging, located on lateral scutal angles at the level of scutal mid-length. Cervical grooves long and deep reaching the scutal posterior third. Basis capituli dorsally broadly hexagonal, without cornua. Hypostome spatulate, dental formula 2/2. Legs: coxa I with two pointed spurs, the external slightly longer than the internal; coxae II–IV with a small triangular spur each.

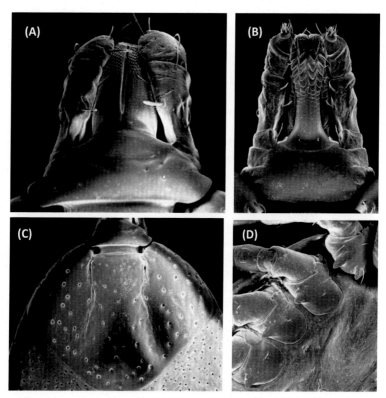

FIGURE 2.30   *A. dubitatum* nymph: (A) capitulum dorsal view, (B) capitulum ventral view, (C) scutum, (D) coxae. *Figures (A) to (D) are reproductions of figure 23:* A. dubitatum *nymph. Capitulum, dorsum; Capitulum, venter; Scutum; Coxae I—IV. From Martins TF, Labruna MB, Mangold AJ, Cafrune MM, Guglielmone AA, Nava S. Taxonomic key to nymphs of the genus* Amblyomma *(Acari: Ixodidae) in Argentina, with description and redescription of the nymphal stage of four* Amblyomma *species. Ticks Tick-borne Dis 2014;5:753—70, with permission of Elsevier.*

Taxonomic notes: the presence of carena is described for the males of *A. dubitatum*, but in some specimens this morphological feature is almost imperceptible. *A. dubitatum* has been frequently named in the 20th century scientific literature as *A. cooperi* Nuttall and Warburton, 1908, but Estrada-Peña et al.[14] demonstrated that the name *A. cooperi* is a synonym of *A. dubitatum*. See also Taxonomic notes of *A. coelebs* Neumann, 1899.

DNA sequences with relevance for tick identification and phylogeny: the following mitochondrial DNA sequences of *A. dubitatum* are available in the Gen Bank: mitochondrial genes 12S rRNA (accession numbers AY342256— AY342258) and 16S rRNA (accession numbers GU301910—GU301914, DQ858954, DQ858955). There also are sequences of the nuclear genome: the fragment 5.8S ribosomal RNA gene-internal transcribed spacer 2-28S ribosomal RNA gene (accession number AY887116) and 18S rDNA sequences (accession number FJ464425).

REFERENCES

1. Nava S, Venzal JM, Labruna MB, Mastropaolo M, González EM, Mangold AJ, et al. Hosts, distribution and genetic divergence (16S rDNA) of *Amblyomma dubitatum* (Acari: Ixodidae). *Exp Appl Acarol* 2010;**51**:335−51.
2. Zerpa C, Venzal JM, López N, Mangold AJ, Guglielmone AA. Garrapatas de Catamarca y Tucumán: estudio de una colección de hospedadores silvestres y domésticos. *Rev FAVE Cienc Vet* 2003;**2**:167−71.
3. Debárbora VN, Nava S, Cirignoli S, Guglielmone AA, Poi ASG. Ticks (Acari: Ixodidae) parasitizing endemic and exotic wild mammals in the Esteros del Iberá wetlands, Argentina. *Syst Appl Acarol* 2012;**17**:243−50.
4. Debárbora VN, Mangold AJ, Oscherov EB, Guglielmone AA, Nava S. Study of the life cycle of *Amblyomma dubitatum* (Acari: Ixodidae) based on field and laboratory data. *Exp Appl Acarol* 2014;**63**:93−105.
5. Coelho MG, Ramos VN, Limongi JE, Lemos ERS, Guterres A, Neto SFC, et al. Serologic evidence of the exposure of small mammals to spotted-fever *Rickettsia* and *Ricekttsia bellii* in Minas Gerais, Brazil. *J Infect Dev Countries* 2016;**10**:275−82.
6. Sponchiado J, Melo GL, Martins TF, Krawczak FS, Labruna MB, Cáceres NC. Association patterns of ticks (Acari: Ixodida: Ixodidae, Argasidae) of small mammals in Cerrado fragments, western Brazil. *Exp Appl Acarol* 2015;**65**:389−91.
7. Witter R, Martins TF, Campos AK, Melo ALT, Correa SHR, Morgado TO, et al. Rickettsial infection in ticks (Acari: Ixodidae) of wild animals in Midwestern Brazil. *Ticks Tick-borne Dis* 2016;**7**:415−23.
8. Flores FS, Nava S, Batallán G, Tauro LB, Contigiani MS, Diaz LA, et al. Ticks (Acari: Ixodidae) on wild birds in north-central Argentina. *Tick Tick-borne Dis* 2014;**5**:715−21.
9. Labruna MB, Pacheco RC, Ataliba AC, Szabó MJP. Human parasitism by the capibara tick, *Amblyomma dubitatum* (Acari: Ixodida). *Entomol News* 2007;**118**:77−80.
10. Lado P, Castro O, Labruna MB, Venzal JM. First molecular detection of *Rickettsia parkeri* in *Amblyomma tigrinum* and *Amblyomma dubitatum* ticks from Uruguay. *Ticks Tick-borne Dis* 2014;**5**:660−2.
11. Monje LD, Nava S, Eberhardt AT, Correa AI, Guglielmone AA, Beldoménico PM. Molecular detection of the human pathogenic *Rickettsia* sp. strain Atlantic Rainforest in *Amblyomma dubitatum* ticks from Argentina. *Vector-borne Zoon Dis* 2015; **15**:167−9.
12. Labruna MB, Whitworth T, Horta MC, Bouyer DH, McBride J, Pinter A, et al. *Rickettsia* species infecting *Amblyomma cooperi* ticks from an area in the state of São Paulo, Brazil, where Brazilian spotted fever is endemic. *J Clin Microbiol* 2007;**42**:90−8.
13. Parola P, Paddock CD, Scolovschi C, Labruna MB, Mediannokov O, Kernif T, et al. Update on tick-borne rickettsioses around the world: a geographic approach. *Clin Microbiol Rev* 2013;**26**:657−702.
14. Estrada-Peña A, Venzal JM, Guglielmone AA. *Amblyomma dubitatum* Neumann: description of nymph and redescription of adults, together with the description of the immature stages of *A. triste* Koch. *Acarologia* 2002;**42**:323−33.

Amblyomma hadanii
Nava, Mastropaolo, Mangold, Martins, Venzal, and Guglielmone, 2014

Nava, S., Mastropaolo, M., Mangold, A.J., Martins, T.F., Venzal, J.M., & Guglielmone, A.A. (2014) *Amblyomma hadanii* n. sp. (Acari: Ixodidae), a tick from northwestern Argentina previously confused with *Amblyomma coelebs* Neumann, 1899. *Systematic Parasitology*, 88, 261–272.

DISTRIBUTION

Argentina.[1]

Biogeographic distribution in the Southern Cone of America: Yungas (Neotropical Region) (see Fig. 1 in the introduction).

HOSTS

Adults of *A. hadanii* were collected on *Tapirus terrestris*, cattle and horses, nymphs were found on cattle, horses, dogs, *Cerdocyon thous*, and humans,[1] and larvae have been detected attached to humans by one of the authors (Santiago Nava). However, records of *A. hadanii* are few to reach well-supported conclusions on its host range and host specificity.[1]

ECOLOGY

With the exception of the reports on distribution and host association presented in the original description of *A. hadanii*, there is no information on ecological aspects of this species.

SANITARY IMPORTANCE

Nymphs and larvae of *A. hadanii* were found parasitizing humans, adults were recorded on cattle and horses, and nymphs on cattle, horses, and dogs. Specimens of *A. hadanii* were found to be infected with "*Candidatus* Rickettsia amblyommii," bacteria of unknown pathogenicity belonging to the spotted fever group *Rickettsia*.[2]

DIAGNOSIS

Male (Fig. 2.31): large body size, total length 4.3 mm (3.9–4.7), breadth 3.2 mm (3.0–3.5). Body outline oval, scapulae rounded; cervical grooves deep and short, comma-shaped; marginal groove complete. Eyes flat. Scutum

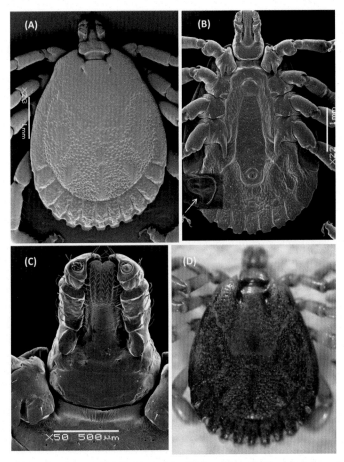

FIGURE 2.31 *A. hadanii* male: (A) dorsal view, (B) ventral view, (C) capitulum ventral view, (D) scutum ornamentation. *White arrow* in (B) shows the shape of the spiracular plate.

ornate, limiting spots converging posteriorly toward the median line forming the outline of a pseudoscutum, lateral spots, postero-accessory spots and postero-median spot small and barely perceptible; punctuations numerous, uniformly distributed. Carena present, irregular in shape, larger and sometimes with a small incision on festoons 4, 5, and 6. Basis capituli dorsally rectangular, cornua short. Hypostome spatulate, dental formula 3/3. Genital aperture located at level of coxa II, U-shaped. Legs: coxa I with two distinct, triangular blunt spurs, the external spur slightly longer than the internal spur; coxae II–III with a triangular short spur each; coxa IV with a triangular sharp spur, longer than spurs on coxae II–III; trochanters without spurs. Spiracular plates comma-shaped.

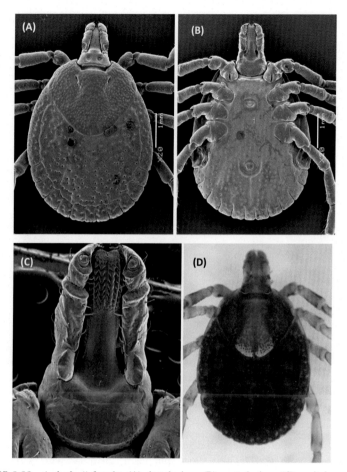

FIGURE 2.32   *A. hadanii* female: (A) dorsal view, (B) ventral view, (C) capitulum ventral view, (D) scutum ornamentation.

Female (Fig. 2.32): large body size, total length 4.6 mm (4.0−5.0), breadth 3.6 mm (3.1−4.0). Body outline oval. Scutum length 2.2 mm (1.9−2.4), breadth 2.5 mm (2.2−2.7); scapulae rounded, cervical grooves deep anteriorly, shallow posteriorly, sigmoid in shape. Eyes flat. Chitinous tubercles at posterior body margin absent. Scutum ornate, central area characterized by a patchy yellow-whitish enameled posterior spot, with a central stripe reaching posterior scutal margin, central stripe become narrower toward posterior scutal margin; punctuations numerous, uniformly distributed. Notum glabrous. Basis capituli dorsally subrectangular, cornua short, porose areas rounded. Hypostome spatulate, dental formula 3/3. Genital aperture located at level of coxa II, U-shaped. Legs: coxa I with two distinct,

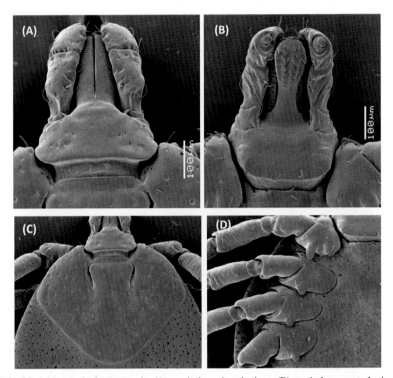

FIGURE 2.33 *A. hadanii* nymph: (A) capitulum dorsal view, (B) capitulum ventral view, (C) scutum, (D) coxae.

triangular blunt spurs, the external spur slightly longer than the internal spur; coxae II−III with a triangular short spur each; coxa IV with a triangular short spur, longer than spurs on coxae II−III; trochanters without spurs. Spiracular plates comma-shaped.

Nymph (Fig. 2.33): large body size, total length 1.9 mm (1.8−2.1), breadth 1.6 mm (1.4−1.7). Body outline oval. Chitinous tubercles at posterior body margin absent. Scutum with punctuations evenly distributed, larger laterally, smaller centrally; eyes slightly bulging, located on lateral scutal angles at the level of scutal mid-length; cervical grooves deep and short, ending as small shallow depressions at the level of the posterior margin of the eyes. Basis capituli dorsally rectangular, without cornua. Hypostome spatulate, dental formula 2/2. Legs: coxa I with two pointed spurs, the internal spur broader than the external, and the external longer than the internal spur; coxae II−IV with a small triangular spur each.

Taxonomic notes: the presence of carena is described for the males of *A. hadanii*, but in some specimens this morphological feature is almost imperceptible. See also Taxonomic notes of *A. coelebs* Neumann, 1899.

DNA sequences with relevance for tick identification and phylogeny: DNA sequences of genes of *A. hadanii* are available in the Gen Bank as follows: 16S rDNA (accession numbers KJ584370−KJ584372) and 18S rDNA (accession number KJ584368).

REFERENCES

1. Nava S, Mastropaolo M, Mangold AJ, Martins TF, Venzal JM, Guglielmone AA. *Amblyomma hadanii* n. sp. (Acari: Ixodidae), a tick from northwestern Argentina previously confused with *Amblyomma coelebs* Neumann, 1899. *Syst Parasitol* 2014; **88**:261−72.
2. Mastropaolo M, Tarragona EL, Silaghi C, Pfister K, Nava S. High prevalence of "*Candidatus* Rickettsia amblyommii" in *Amblyomma* ticks from a Spotted Fever Endemic Region in North Argentina. *Comp Immunol Microbiol Infect Dis* 2016; **46**:73−6.

Amblyomma incisum Neumann, 1906

Neumann, L.G. (1906) Notes sur les Ixodidés. IV. *Archives de Parasitologie*, 10, 195−219.

DISTRIBUTION

Argentina, Bolivia, Brazil, Paraguay, and Peru.[1] The specimens of *A. incisum* from Guyana and Venezuela reported by Jones et al.[2] were redetermined as *A. latepunctatum* by Labruna et al.[1] Records of *A. incisum* from Ecuador[3,4] and French Guiana[5] require confirmation.

Biogeographic distribution in the Southern Cone of America: Chaco and Parana Forest (Neotropical Region) (see Fig. 1 in the introduction).

HOSTS

Tapirs are the principal hosts for adults of *A. incisum*.[1,6] There are records of nymphs of this tick from artiodactyls and carnivorous (Table 2.9).

ECOLOGY

A. incisum has a 1-year life cycle. In a study performed in the Atlantic rain-forest in south−eastern Brazil, Szabó et al.[7] have found that larvae were more abundant in autumn and winter, nymphs in spring, while adults have the peak of abundance in summer.

TABLE 2.9 Hosts of Adults (A) and Nymphs (N) of *A. incisum*

MAMMALIA		*Puma concolor*	N
ARTIODACTYLA		**Procyonidae**	
Cervidae		*Nasua nasua*	N
Mazama bororo	N	**PERISSODACTYLA**	
M. gouazoubira	N	**Tapiridae**	
CARNIVORA		<u>*Tapirus terrestris*</u>	A
Canidae		**PRIMATES**	
C. thous	N	**Hominidae**	
Felidae		Human	AN
Panthera onca	N	**RODENTIA**	
		Caviidae	
		Hydrochoerus hydrochaeris	

Main host for adult ticks underlined.

SANITARY IMPORTANCE

Nymphs of *A. incisum* have been recorded parasitizing humans.[8,9] *Rickettsia monteiroi*, a species of unknown pathogenicity, has been isolated from *A. incisum* ticks.[10]

DIAGNOSIS

Male (Fig. 2.34): large body size, total length 4.5 mm (4.0−4.9), breadth 3.6 mm (3.2−4.0). Body outline oval, scapulae rounded; cervical grooves deep and short, comma-shaped; marginal groove absent. Eyes slightly bulging. Scutum ornate, postero-accessory and postero-median spots conspicuous; lateral spots distinct, second and third lateral spots usually conjoined; punctuations numerous, large and deep in lateral and posterior fields, fewer and shallower punctuations in antero-central field. Carena present, all ventral plates incised, except the plate corresponding to the central festoon. Basis capituli dorsally rectangular, with cornua. Hypostome spatulate, dental formula 3/3. Genital aperture located at level of coxa II, U-shaped. Legs: coxa I with two distinct, long, triangular sharp spurs, the internal spur wider than the external spur; coxae II−IV with a triangular short spur each; trochanters without spurs. Spiracular plates comma-shaped.

Female (Fig. 2.35): large body size, total length 5.9 mm (4.8−6.5), breadth 4.5 mm (3.9−4.9). Body outline oval. Scutum length 2.7 mm

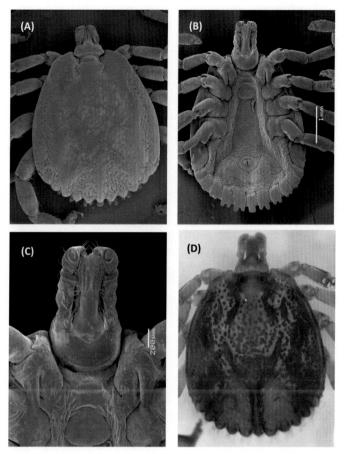

FIGURE 2.34   *A. incisum* male: (A) dorsal view, (B) ventral view, (C) capitulum ventral view, (D) scutum ornamentation.

(2.5−3.0), breadth 3.1 mm (2.7−3.3); scapulae rounded; cervical grooves short and deep, sigmoid in shape. Eyes slightly bulging. Chitinous tubercles at posterior body margin present. Scutum ornate, extensively pale yellowish, cervical spots narrow and divergent posteriorly; punctuations numerous, deep, larger in the antero-lateral fields. Notum glabrous. Basis capituli dorsally rectangular, cornua short, porose areas rounded. Hypostome spatulate, dental formula 4/4. Genital aperture located at level of coxa II, U-shaped. Legs: coxa I with two distinct, long, triangular sharp spurs, the internal spur wider than the external spur; coxae II−III each with a triangular, short spur, with a small salient ridge extending from the spur to the postero-internal angle of the coxa; coxa IV with a triangular, short spur, longer than spurs on coxae II−III; trochanters without spurs. Spiracular plates comma-shaped.

Nymph (Fig. 2.36): large body size, total length 2.0 mm (1.9−2.1), breadth 1.6 mm (1.5−1.7). Body outline oval. Chitinous tubercles at

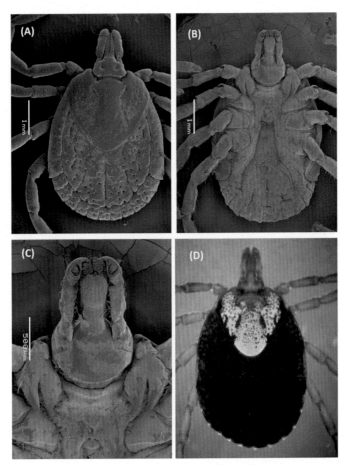

FIGURE 2.35 *A. incisum* female: (A) dorsal view, (B) ventral view, (C) capitulum ventral view, (D) scutum ornamentation.

posterior body margin present. Scutum with few medium and shallow punctuations evenly distributed, deeper laterally. Eyes slightly bulging, located on lateral scutal angles at the level of scutal mid-length. Cervical grooves short and deep at the anterior third, then diverging posteriorly as shallow depressions. Basis capituli dorsally rectangular, with cornua. Hypostome spatulate, dental formula 2/2. Legs: coxa I with two distinct, narrow spurs, the external longer than the internal; coxae II−IV with a small triangular spur.

Taxonomic notes: the presence of carena is described for the males of *A. incisum*, but in some specimens this morphological feature is almost imperceptible. *A. incisum* is morphologically and phylogenetically close to *A. latepunctatum* Tonelli-Rondelli, 1931, and *A. scalpturatum* Neumann, 1906,[1] and confused with each other due to their similarities. A morphological

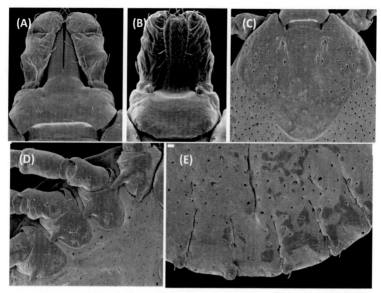

FIGURE 2.36   *A. incisum* nymph: (A) capitulum dorsal view, (B) capitulum ventral view, (C) scutum, (D) coxae, (E) chitinous tubercles at the posterior body margin. *Figures (A) to (E) are reproductions of figure 10:* A. incisum *nymph. Capitulum, dorsum; Capitulum, venter; Scutum; Coxae I−IV; Ventral festoons. From Martins TF, Labruna MB, Mangold AJ, Cafrune MM, Guglielmone AA, Nava S. Taxonomic key to nymphs of the genus* Amblyomma *(Acari: Ixodidae) in Argentina, with description and redescription of the nymphal stage of four* Amblyomma *species.* Ticks Tick-borne Dis *2014;5:753−70, with permission of Elsevier.*

and molecular (sequences of the internal transcribed spacer 2) comparison of *A. incisum* ticks from different populations has shown the existence of two groups, one represented by ticks from the Amazon region (the north group) and another represented by ticks from the southern part of South America (the south group).[1] Phenotypically, these two groups have differences in the density of the punctuations in the scutum, in the size of the spurs on coxae II−IV, and in the size of the cornua of males[1] being uncertain if these groups represent variation of *A. incisum* or different species. Labruna et al.[1] and most tick workers defined the eyes of the adults of *A. incisum* as flat, but Figs. 2.34 and 2.35 show slightly bulging eyes for this species being uncertain if this small difference is due to intraspecific morphological variation or technicalities.

DNA sequences with relevance for tick identification and phylogeny: DNA sequences of *A. incisum* are available in the Gen Bank as follows: 16S rDNA gene (accession numbers KM519939, FJ424405), and of the internal transcribed spacer 2 (accession numbers AY619577, AY619578, AY619576, KM519939).

REFERENCES

1. Labruna MB, Keirans JE, Camargo LMA, Ribeiro AF, Soares RM, Camargo EP. *Amblyomma latepunctatum*, a valid tick species (Acari: Ixodidae) long misidentified with both *Amblyomma incisum* and *Amblyomma scalpturatum*. *J Parasitol* 2005;**91**:527−41.
2. Jones EK, Clifford CM, Keirans JE, Kohls GM. The ticks of Venezuela (Acarina: Ixodoidea) with a key to the species of *Amblyomma* in the Western Hemisphere. *Brigham Young Univ Sci Bull Biol Ser* 1972;**17**(4):1−40.
3. Neumann LG. Notes sur les Ixodidés. IV. *Arch Parasitol* 1906;**10**:195−219.
4. Tonelli Rondelli M. Ixodoidea del Museo di Torino. *Boll Mus Zool Anat Comp R Univ Torino Ser III* 1931;**41**(6):1−10.
5. Floch H, Fauran P. Ixodidés de la Guyane et des Antilles Françaises. *Arch Inst Pasteur Guyane Inini* 1958;**446**:94.
6. Guglielmone AA, Robbins RG, Apanaskevich DA, Petney TN, Estrada-Peña A, Horak IG. *The hard ticks of the world (Acari: Ixodida: Ixodidae)*. London: Springer; 2014.
7. Szabó MPJ, Labruna MB, García MV, Pinter A, Castagnolli KC, Pacheco RP, et al. Ecological aspects of the free-living ticks (Acari: Ixodidae) on animal trails within Atlantic rainforest in South-eastern Brazil. *Ann Trop Med Parasitol* 2009;**103**:57−72.
8. Szabó MPJ, Labruna MB, Castagnolli KC, García MV, Pinter A, Veronez VA, et al. Ticks (Acari: Ixodidae) parasitizing humans in an Atlantic rainforest reserve of Southeastern Brazil with notes on host suitability. *Exp Appl Acarol* 2006;**39**:339−46.
9. Guglielmone AA, Beati L, Barros-Battesti DM, Labruna MB, Nava S, Venzal JM, et al. Ticks (Ixodidae) on humans in South America. *Exp Appl Acarol* 2006;**40**:83−100.
10. Pacheco R, Moraes-Filho J, Marcili A, Richtzenhain LJ, Szabó MPJ, Catroxo MHB, et al. *Rickettsia monteiroi* sp. nov. infecting the tick *Amblyomma incisum* in Brazil. *Appl Environ Microbiol* 2011;**77**:5207−11.

Amblyomma longirostre (Koch, 1844)

Haemalastor longirostris. Koch, C.L. (1844) Systematische Übersicht über die Ordnung der Zecken. *Archiv für Naturgeschichte*, 10, 217−239.

DISTRIBUTION

Argentina, Bolivia, Brazil, Colombia, Costa Rica, French Guiana, Honduras, Mexico, Panama, Paraguay, Peru, Trinidad and Tobago, Uruguay, and Venezuela.[1−4] Immature stages of *A. longirostre* have also been found on migratory birds reaching the Nearctic Region,[5,6] but this tick is not established in that region probably due to the lack of suitable environmental conditions or appropriate hosts for adults.[1,7]

Biogeographic distribution in the Southern Cone of America: Pampa and Parana Forest (Neotropical Region) (see Fig. 1 of the introduction).

HOSTS

Porcupines (Rodentia: Erethizontidae) and passeriform birds are the principal hosts for adults and immature stages of *A. longirostre*, respectively (Table 2.10).[1,2,4,7,8] Adults and nymphs of *A. longirostre* were also found on

TABLE 2.10 Hosts of Adults (A), Nymphs (N), and Larvae (L) of *A. longirostre*

MAMMALIA		*Basileuterus leucoblepharus*	N
ARTIODACTYLA		*B. rufifrons*	A
Cervidae		*Geothlypis aequinoctialis*	N
Blastoceros dichotomus	A	*Myiothlypis flaveola*	L
CARNIVORA		**Pipridae**	
Canidae		*Antilophia galeata*	NL
Cerdocyon thous	A	*Chiroxiphia caudata*	NL
Domestic dog	A	*C. linearis*	L
Felidae		*C. pareola*	L
Leopardus geoffroyi	N	*Dixiphia pipra*	NL
Mustelidae		*Ilicura militaris*	NL
Eira barbara	A	*Manacus candei*	NL
CHIROPTERA		*M. manacus*	NL
Phyllostomidae		*Neopelma pallescens*	NL
Artibeus literatus	N	*Pipra fasciicauda*	N
Platyrrhinchus umbratus	A	**Thamnophilidae**	
PERISSODACTYLA		*Cercomacra tyrannina*	L
Equidae		*Dysithamnus mentalis*	NL
Horse	A	*Euphonia hirundinacea*	L
PILOSA		*E. pectoralis*	N
Bradypodidae		*E. violacea*	N
Bradypus tridactylus	AN	*E. xanthogaster*	N
PRIMATES		*Epinecrophylla leucophthalma*	N
Hominidae		*Formicivora melanogaster*	N
Human	AN	*Hylophylax naevioides*	N

(Continued)

TABLE 2.10 (Continued)

RODENTIA		H. naevius	L
Erethizontidae		Hypocnemoides maculicauda	NL
Chaetomys subspinosus	A	Mackenziaena severa	N
Coendou bicolor	A	Myrmotherula longipennis	L
C. prehensilis	AN	Phlegopsis nigromaculata	L
C. rothschildi	AN	Pyriglena leuconota	NL
Sphiggurus insidiosus	A	P. leucoptera	NL
S. mexicanus	A	Taraba major	L
S. spinosus	AN	Thamnomanes caesius	L
S. vestitus	A	T. aethiops	N
S. villosus	AN	T. bridgesi	N
Sciuridae		T. caerulescens	ANL
Sciurus granatensis	N	T. pelzelni	NL
AVES		Willisornis poecilonotus	L
CAPRIMULGIFORMES		Thraupidae	
Caprimulgidae		Dacnis cayana	NL
Nyctidromus albicollis	AN	Eucometis penicillata	NL
Trochilidae		Habia rubica	NL
Amazilia tzacatl	N	Leptopogon amaurocephalus	L
Chlorostibon lucidus	N	Nemosia pileata	N
Eupetomena macroura	N	Pipraeidea melanonota	N
Phaethornis superciliosus	N	Rhampocelus carbo	N
Thalurania glaucopis	N	R. sanguinolentus	N
CORACIIFORMES		Tachyphonus coronatus	NL
Alcedinidae		T. cristatus	NL
Chloroceryle aenea	N	T. luctuosus	N
Momotidae		T. rufus	NL
Baryphthengus ruficapillus	NL	T. surinamus	L
CUCULIFORMES		Tangara cayana	N
Cuculidae		T. palmarum	N
Crotophaga ani	N	T. seledon	N

(Continued)

TABLE 2.10 (Continued)

COLUMBIFORMES		Tersinia viridis	N
Columbidae		Thraupis sayaca	L
Leptotila rufaxilla	N	Trichothraupis melanops	NL
L. verrauxi	N	**Troglodytidae**	
FALCONIFORMES		Henicorhina leucosticta	L
Accipitridae		Thryothorus atrogularis	L
Buteo platypterus	L	T. longirostris	N
Falconidae		T. modestus	L
Micrastur ruficollis	N	T. nigricapillus	NL
GALLIFORMES		T. rufalbus	L
Cracidae		Troglodytes musculus	N
Penelope obscura	A	**Turdidae**	
GRUIFORMES		Turdus albicollis	NL
Gruidae		T. amaurochalinus	NL
Laterallus albigularis	L	T. flavipes	N
PASSERIFORMES		T. fumigatus	N
Cardinalidae		T. grayi	N
Cyanocompsa brissonii	N	T. leucomelas	NL
C. cyanoides	L	T. nigriceps	N
Piranga flava	NL	T. nudigensis	N
Saltator atriceps	L	T. rufiventris	NL
S. maximus	N	T. subalaris	N
S. similis	NL	**Tyrannidae**	
Conopophagidae		Attila rufus	N
Conopophaga lineata	NL	Casiornis rufus	N
C. melanops	N	Cnemotriccus fuscatus	N
Corvidae		Cyclarhis guyanensis	N
Cyanocorax cristatellus	N	Elaenia cristata	AN
Cotingidae		E. flavogaster	N
Pachyramphus polychopterus	L	E. mesoleuca	N
Schiffornis turdina	L	E. parvirostris	N
S. virescens	L	Empidonax traillii	N

(Continued)

TABLE 2.10 (Continued)

Dendrocolaptidae		*Hemitriccus margaritaceiventer*	N
Campylorhamphus falcularius	NL	*H. minor*	L
Deconychura longicauda	L	*H. nidipendulus*	N
D. stictolaema	L	*Lathroticcus euleri*	N
Dendrocincla anabatina	N	*Leptopogon amaurocephalus*	NL
D. fuliginosa	L	*L. superciliaris*	L
D. merula	NL	*Megarynchus pitangua*	N
Dendrocolaptes certhia	N	*Mionectes macconelli*	L
D. hoffmansi	NL	*M. oleagineus*	NL
Glyphorynchus spirurus	NL	*M. rufiventris*	NL
Hylexetastes brigidai	NL	*Myiarchus ferox*	L
Lepidocolaptes angustirostris	N	*M. tuberculifer*	N
L. souleyetti	NL	*Myiobius barbatus*	L
L. spirurus	N	*Myiopagis viridicata*	L
L. squamatus	L	*Myiophobus fasciatus*	L
Sittasomus griseicapillus	NL	*Myiozetetes similis*	N
Xiphorhynchus elegans	NL	*Myodynastes maculatus*	N
X. fuscus	L	*Onychorhynchus coronatus*	L
X. guttatus	NL	*Phylloscartes ventralis*	N
X. lachrymosus	N	*Platyrinchus cancrominus*	L
Emberizidae		*P. leucoryphus*	L
Haplospiza unicolor	N	*P. mystaceus*	NL
O. angolensis	N	*Poecilotriccus latirostris*	L
Sporophila caerulescens	N	*Tolmomyias poliocephalus*	L
S. corvina	N	*T. sulphurescens*	NL
S. leucoptera	N	*Tyrannus melancholicus*	N
S. lineola	N	*T. savana*	N
S. nigricollis	N	*Xolmis cinereus*	N
Furnariidae		*X. velatus*	N

(Continued)

TABLE 2.10 (Continued)

Anabacerthia amaurotis	N	**Vireonidae**	
Anabazenops fuscus	NL	Vireo chivi	L
Automolus leucophtalmus	NL	V. griseus	N
A. ochrolaemus	L	V. olivaceus	N
Certhiaxis cinnamomeus	N	**PICIFORMES**	
Philydor atricapillus	N	**Bucconidae**	
P. rufus	NL	Malacoptila striata	NL
Synallaxis ruficapilla	NL	**Galbulidae**	
S. spixi	A	Galbula ruficauda	L
Syndactyla rufosuperciliata	N	**Picidae**	
Xenops minutus	L	Celeus flavescens	N
Icteridae		**Ramphastidae**	
Cacicus cela	N	Pteroglossus bitorquatus	N
C. haemorrohous	N	Ramphastos dicolorus	N
Icterus icterus	N	R. vitellinus	N
Parulidae		**STRIGIFORMES**	
Basileuterus culicivorus	NL	**Strigidae**	
B. flaveolus	N	Pulsatrix koeniswaldiana	N
B. hypoleucus	NL		

mammals belonging to different orders, and there are records of adults, nymphs, and larvae on non-passeriform birds (Table 2.10).

ECOLOGY

Besides the data on host association and geographical distribution, there is little information on the ecology of this tick. Labruna et al.[7] hypothesized that the free-living stages of A. longirostre complete their development in the arboreal canopy, due to the fact that its main hosts have arboreal habits, but additional evidence is needed to confirm this hypothesis.

SANITARY IMPORTANCE

There are a few records of A. longirostre parasitizing humans; these records include some adult ticks and just one nymph[1,9] and records on domestic animals are exceptional. The infection with "Candidatus Rickettsia *amblyommii,*" an organism with unknown pathogenicity, in *A. longirostre* appears to be a ubiquitous phenomenon because it was reported in several countries of the Neotropical Region.[3,4,8,10,11]

DIAGNOSIS

Male (Fig. 2.37): large body size, total length 5.6 mm (5.5−5.8), breadth 3.7 mm (3.5−3.8). Body outline oval elongate, scapulae rounded; cervical grooves short, comma-shaped; marginal groove incomplete. Eyes flat. Scutum ornate, with irregular pale spots extending from the scapular fields to the posterior third of the scutum, and small pale spots on the posterior field of the scutum; punctuations numerous, small, uniformly distributed.

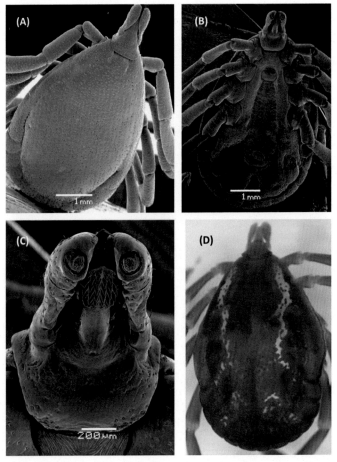

FIGURE 2.37 *A. longirostre* male: (A) dorsal view, (B) ventral view, (C) capitulum ventral view, (D) scutum ornamentation.

Carena absent. Presence of five ventral plates in the posterior field of the ventral surface. Basis capituli dorsally subtriangular, without cornua. Hypostome spatulate, dental formula 3/3. Genital aperture located at level of coxa II, U-shaped. Legs: coxa I with two distinct, short triangular spurs, the external spur longer than the internal spur; coxae II−IV with a triangular short spur each; trochanters without spurs. Spiracular plates comma-shaped.

Female (Fig. 2.38): large body size, total length 5.7 mm (5.1−6.1), breadth 3.7 mm (3.6−4.4). Body outline oval elongated. Scutum length 3.3 mm (2.8−3.5), breadth 2.6 mm (2.3−2.6); scapulae rounded; cervical grooves cervical grooves sigmoid, deeper anteriorly. Eyes flat. Chitinous tubercles at posterior body margin absent. Scutum ornate, with an irregular

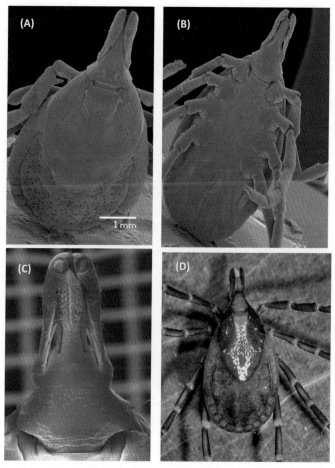

FIGURE 2.38 *A. longirostre* female: (A) dorsal view, (B) ventral view, (C) capitulum ventral view, (D) scutum ornamentation.

longitudinal pale patch in the median field; punctuations numerous, small, uniformly distributed. Notum glabrous. Basis capituli dorsally triangular, without cornua, porose areas oval. Hypostome pointed, lanceolate, dental formula 3/3. Genital aperture located at level of coxa II, U-shaped. Legs: coxa I with two distinct, short triangular spurs, the external spur longer than the internal spur; coxae II–IV with a triangular short spur each; trochanters without spurs. Spiracular plates comma-shaped.

Nymph (Fig. 2.39): large body size, total length 1.9 mm (1.5–2.0), breadth 1.4 mm (1.2–1.6). Body outline oval. Chitinous tubercles at posterior body margin absent. Scutum with few punctuations, larger and deeper laterally. Eyes flat, located on lateral scutal angles at the level of scutal

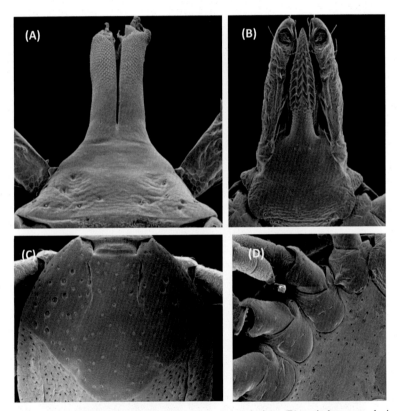

FIGURE 2.39   *A. longirostre* nymph: (A) capitulum dorsal view, (B) capitulum ventral view, (C) scutum, (D) coxae. *Figures (A) to (D) are reproductions of figure 7: A. longirostre nymph. Capitulum, dorsum; Capitulum, venter; Scutum; Coxae I–IV. From Martins TF, Labruna MB, Mangold AJ, Cafrune MM, Guglielmone AA, Nava S. Taxonomic key to nymphs of the genus* Amblyomma *(Acari: Ixodidae) in Argentina, with description and redescription of the nymphal stage of four* Amblyomma *species. Ticks Tick-borne Dis 2014;5:753–70, with permission of Elsevier.*

mid-length. Cervical grooves deep anteriorly continuing as a shallow depression. Basis capituli dorsally triangular, without cornua; pseudoauricula present; ventral cornua present. Hypostome pointed, lanceolate, dental formula 2/2. Legs: coxa I with two short spurs, the external broader and longer than the internal; coxae II–IV each with a ridge-like spur.

Taxonomic notes: *A. longirostre*, *A. geayi* Neumann, 1899, and *A. parkeri* Fonseca and Aragão, 1952, constitute a phylogenetically closely related group.[12] These three tick species share a combination of some morphological characters unique among Neotropical ticks: presence of five ventral plates in the posterior field of the ventral surface of males, basis capituli dorsally triangular in females, and coxa I with two short triangular spurs, unequal in size.

DNA sequences with relevance for tick identification and phylogeny: DNA sequences of *A. longirostre* genes are available in the Gen Bank as follows: mitochondrial 16S rDNA (accession numbers KP762568–KP762573, KP835784–KP835790, KM262205–KM262207, KF702347, JN800424–JN800430, EU805563–EU805565, AY342264–AY342266), mitochondrial 12S rDNA (accession numbers KT386304, KF702337, JX192899–JX192920, EU805557–EU805562), and of the nuclear genome (the fragment 5.8S ribosomal RNA gene-internal transcribed spacer 2-28S ribosomal RNA gene) (accession number AY887120).

REFERENCES

1. Guglielmone AA, Estrada-Peña A, Keirans JE, Robbins RG. Ticks (Acari: Ixodida) of the neotropical zoogeographic region. Special publication of the integrated consortium on ticks and tick-borne diseases-2. Houten (The Netherlands): Atalanta; 2003.

2. Nava S, Velazco PM, Guglielmone AA. First record of *Amblyomma longirostre* (Koch, 1844) (Acari: Ixodidae) from Peru, with a review of this tick's host relationships. *Syst Appl Acarol* 2010;**15**:21−30.

3. Novakova M, Literak I, Chevez L, Martins TF, Ogrzewalska M, Labruna MB. Rickettsial infections in ticks from reptiles, birds and humans in Honduras. *Ticks Tick-borne Dis* 2015;**6**:737−42.

4. Ogrzewalska M, Literak I, Capek M, Sychra O, Álvarez Calderón V, Rodríguez BC, et al. Bacteria of the genus *Rickettsia* in ticks (Acari: Ixodidae) collected from birds in Costa Rica. *Ticks Tick-borne Dis* 2015;**6**:478−82.

5. Fairchild GB, Kohls GM, Tipton VJ. The ticks of Panama (Acarina: Ixodoidea). In: Wenzel WR, Tipton VJ, editors. *Ectoparasites of Panama*. Chicago (IL): Field Museum of Natural History; 1966. pp. 167−219.

6. Scott JD, Fernando K, Banerjee SN, Durden L, Byrne SK, Banerjee M, et al. Birds disperse ixodid (Acari: Ixodidae) and *Borrelia burgdorferi* infected ticks in Canada. *J Med Entomol* 2001;**38**:493−500.

7. Labruna MB, Sanfilippo LE, Demetrio C, Menezes AC, Pinter A, Guglielmone AA, et al. Ticks collected on birds in the state of São Paulo, Brazil. *Exp Appl Acarol* 2007;**43**:147−60.

8. Ogrzewalska M, Literak I, Martins TF, Labruna MB. Rickettsial infections in ticks from wild birds in Paraguay. *Ticks Tick-borne Dis* 2014;**5**:83−9.

9. Rodríguez-Peraza JL, Forlano MD, Meléndez RD, Carrero AA, Sánchez H, Sira E, et al. Identificación morfológica de *Amblyomma longirostre* (Koch, 1844) (Acari: Ixodidae) en Venezuela. *Gac Cienc Vet* 2014;**19**(2):40−5.

10. Labruna MB, Mcbride JW, Bouyer DH, Camargo LMA, Camargo EP, Walker DH. Molecular evidence for a spotted fever group *Rickettsia* species in the tick *Amblyomma longirostre* in Brazil. *J Med Entomol* 2004;**41**:533−7.

11. Ogrzewalska M, Pacheco R, Uezu A, Ferreira F, Labruna MB. Ticks (Acari: Ixodidae) infesting wild birds in an Atlantic Forest area in the state of São Paulo, Brazil, with isolation of *Rickettsia* from the tick *Amblyomma longirostre*. *J Med Entomol* 2008;**45**:770−4.

12. Labruna MB, Onofrio VC, Beati L, Arzua M, Bertola PB, Ribeiro AF, et al. Redescription of the female, description of the male, and several new records of *Amblyomma parkeri* (Acari: Ixodidae), a South American tick species. *Exp Appl Acarol* 2009;**49**:243−60.

Amblyomma neumanni
Ribaga, 1902

Ribaga, C. (1902) Acari sudamericani. *Zoologischer Anzeiger*, 25, 502−508.

DISTRIBUTION

Argentina and Colombia.[1] The presence of this tick in Uruguay has been mentioned by Vogelsang,[2] but this record was considered doubtful by Venzal et al.[3]

Biogeographic distribution in the Southern Cone of America: Chaco and Yungas (Neotropical Region) (see Fig. 1 in the introduction).

HOSTS

A. neumanni is a tick species with low host specificity. All parasitic stages of *A. neumanni* feed on large wild and domestic mammals of different orders[4,5] (Table 2.11).

ECOLOGY

The life cycle of *A. neumanni* under natural conditions has been studied by Guglielmone et al.[6] and Nava et al.[7] *A. neumanni* presents a complex life cycle which is regulated by diapause, and the seasonal distribution of all stages is characterized by the absence of parasitic activity during summer and differences in abundance based on the latitude. Larvae of *A. neumanni* are active from late summer to early winter, with the peak of abundance in autumn, nymphs from late autumn to early spring, with the peak of abundance in winter, and adults are found from autumn to late spring. The seasonal distribution pattern of adults may show a bimodal curve, with a peak in autumn and other during early and middle spring.[7] The females of *A. neumanni* that feed, copulated and engorged in autumn overwinter in

TABLE 2.11 Hosts of Adults (A), Nymphs (N), and Larvae (L) of *A. neumanni*

MAMMALIA		CARNIVORA	
ARTIODACTYLA		**Canidae**	
Bovidae		*Cerdocyon thous*	A
Cattle	ANL	Domestic dog	AN
Goat	AN	*Lycalopex gymnocercus*	ANL
Sheep	AN	**PERISSODACTYLA**	
Cervidae		**Equidae**	
Mazama americana	ANL	Donkey	AN
M. gouazoubira	ANL	Horse	ANL
Suidae		Mule	AN
Domestic pig	AN	**PILOSA**	
Tayassuidae		**Myrmecophagidae**	
Catagonus wagneri	A	*Myrmecophaga tridactyla*	AN
Pecari tajacu	AN	**PRIMATES**	
		Hominidae	
		Human	ANL

morphogenetic diapause, and they will lay eggs in spring, simultaneously with the females that feed and copulate in this season. Consequently, oviposition is synchronized to coincide with the onset of rainy season in the spring. It is speculated that the females that conform the peak during autumn are formed by specimens that have been unable to find a host in the previous spring and undergo behavioral diapause in the summer months. Even though *A. neumanni* has a wide range of hosts, it is important to consider that this species has the ability to survive with just the presence of cattle in the environment, since all their parasitic stages are adapted to feed successfully on this host.[7,8] *A. neumanni* shows a strong preference for hilly areas with an altitude range of 500−1600 m in the Chaco Biogeographic Province.

SANITARY IMPORTANCE

A. neumanni is the most common tick species biting humans in Argentina.[9−11] So far *Rickettsia bellii* and "*Candidatus Rickettsia amblyommii*" were detected in this tick,[12,13] but these *Rickettsia* species are currently considered of unknown pathogenicity. *A. neumanni* is a common parasite of cattle along its geographic distribution,[7,8,14] and Gaido et al.[15] have demonstrated the potential capacity of this tick to transmit *Anaplasma marginale*, an intraerythrocytic bacteria causing a severe cattle disease.

DIAGNOSIS

Male (Fig. 2.40): small body size, total length 2.7 mm (2.7−3.5), breadth 2.1 mm (1.8−2.3). Body outline oval elongated, narrower in the anterior part; scapulae rounded; cervical grooves deep and short, comma-shaped; marginal groove incomplete. Eyes flat. Scutum ornate, with reddish brown

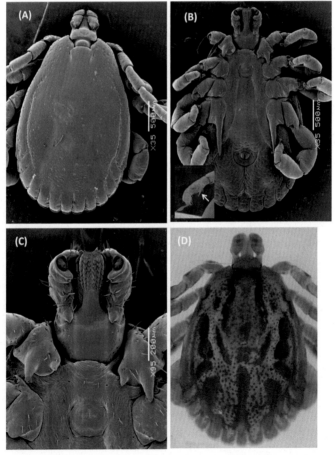

FIGURE 2.40 *A. neumanni* male: (A) dorsal view, (B) ventral view and image of the spine present on the tibia of legs II−IV in both male and female, (C) capitulum ventral view, (D) scutum ornamentation.

spots on a pale ground; postero-median spot extending to the level of spiracular plates; postero-accessory spots short; lateral spots small, distinct or conjoined; cervical spots narrow and divergent posteriorly; punctuations small, numerous, larger in the anterior and posterior margins of the scutum. Carena absent. Basis capituli dorsally subrectangular, cornua long. Hypostome spatulate, dental formula 3/3. Genital aperture located at level of coxa II, U-shaped. Legs: coxa I with two distinct, triangular spurs, the external longer than the internal; coxae II—III with a triangular, short blunt spur each; coxa IV with a narrow, sharp long spur, reaching the anus level; trochanters without spurs; presence of one spine on the tibia of legs II—IV. Spiracular plates comma-shaped.

Female (Fig. 2.41): medium body size, total length 3.2 mm (3.2—3.4), breadth 2.2 mm (2.1—2.3). Body outline oval. Scutum length 1.6 mm

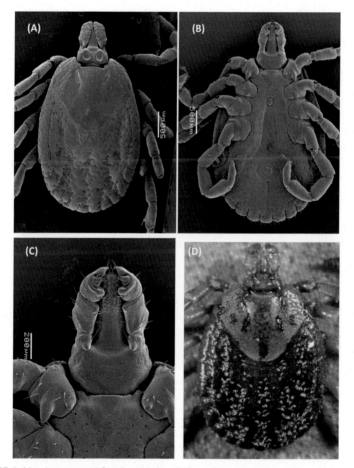

FIGURE 2.41   *A. neumanni* female: (A) dorsal view, (B) ventral view, (C) capitulum ventral view, (D) scutum ornamentation.

(1.5−1.7), breadth 1.7 mm (1.6−1.8); scapulae rounded, cervical grooves deep anteriorly, shallow posteriorly, sigmoid in shape. Eyes flat. Chitinous tubercles at posterior body margin absent. Scutum ornate, extensively pale yellowish, cervical spots narrow and divergent posteriorly, central area short and narrow; punctuations numerous, small, larger in the antero-lateral fields. Notum with short, coarse setae. Basis capituli dorsally rectangular, cornua absent, porose areas oval. Hypostome spatulate, dental formula 3/3. Genital aperture located at level of coxa II, U-shaped. Legs: coxa I with two distinct, triangular short spurs, the external longer than the internal; coxae II−IV each with a triangular, short blunt spur; trochanters without spurs; presence of one spine on the tibia of legs II−IV. Spiracular plates comma-shaped.

Nymph (Fig. 2.42): medium body size, total length 1.5 mm (1.4−1.6), breadth 1.1 mm (1.0−1.2). Body outline oval. Chitinous tubercles at posterior body margin absent. Scutum with punctuations evenly distributed, larger laterally, smaller centrally. Eyes bulging, located on lateral scutal angles at

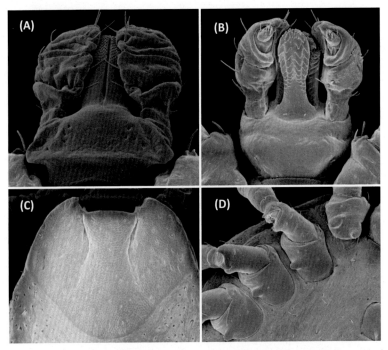

FIGURE 2.42 *A. neumanni* nymph: (A) capitulum dorsal view, (B) capitulum ventral view, (C) scutum, (D) coxae. *Figures (A) to (D) are reproductions of figure 17:* A. neumanni *nymph. Capitulum, dorsum; Capitulum, venter; Scutum; Coxae I−IV. From Martins TF, Labruna MB, Mangold AJ, Cafrune MM, Guglielmone AA, Nava S. Taxonomic key to nymphs of the genus* Amblyomma *(Acari: Ixodidae) in Argentina, with description and redescription of the nymphal stage of four* Amblyomma *species. Ticks Tick-borne Dis 2014;5:753−70, with permission of Elsevier.*

the level of scutal mid-length. Cervical grooves long and deep reaching the scutal posterior third. Basis capituli dorsally rectangular, without cornua. Hypostome spatulate, dental formula 2/2. Legs: coxa I with two short, triangular spurs, the external longer than internal; coxae II–IV with a small triangular spur each.

Taxonomic notes: *A. neumanni* has been included in the subgenus *Anastosiella* by Santos Dias[16] with *A. maculatum* Koch, 1844, *A. tigrinum* Koch, 1844, *A. triste* Koch, 1844, *A. parvitarsum* Neumann, 1901, *A. pecarium* Dunn, 1933 and *A. brasiliense* Aragão, 1908, while Camicas et al.[17] include the following species within the subgenus *Anastosiella*: *A. neumanni*, *A. maculatum*, *A. tigrinum*, *A. triste*, *A. parvitarsum*, *A. aureolatum* (Pallas, 1772), and *A. ovale* (Koch, 1884). However, Estrada-Peña et al.[1] have shown that these species do not constitute a natural group. The statement of Estrada-Peña et al.[1] is supported by a phylogenetic analysis constructed with mitochondrial DNA sequences (Fig. 2.7), which shows that *A. neumanni* is not phylogenetically related to *A. maculatum*, *A. tigrinum*, and *A. triste*. Jones et al.[18] incorrectly described *A. neumanni* without the spine on the tibia of legs II–IV and with marginal groove complete. The description of *A. neumanni* in Voltzit[19] corresponds to *A. parvitarsum* as stated in Nava et al.[7]

DNA sequences with relevance for tick identification and phylogeny: the following DNA sequences of *A. neumanni* genes are available in the Gen Bank: 16S rDNA (accession numbers AY498560; FJ965564–FJ965593), 12S rDNA (accession number AY342272), and 18S rDNA (accession number FJ464424).

REFERENCES

1. Estrada-Peña A, Venzal JM, Mangold AJ, Cafrune MM, Guglielmone AA. The *Amblyomma maculatum* Koch, 1844 (Acari: Ixodidae: Amblyomminae) tick group: diagnostic characters, description of the larva of *A. parvitarsum* Neumann, 1901, 16S rDNA sequences, distribution and hosts. *Syst Parasitol* 2005;**60**:99–112.
2. Vogelsang EG. Notas parasitológicas. II. *Rev Med Vet (Montevideo)* 1927;**10**:474–5.
3. Venzal JM, Castro O, Cabrera PA, De Souza CG, Guglielmone AA. Las garrapatas de Uruguay: especies, hospedadores, distribución e importancia sanitaria. *Veterinaria (Montevideo)* 2003;**38**:17–28.
4. Guglielmone AA, Nava S. Las garrapatas argentinas del género *Amblyomma* (Acari: Ixodidae): distribución y hospedadores. *Rev Invest Agropec* 2006;**35**(3):135–55.
5. Nava S, Guglielmone AA. A meta-analysis of host specificity in Neotropical hard ticks (Acari: Ixodidae). *Bull Entomol Res* 2013;**103**:216–24.
6. Guglielmone AA, Mangold AJ, Aguirre DH, Gaido AB. Ecological aspects of four species of ticks found on cattle in Salta, northwest Argentina. *Vet Parasitol* 1990;**35**:93–101.
7. Nava S, Estrada-Peña A, Mangold AJ, Guglielmone AA. Ecology of *Amblyomma neumanni* (Acari: Ixodidae). *Acta Trop* 2009;**111**:226–36.
8. Nava S, Mangold AJ, Guglielmone AA. The natural hosts for larvae and nymphs of *Amblyomma neumanni* and *Amblyomma parvum* (Acari: Ixodidae). *Exp Appl Acarol* 2006;**40**:123–31.

9. Guglielmone AA, Mangold AJ, Viñabal AE. Ticks (Ixodidae) parasitizing humans in four provinces of northwestern Argentina. *Ann Trop Med Parasitol* 1991;**85**:539–42.

10. Guglielmone AA, Beati L, Barros-Battesti DM, Labruna MB, Nava S, Venzal JM, et al. Ticks (Ixodidae) on humans in South America. *Exp Appl Acarol* 2006;**40**:83–100.

11. Nava S, Caparrós JA, Mangold AJ, Guglielmone AA. Ticks (Acari: Ixodida: Argasidae, Ixodidae) infesting humans in northwestern Córdoba Province, Argentina. *Medicina (Buenos Aires)* 2006;**66**:225–8.

12. Labruna MB, Pacheco RC, Nava S, Brandao PE, Richtzenhain LJ, Guglielmone AA. Infection by *Rickettsia bellii* and *"Rickettsia amblyommii"* in *Amblyomma neumanni* ticks from Argentina. *Microb Ecol* 2007;**54**:126–33.

13. Saracho Bottero MN, Tarragona EL, Nava S. Spotted fever group rickettsiae in *Amblyomma* ticks likely to infest humans in rural areas from northwestern Argentina. *Medicina (Buenos Aires)* 2015;**74**:391–5.

14. Guglielmone AA, Hadani A. *Amblyomma* ticks found on cattle in the Northwest of Argentina. *Ann Parasitol Hum Comp* 1982;**57**:91–7.

15. Gaido AG, Viñabal AE, Aguirre DH, Echaide S, Guglielmone AA. Transmission of *Anaplasma marginale* by the three-host tick *Amblyomma neumanni* under laboratory conditions. *Folia Parasitol* 1995;**42**:72.

16. Santos Dias JAT. Nova contribuição para o estudo da sistemática do género *Amblyomma* Koch, 1844 (Acarina-Ixodoidea). *García de Orta Sér Zool* 1993;**19**:11–19.

17. Camicas JL, Hervy JP, Adam F, Morel PC. *Les tiques du monde. Nomenclature, stades décrits, hôtes, répartition (Acarida, Ixodida).* Paris: ORSTOM; 1998.

18. Jones EK, Clifford CM, Keirans JE, Kohls GM. The ticks of Venezuela (Acarina: Ixodoidea) with a key to the species of *Amblyomma* in the Western Hemisphere. *Brigham Young Univ Sci Bull Biol Ser* 1972;**17**(4):1–40.

19. Voltzit OV. A review of Neotropical *Amblyomma* species (Acari: Ixodidae). *Acarina* 2007;**15**:3–134.

Amblyomma nodosum Neumann, 1899

Neumann, L.G. (1899) Révision de la famille des ixodidés (3e mémoire). *Mémoires de la Société Zoologique de France*, 12, 107–294.

DISTRIBUTION

Argentina, Belize, Bolivia, Brazil, Colombia, Costa Rica, Guatemala, Honduras, Mexico, Nicaragua, Panama, Paraguay, Trinidad and Tobago, and Venezuela.[1] Keirans and Durden[2] have presented records of *A. nodosum* in the Nearctic Region, but there is no evidence of established populations of this tick in that region.[3]

Biogeographic distribution in the Southern Cone of America: Chaco (Neotropical Region) (see Fig. 1 in the introduction).

HOSTS

Anteaters (Pilosa: Myrmecophagidae) and passerine birds are the principal hosts for adults and immature stages of *A. nodosum*, respectively.[3–13] The host range for adults of *A. nodosum* also includes species belonging to different orders of mammals, and immature stages were found in other three orders of birds besides Passeriformes (Table 2.12).

TABLE 2.12 Hosts of Adults (A), Nymphs (N), and Larvae (L) of *A. nodosum*

MAMMALIA			*Automolus leucophtalmus*		N
CARNIVORA			*Cranioleuca vulpine*		N
Canidae			*Xenops minutus*		N
Domestic dog	A		**Parulidae**		
CINGULATA			*Basileutetrus culicivorus*		NL
Dasypodidae			*B. flaveolus*		N
Euphractus sexcinctus	A		**Pipridae**		
PILOSA			*Chiroxiphia caudata*		NL
Bradypodidae			*Dixiphia pipra*		N
Bradypus sp.	A		*Manacus manacus*		N
Myrmecophagidae			*Pipra fasciicauda*		N
Myrmecophaga tridactyla	A		**Thamnophilidae**		
Tamandua mexicana	A		*Drymophila squamata*		N
T. tetradactyla	A		*Dysithamnus mentalis*		NL
RODENTIA			*Formicivora grisea*		N
Erethizontidae			*F. rufus*		N
Coendu prehenselis	A		*Herpsilochmus atricapillus*		N
AVES			*Hypocnemoides maculicauda*		N
CAPRIMULGIFORMES			*Taraba major*		N
Caprimulgidae			*Thamnophilus caerulescens*		N
Eupetomena macroura	N		*T. torquatus*		N
CORACIIFORMES			*T. doliatus*		N
Momotidae			*T. pelzeni*		NL

(Continued)

TABLE 2.12 (Continued)

Baryphthengus ruficapillus	N	**Thraupidae**	
COLUMBIFORMES		*Ramphocelus carbo*	N
Columbidae		*Tachyphonus cristatus*	N
Leptotila verrauxi	N	*T. luctuosus*	N
PASSERIFORMES		*T. phoenicius*	N
Cardinalidae		*T. rufus*	N
Cyanacompsa cyanoides	N	*Tricothraupis melanops*	NL
Saltator maximus	N	**Turdidae**	
S. similis	N	*Turdus amaurochalinus*	N
Conopophagidae		*T. leucomelas*	NL
Conopophaga lineata	NL	**Tyrannidae**	
C. melanops	N	*Casiornis rufus*	N
Cotingidae		*Cnemotriccus fuscatus*	N
Pachyramphus polychopterus	N	*Elaenaia chiriquensis*	N
Dendrocolaptidae		*E. cristata*	N
Dendrocolaptes platyrostris	N	*E. obscura*	N
Glyphorynchus spirurus	N	*E. mesoleuca*	N
Lepidocolaptes squamatus	N	*Hemitriccus margaritaceiventer*	N
Emberizidae		*Leptopogon amaurocephalus*	N
Arremon flavirostris	L	*Myiarchus tyranullus*	N
Coryphospingus cucullatus	N	*Platyrinchus mystaceus*	NL
Sporophila caerulescens	N	*Poecilotriccus latirostris*	N
Furnariidae		*Tolmomyas sulphurescens*	N
Anabazenops fuscus	NL		

Main hosts for adult ticks underlined.

ECOLOGY

Ecological information about A. *nodosum* refers just to distribution and host association.

SANITARY IMPORTANCE

A. *nodosum* ticks were found to be infected with *Rickettsia bellii* and *Rickettsia parkeri* strain NOD,[14] but there are no records of this tick on humans. With the exception of one report of an adult on a dog,[15] *A. nodosum* was never found on domestic animals.

DIAGNOSIS

Male (Fig. 2.43): body size medium, total length 3.9 mm (3.7−4.1), breadth 3.0 mm (2.9−3.2). Body outline oval, scapulae rounded; cervical grooves deep and short, comma-shaped; marginal groove absent. Eyes flat. Scutum ornate, with irregular pale spots J-shaped in the antero-lateral fields, postero-accessory and postero-median spots narrow, lateral spots conspicuous and small; punctuations numerous, uniformly distributed. Carena absent. Basis capituli dorsally rectangular, cornua long, article II of palps with a

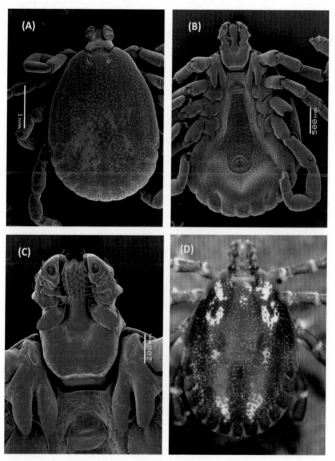

FIGURE 2.43 *A. nodosum* male: (A) dorsal view, (B) ventral view, (C) capitulum ventral view, (D) scutum ornamentation.

postero-dorsal projection. Hypostome spatulate, dental formula 3/3. Genital aperture located at level of coxa II, U-shaped. Legs: coxa I with two distinct, long, triangular blunt spurs, subequal in size; coxae II−IV with one triangular short spur each; trochanters without spurs. Spiracular plates comma-shaped.

Female (Fig. 2.44): large body size, total length 5.5 mm (5.4−5.6), breadth 3.5 mm (3.4−3.6). Body outline oval. Scutum length 2.5 mm (2.4−2.6), breadth 2.6 mm (2.5−2.7); scapulae rounded, cervical grooves deep and short, comma-shaped. Eyes flat. Chitinous tubercles at posterior body margin absent. Scutum ornate, with a Y-shaped pale spot in the antero-lateral field, and a small and irregular pale spot in the posterior field; punctuations numerous, uniformly distributed. Notum with small setae, barely perceptible. Basis capituli dorsally subtriangular, cornua short, porose areas oval. Hypostome spatulate, dental formula 3/3. Genital aperture located at level of coxa II, V-shaped. Legs: coxa I with two distinct, triangular blunt

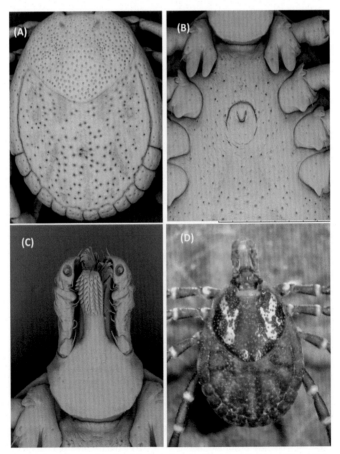

FIGURE 2.44 *A. nodosum* female: (A) dorsal view, (B) ventral view, (C) capitulum ventral view, (D) scutum ornamentation.

spurs, equal in size; coxae II–IV with one triangular short spur each; trochanters without spurs. Spiracular plates comma-shaped.

Nymph (Fig. 2.45): medium body size, total length 1.3 mm (1.2–1.4), breadth 1.1 mm (1.0–1.2). Body outline oval. Chitinous tubercles at posterior body margin absent. Scutum with surface rugose; few medium punctuations, more numerous on lateral fields. Eyes slightly bulging, located on lateral scutal angles at the level of scutal mid-length. Cervical grooves long, reaching the scutal posterior third, deeper anteriorly. Basis capituli dorsally subtriangular, without cornua. Hypostome spatulate, dental formula 2/2. Legs: coxa I with two triangular spurs, the external pointed and longer than the internal; coxae II–IV with one small triangular spur each.

Taxonomic notes: see Taxonomic notes of *A. calcaratum* Neumann, 1899.

DNA sequences with relevance for tick identification and phylogeny: the following DNA sequences of *A. nodosum* are available in the

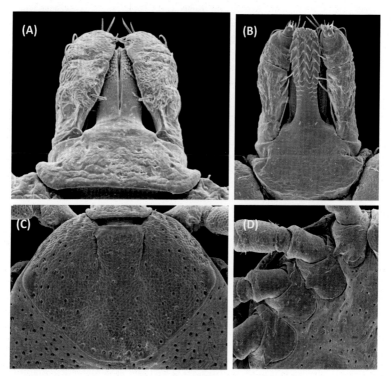

FIGURE 2.45   *A. nodosum* nymph: (A) capitulum dorsal view, (B) capitulum ventral view, (C) scutum, (D) coxae. *Figures (A) to (D) are reproductions of figure 19:* A. nodosum *nymph. Capitulum, dorsum; Capitulum, venter; Scutum; Coxae I–IV. From Martins TF, Labruna MB, Mangold AJ, Cafrune MM, Guglielmone AA, Nava S. Taxonomic key to nymphs of the genus* Amblyomma *(Acari: Ixodidae) in Argentina, with description and redescription of the nymphal stage of four* Amblyomma *species. Ticks Tick-borne Dis 2014;5:753–70, with permission of Elsevier.*

Gen Bank: mitochondrial genes 12S rrna (accession numbers AY225321; KT386303; JX192882—JX192898) and 16S rrna (accession numbers KP686064, KP835782, KP835783, FJ424402, FJ424403, KM262201), cytochrome oxidase I (accession numbers KF200138, KF200131), and sequences of the nuclear marker 28S rrna (accession number AY225323).

REFERENCES

1. Guglielmone AA, Estrada-Peña A, Keirans JE, Robbins RG. Ticks (Acari: Ixodida) of the neotropical zoogeographic region. Special publication of the integrated consortium on ticks and tick-borne diseases-2. Houten (The Netherlands): Atalanta; 2003.

2. Keirans JE, Durden LA. Invasion: exotic ticks (Acari: Argasidae, Ixodidae) imported into the United States. A review and new records. *J Med Entomol* 2001;**38**:850—61.

3. Guglielmone AA, Robbins RG, Apanaskevich DA, Petney TN, Estrada-Peña A, Horak IG. *The hard ticks of the world (Acari: Ixodida: Ixodidae)*. London: Springer; 2014.

4. Jones EK, Clifford CM, Keirans JE, Kohls GM. The ticks of Venezuela (Acarina: Ixodoidea) with a key to the species of *Amblyomma* in the Western Hemisphere. *Brigham Young Univ Sci Bull Biol Ser* 1972;**17**(4):1—40.

5. Labruna MB, Sanfilippo LE, Demetrio C, Menezes AC, Pinter A, Guglielmone AA, et al. Ticks collected on birds in the state of São Paulo, Brazil. *Exp Appl Acarol* 2007;**43**:147—60.

6. Ogrzewalska M, Pacheco R, Uezu A, Richtzenhain LJ, Ferreira F, Labruna MB. Ticks (Acari: Ixodidae) infesting birds in an Atlantic rain forest region of Brazil. *J Med Entomol* 2009;**46**:1225—9.

7. Ogrzewalska M, Saraiva DG, Moraes-Filho J, Martins TF, Costa FB, Pinter A, et al. Epidemiology of Brazilian spotted fever in the Atlantic Forest, state of São Paulo, Brazil. *Parasitology* 2012;**139**:1283—300.

8. Tolesano-Pascoli GV, Torga K, Franchin AG, Ogrewalska M, Gerardi M, Olegário MMM, et al. Ticks on birds in a forest fragment of Brazilian cerrado (savanna) in the municipality of Uberlândia, State of Minas Gerais, Brazil. *Rev Bras Parasitol Vet* 2010;**19**:244—8.

9. Luz H, Faccini JLH, Landulfo GA, Berto BP, Ferreira I. Bird ticks in an area of the Cerrado of Minas Gerais State, southeast Brazil. *Exp Appl Acarol* 2012;**58**:88—99.

10. Lugarini C, Martins TF, Ogrzewalska M, Vasconcelos NCT, Ellis VA, Oliveira JB. Rickettsial agents in avian ixodid ticks in northeast Brazil. *Ticks Tick-borne Dis* 2015;**6**:364—75.

11. Maturano R, Faccini JLH, Daemon E, Fazza POC, Bastos RR. Additional information about tick parasitism in Passeriformes birds in an Atlantic Forest in southeastern Brazil. *Parasitol Res* 2015;**114**:4181—93.

12. Ramos DG, Melo ALT, Martins TF, Alves AS, Pacheco TA, Pinto LB, et al. Rickettsial infection in ticks from wild birds from Cerrado and the Pantanal region of Mato Grosso, Midwestern, Brazil. *Ticks Tick-borne Dis* 2015;**6**:836—42.

13. Witter R, Martins TF, Campos AK, Melo ALT, Correa SHR, Morgado TO, et al. Rickettsial infection in ticks (Acari: Ixodidae) of wild animals in Midwestern Brazil. *Ticks Tick-borne Dis* 2016;**7**:415—23.

14. Ogrzewalska M, Pacheco R, Uezu A, Richtzenhain LJ, Ferreira F, Labruna MB. Rickettsial infection in *Amblyomma nodosum* ticks (Acari: Ixodidae) from Brazil. *Ann Trop Med Parasitol* 2009;**103**:413—25.

15. Mazioli R, Szabó M, Mafra C. *Amblyomma nodosum* (Acari: Ixodidae) parasitizing a domestic dog in Colatina, Espírito Santo, Brazil. *Rev Bras Parasitol Vet* 2012;**21**:428—9.

Amblyomma ovale
Koch, 1844

Koch, C.L. (1844) Systematische Übersicht über die Ordnung der Zecken. *Archiv für Naturgeschichte*, 10, 217–239.

DISTRIBUTION

Argentina, Belize, Bolivia, Brazil, Colombia, Costa Rica, Ecuador, El Salvador, French Guiana, Guatemala, Guyana, Mexico, Nicaragua, Panama, Paraguay, Peru, Surinam, Trinidad and Tobago, and Venezuela.[1] *A. ovale* was also recorded in localities from the Nearctic Region in Mexico and United States.[1]

Biogeographic distribution in the Southern Cone of America: Chaco, Parana Forest, Yungas (Neotropical Region), and Monte (South American transition zone) (see Fig. 1 in the introduction).

HOSTS

The host range of *A. ovale* is ample as detailed in Table 2.13 but the principal hosts for adult ticks are carnivorous mammals, while rodents of the families Cricetidae and Echimyidae are the most common hosts for nymphs and larvae,[1–5] although birds have been found increasingly relevant as hosts for immature stages.[6–13] It is also considerable the contribution of tapirs and carnivorous animals as hosts for adults and immatures, respectively[1,14] and in a lesser extent other types of hosts.[15] It is uncertain if the record of *A. ovale* on *Dasypus novemcinctus* by Gomes et al.[16] belongs to this tick species and this host is not included in Table 2.13. There are some vague records of *A. ovale* from Anura and Testudines in Guglielmone et al.[17] but they are not included in the host range for *A. ovale* presented in Table 2.13.

ECOLOGY

Although the life cycle of *A. ovale* was studied under laboratory conditions,[18] investigations on its ecological preferences and life cycle under field conditions are lacking. All ecological information about *A. ovale* refers to distribution and host association.

SANITARY IMPORTANCE

Adults of *A. ovale* are common parasites of dogs in rural and forested areas, and the records of *A. ovale* adults biting humans in South America are numerous.[1,14,19,20] This tick species has capacity to transmit *Hepatozoon canis*,[21] the causative agent of a serious dog disease; it is also a potential vector of

TABLE 2.13 Hosts of Adults (A), Nymphs (N), and Larvae (L) of *A. ovale*

MAMMALIA		Cricetidae	
ARTIODACTYLA		*Euryoryzomys russatus*	N
Cervidae		*Holochilus brasiliensis*	N
Blastocerus dichotomus	A	*Hylaemys megacephalus*	N
Mazama americana	A	*Nectomys squamipes*	N
M. gouazoubira	A	*Oligoryzomys flavescens*	NL
Suidae		*O. nigripes*	NL
Pigs (domestic and feral)	A	*Sigmodon hispidus*	AN
Tayassuidae		*Sooretamys agouya*	N
Pecari tajacu	A	*Transandinomys bolivaris*	N
CARNIVORA		*T. talamancae*	AN
Canidae		*Zygodontomys brevicauda*	ANL
Canis latrans	ANL	**Echimyidae**	
Cerdocyon thous	AN	*Proechimys canicollis*	NL
Chrysocyon brachyurus	ANL	*P. guyannensis*	N
Domestic dog	ANL	*P. semispinosus*	N
Lycalopex gymnocercus	A	*Thrichomys apereoides*	N
Speothos venaticus	A	*T. fosteri*	N
Urocyon cinereoargenteus	AN	**Erethizontidae**	
Felidae		*Sphiggurus villosus*	A
Domestic cat	AN	**Heteromyidae**	
Herpailurus yagouaroundi	A	*Heteromys anomalus*	N
Leopardus geoffroyi	A	**Muridae**	
L. pardalis	AN	*Rattus rattus*	ANL
L. tigrinus	A	**AVES**	
L. wiedii	A	**CORACIIFORMES**	
Panthera onca	AN	**Momotidae**	
Puma concolor	AN	*Baryphthengus martii*	L
Mephitidae		**GALLIFORMES**	
Conepatus semistriatus	A	**Cracidae**	
Mustelidae		*Mitu tuberosum*	L
Eira barbara	AN	*Penelope superciliaris*	A

(Continued)

TABLE 2.13 (Continued)

Galictis cuja	A	**PASSERIFORMES**	
G. vittata	A	**Cardinalidae**	
Lontra longicaudis	A	Cyanacompsa brissonii	N
Procyonidae		**Conopophagidae**	
Nasua nasua	AN	Conopophaga lineata	L
N. narica	A	**Emberizidae**	
Procyon cancrivorus	AN	Arremon flavirostris	NL
P. lotor	AN	Arremonops conirostris	A
DIDELPHIMORPHIA		Sporophila nigricollis	AN
Didelphidae		**Formicariidae**	
Didelphis sp.	NL	Formicarius analis	N
D. albiventris	AN	**Furnariidae**	
D. marsupialis	N	Sittasomus griseicapillus	L
Marmosa robinsoni	N	**Thamnophilidae**	
PERISSODACTYLA		Mackenziana severa	N
Equidae		Thamnophilus pelezni	N
Horse	A	**Thraupidae**	
Mule	A	Trichothraupis melanops	L
Tapiridae		**Tityridae**	
Tapirus bairdii	A	Schiffornis turdina	L
T. terrestris	AN	**Trogloditydae**	
PRIMATES		Troglodytes musculus	N
Atelidae		**Turdidae**	
A. guariba	A	Turdus albicollis	NL
Cebidae		T. amaurochalinus	N
Sapajus apella	A	T. leucomelas	N
Hominidae		T. rufiventris	N
Human	AN	**Tyrannidae**	
RODENTIA		Myiarchus ferox	N
Caviidae		Tyrannus melancholicus	N
Hydrochoerus hydrochaeris	A		

the human pathogen *Rickettsia* sp. strain Atlantic rainforest.[20] A case of human paralysis due to the bite of *A. ovale* was reported in Panama.[22]

DIAGNOSIS

Male (Fig. 2.46): medium body size, total length 3.5 mm (2.7−4.2), breadth 2.3 mm (1.8−2.7). Body outline oval elongated, scapulae small and rounded; cervical grooves sigmoid, deeper anteriorly; marginal groove complete. Eyes flat. Scutum ornate, with two narrow yellow stripes in the lateral fields; cervical spot long; postero-accessory and postero-median spots short and broad, lateral spots conspicuous and small; punctuations numerous, uniformly

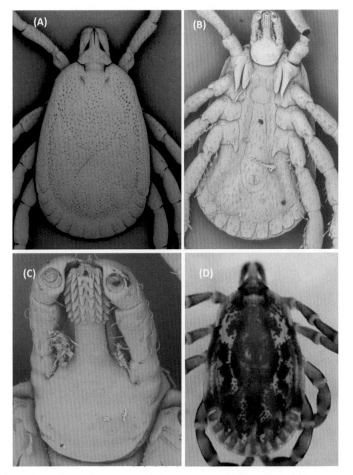

FIGURE 2.46 *A. ovale* male: (A) dorsal view, (B) ventral view, (C) capitulum ventral view, (D) scutum ornamentation.

distributed. Carena present. Basis capituli dorsally subtriangular, cornua short. Hypostome spatulate, dental formula 3/3. Genital aperture located at level of coxa II, U-shaped. Legs: coxa I with two distinct, long, triangular sharp spurs, the external spur slightly longer than the internal spur, tip of the external spur with a slight curve outward; coxae II–III each with a short triangular spur; coxa IV with a triangular sharp spur, longer than spurs on coxae II–III; trochanters without spurs. Spiracular plates comma-shaped.

Female (Fig. 2.47): medium body size, total length 3.7 mm (3.5–4.0), breadth 2.6 mm (2.5–2.6). Body outline oval. Scutum length 2.5 mm (2.4–2.6), breadth 2.5 mm (2.4–2.6); scapulae rounded; cervical grooves sigmoid, deeper anteriorly. Eyes flat. Chitinous tubercles at posterior body

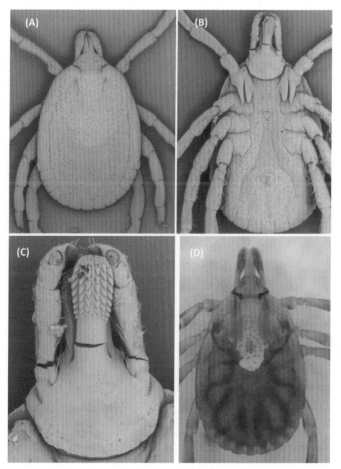

FIGURE 2.47   *A. ovale* female: (A) dorsal view, (B) ventral view, (C) capitulum ventral view, (D) scutum ornamentation.

margin absent. Scutum ornate, with a yellowish posterior spot without central stripe, cervical spots narrow and elongated; punctuations numerous, uniformly distributed. Notum glabrous. Basis capituli dorsally subtriangular, cornua short, porose areas oval. Hypostome spatulate, dental formula 3/3. Genital aperture located at level of coxa II, U-shaped. Legs: coxa I with two distinct, long, triangular sharp spurs, the external spur slightly longer than the internal spur, tip of the external spur with a slight curve outward; coxae II−IV with a short triangular spur each; trochanters without spurs. Spiracular plates comma-shaped.

Nymph (Fig. 2.48): body size small, total length 1.4 mm (1.3−1.5), breadth 0.9 mm (0.9−1.0). Body outline oval. Chitinous tubercles at posterior body margin absent. Scutum with few punctuations, larger and deeper laterally. Eyes flat, located on lateral scutal angles at the level of the posterior third of the scutum. Cervical grooves long, reaching the scutal posterior

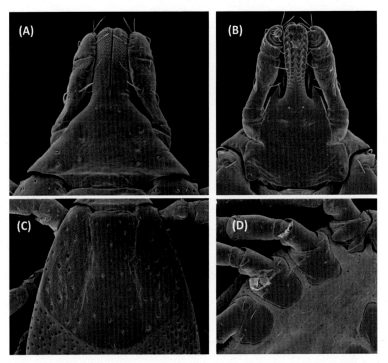

FIGURE 2.48 *A. ovale* nymph: (A) capitulum dorsal view, (B) capitulum ventral view, (C) scutum, (D) coxae. *Figures (A) to (D) are reproductions of figure 8: A. ovale nymph. Capitulum, dorsum; Capitulum, venter; Scutum; Coxae I−IV. From Martins TF, Labruna MB, Mangold AJ, Cafrune MM, Guglielmone AA, Nava S. Taxonomic key to nymphs of the genus* Amblyomma *(Acari: Ixodidae) in Argentina, with description and redescription of the nymphal stage of four* Amblyomma *species. Ticks Tick-borne Dis 2014;5:753−70, with permission of Elsevier.*

fourth, deeper in the anterior fourth. Basis capituli dorsally triangular, without cornua, pseudoauricula present, ventral cornua present. Hypostome spatulate, dental formula 2/2. Legs: coxa I with two triangular blunt spurs, the external longer than the internal; coxae II–IV with a small triangular spur each.

Taxonomic notes: the presence of carena is described for the males of *A. ovale*, but in some specimens this morphological structure is almost imperceptible. See also Taxonomic notes of *A. aureolatum* (Pallas, 1772).

DNA sequences with relevance for tick identification and phylogeny: the following DNA sequences of *A. ovale* are available in the Gen Bank: mitochondrial genes 12S rRNA (accession number AY342273), 16S rRNA (accession numbers JN573304, KF179347, KM042852, FJ424409, KR605466–KR605469, AF541255) and cytochrome oxidase I (accession numbers KF200079, KF200080, KF200129, KF200143, KF200158), and sequences of the nuclear genome (the fragment 5.8S ribosomal RNA gene-internal transcribed spacer 2-28S ribosomal RNA gene) (accession number AY887119).

REFERENCES

1. Guglielmone AA, Estrada-Peña A, Mangold AJ, Barros-Battesti DM, Labruna MB, Martins JR, et al. *Amblyomma aureolatum* (Pallas, 1772) and *Amblyomma ovale* Koch, 1844 (Acari: Ixodidae): DNA sequences, hosts and distribution. *Vet Parasitol* 2003;113:273–88.
2. Guglielmone AA, Nava S. Rodents of the subfamily Sigmodontinae as hosts for South American hard ticks (Acari: Ixodidae) with hypothesis on life history. *Zootaxa* 2011;2904:45–65.
3. Guglielmone AA, Nava S. Distribución geográfica, hospedadores y variabilidad genética de *Amblyomma ovale* y *Amblyomma aureolatum* (Acari: Ixodidae), dos vectores potenciales de rickettsias en la Argentina. In: Farjat JB, Enría D, Martino P, Rosenvitz M, Seijo A, editors. *Temas de Zoonosis VI*. Buenos Aires: Asociación Argentina de Zoonosis; 2014. pp. 183–91.
4. Murgas IL, Castro AM, Bermúdez SE. Current status of *Amblyomma ovale* (Acari: Ixodidae) in Panama. *Ticks Tick-borne Dis* 2013;4:164–6.
5. Martins TF, Peres MG, Costa FB, Bacchiega TS, Appolinario CM, Antunes JMAP. Ticks infesting wild small rodents in three areas of the state of São Paulo, Brazil. *Ciênc Rural* 2016;46:871–5.
6. Ogrzewalska M, Pacheco RC, Uezu A, Richtzenhain LJ, Ferreira F, Labruna MB. Ticks (Acari: Ixodidae) infesting birds in an Atlantic rainforest region in Brazil. *J Med Entomol* 2009;46:1225–9.
7. Ogrzewalska M, Literák I, Capek M, Sychra O, Álvarez Calderón V, Calvo Rodríguez B. Bacteria of the genus *Rickettsia* in ticks (Acari: Ixodidae) collected from birds in Costa Rica. *Ticks Tick-borne Dis* 2015;6:478–82.
8. Luz HR, Faccini JLH, Landulfo GA, Berto BP, Ferreira I. Bird ticks in an area of the Cerrado of Minas Gerais, southeast Brazil. *Exp Appl Acarol* 2012;58:89–99.
9. Pacheco RC, Arzua M, Nieri-Bastos FA, Moraes-Filho J, Marcili A, Richtzenhain LJ, et al. Rickettsial infection in ticks (Acari: Ixodidae) collected on birds in Southern Brazil. *J Med Entomol* 2012;49:710–16.

10. Flores FS, Nava S, Batallán G, Tauro LB, Contigiani MS, Díaz LA, et al. Ticks (Acari: Ixodidae) on wild birds in north-central Argentina. *Ticks Tick-borne Dis* 2014;**5**:715−21.
11. Maturano R, Faccini JLH, Daemon E, Fazza POC, Bastos RR. Additional information about tick parasitism in Passeriformes birds in an Atlantic Forest in southeastern Brazil. *Parasitol Res* 2015;**114**:4181−93.
12. Soares HS, Barbieri ARM, Martins TF, Minervino AHH, Lima JTR, Marcili A, et al. Ticks and rickettsial infection in the wildlife of two regions of the Brazilian Amazon. *Exp Appl Acarol* 2015;**65**:125−40.
13. Ramos DG, Melo ALT, Martins TF, Alves AS, Pacheco TA, Pinto LB, et al. Rickettsial infection in ticks from wild birds from Cerrado and the Pantanal region of Mato Grosso, Midwestern, Brazil. *Ticks Tick-borne Dis* 2015;**6**:836−42.
14. Szabó MPJ, Martins TF, Nieri-Bastos FA, Spolidorio MG, Labruna MB. A surrogate life cycle of *Amblyomma ovale* Koch, 1844. *Ticks Tick-borne Dis* 2012;**3**:262−4.
15. Lavina MS, Souza AP, Souza JC, Bellato V, Sartor AA, Moura AB. Ocorrência de *Amblyomma aureolatum* (Pallas, 1772) e *A. ovale* (Koch, 1844) [sic] (Acari: Ixodidae) parasitando *Alouatta clamitans* Cabrera, 1940 (Primates: Atelidae) na região norte do estado de Santa Catarina. *Arq Bras Med Vet Zoot* 2016;**63**:266−9.
16. Gomes SN, Pesenti TC, Muller G. Parasitismo por *Amblyomma ovale* e *Amblyomma fuscum* (Acari: Ixodidae) em *Dasypus novemcinctus* (Xenarthra: Dasypodidae) no Brasil. *Arq Inst Biol* 2015;**82**:1−4.
17. Guglielmone AA, Robbins RG, Apanaskevich DA, Petney TN, Estrada-Peña A, Horak IG. *The hard ticks of the world (Acari: Ixodida: Ixodidae).* London: Springer; 2014.
18. Martins TF, Moura MM, Labruna MB. Life-cycle and host preference of *Amblyomma ovale* (Acari: Ixodidae) under laboratory conditions. *Exp Appl Acarol* 2012;**56**:151−8.
19. Guglielmone AA, Beati L, Barros-Battesti DM, Labruna MB, Nava S, Venzal JM, et al. Ticks (Ixodidae) on humans in South America. *Exp Appl Acarol* 2006;**40**:83−100.
20. Szabó MPJ, Nieri-Bastos FA, Spolidorio MG, Martins TF, Barbieri AM, Labruna MB. *In vitro* isolation from *Amblyomma ovale* (Acari: Ixodidae) and ecological aspects of the Atlantic rainforest *Rickettsia*, the causative agent of a novel spotted fever rickettsiosis in Brazil. *Parasitology* 2013;**140**:719−28.
21. Forlano M, Scofield A, Elisei C, Fernandes KR, Ewing SA, Massard CL. Diagnosis of *Hepatozoon* spp. in *Amblyomma ovale* and its experimental transmission in domestic dogs in Brazil. *Vet Parasitol* 2005;**134**:1−7.
22. Baeza CR. Tick paralysis-Canal Zone, Panama. *Morb Mortal Wkly Rep* 1979;**14**:428−33.

Amblyomma pacae
Aragão, 1911

Aragão, H.B. (1911) Notas sobre ixódidas brazileiros. *Memórias do Instituto Oswaldo Cruz*, 3, 145−195.

DISTRIBUTION

Belize, Brazil, Colombia, Costa Rica, Guyana, Mexico, Panama, Paraguay, Surinam, and Venezuela.[1−3]

TABLE 2.14 Hosts of Adults (A), Nymphs (N), and Larvae (L) of *A. pacae*

MAMMALIA		PILOSA	
ARTIODACTYLA		Myrmecophagidae	
Suidae		*Tamandua tetradactyla*	A
Feral pig	A	RODENTIA	
Tayassuidae		Cuniculidae	
Tayassi pecari	A	*Cuniculus paca*	AN
CARNIVORA		Cricetidae	
Canidae		*Rhipidomys macrurus*	N
Domestic dog	AN	Dasyproctidae	
DIDELPHIMORPHIA		*Dasyprocta punctata*	A
Didelphidae		Echimyidae	
Didelphis marsupialis	N	*Proechimys semispinosus*	N
PERISSODACTYLA			
Tapiridae			
Tapirus bairdii	A		

Main host for adult ticks underlined.

Biogeographic distribution in the Southern Cone of America: Chaco (Neotropical Region) (see Fig. 1 in the introduction).

HOSTS

The principal host for adults of *A. pacae* is the rodent *Cuniculus paca*, but other mammals were also recorded as hosts for adults and immature stages of this tick[3–7] (Table 2.14).

ECOLOGY

Besides the few data on host association and geographical distribution, there is no information on the ecology of *A. pacae*.

SANITARY IMPORTANCE

The capacity of *A. pacae* to transmit pathogens has not been investigated to date. There is just one record of an adult of *A. pacae* on humans[5] and, with the exception of a record on domestic dog,[6] *A. pacae* has not been found parasitizing domestic animals.

DIAGNOSIS

Male (Fig. 2.49): medium body size, total length 3.8 mm (3.7−3.9), breadth 2.7 mm (2.6−2.9). Body outline oval, scapulae rounded, cervical grooves short, deep, sigmoid; marginal groove absent. Eyes flat. Scutum ornate, limiting spots converging posteriorly toward the median line forming the outline of a pseudoscutum; punctuations numerous, moderately deep, uniformly distributed. Carena absent. Basis capituli subrectangular dorsally, cornua short. Hypostome spatulate, dental formula 3/3. Genital aperture located

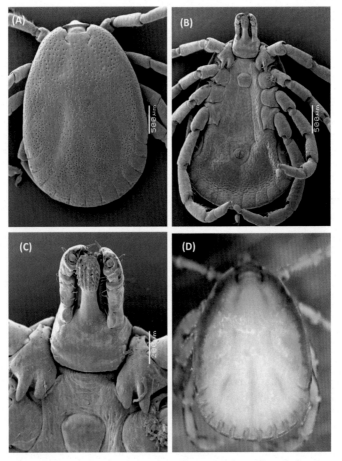

FIGURE 2.49   *A. pacae* male: (A) dorsal view, (B) ventral view, (C) capitulum ventral view, (D) scutum ornamentation.

at level of coxa II, U-shaped. Legs: coxa I with two distinct, short triangular spurs, the external longer than internal; coxae II–IV each with a small triangular spur; trochanters without spurs. Spiracular plates comma-shaped.

Female (Fig. 2.50): medium body size, total length 3.7 mm (3.6–4.0), breadth 2.9 mm (2.7–3.0). Body outline oval. Scutum length 2.0 mm (1.8–2.2), breadth 2.4 mm (2.3–2.5); scapulae rounded, cervical grooves long, deep in the anterior part, continued as shallow and divergent depressions posteriorly. Eyes flat. Chitinous tubercles at the posterior body margin absent. Scutum ornate, with small pale marking in the posterior angler; punctuations numerous, small, uniformly distributed. Notum glabrous. Basis capituli subtriangular dorsally, cornua absent, porose areas oval. Hypostome spatulate, dental formula 3/3. Genital aperture located at level of coxa II, U-shaped. Legs: coxa I with two distinct, short triangular spurs, subequal in

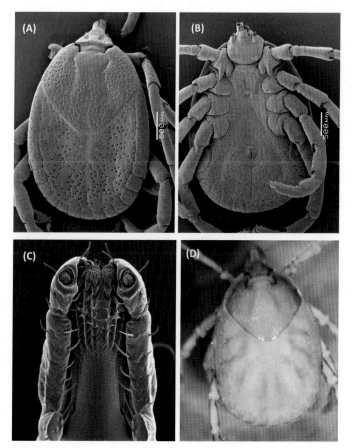

FIGURE 2.50 *A. pacae* female: (A) dorsal view, (B) ventral view, (C) capitulum ventral view, (D) scutum ornamentation.

size; coxae II–IV each with a triangular spur; trochanters without spurs. Spiracular plates comma-shaped.

Nymph (Fig. 2.51): medium body size, total length 1.6 mm (1.3–1.7), breadth 1.3 mm (1.1–1.4). Body outline oval. Chitinous tubercles at the posterior body margin absent. Scutum inornate, numerous large punctuations evenly distributed. Eyes flat, located on lateral scutal angles at the level of scutal mid-length. Cervical grooves long, deeper at the scutal anterior third, followed by a rugose shallow depression surpassing the level of the eyes. Basis capituli broadly hexagonal dorsally, without cornua. Hypostome spatulate, dental formula 2/2. Legs: coxa I with two slender pointed spurs, the external slightly longer than the internal; coxae II–IV with a small triangular spur each.

Taxonomic notes: the phylogenetic analysis performed with the available DNA sequences of *A. pacae* (mitochondrial 16S rRNA and 12S rRNA genes) shows that its evolutionary relationship with the other Neotropical species of *Amblyomma* remains unresolved (Fig. 2.7).

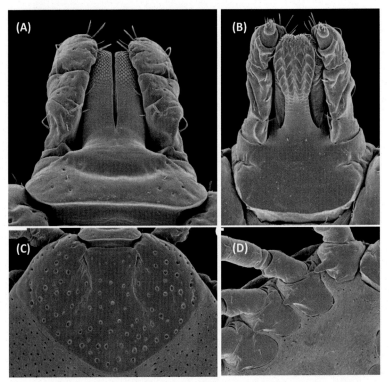

FIGURE 2.51 *A. pacae* nymph: (A) capitulum dorsal view, (B) capitulum ventral view, (C) scutum, (D) coxae. Figures (A)–(D) are reproductions of Figure 12: A. pacae nymph. Capitulum, dorsum; Capitulum, venter; Scutum; Coxae IV. *From Martins TF, Onofrio VC, Barros-Battesti DM, Labruna MB. Nymphs of the genus Amblyomma (Acari: Ixodidae) of Brazil: descriptions, redescriptions, and identification key. Ticks Tick-borne Dis 2010;1:75–99, with permission from Elsevier.*

DNA sequences with relevance for tick identification and phylogeny: the DNA sequences of *A. pacae* available in the Gen Bank correspond to partial sequences of the mitochondrial genes 16S rRNA (accession number JX141384) and 12S rRNA (accession number AY342274).

REFERENCES

1. Guglielmone AA, Estrada-Peña A, Keirans JE, Robbins RG. Ticks (Acari: Ixodida) of the neotropical zoogeographic region. Special publication of the integrated consortium on ticks and tick-borne diseases-2. Houten (The Netherlands): Atalanta; 2003.
2. Álvarez Calderón V, Hernández Fonseca V, Hernández Gamboa J. Catálogo de garrapatas suaves (Acari: Argasidae) y duras (Acari: Ixodidae) de Costa Rica. *Brenesia* 2005;**63−64**:81−8.
3. Guzmán-Cornejo C, Pérez TM, Nava S, Guglielmone AA. First records of the ticks *Amblyomma calcaratum* and *A. pacae* (Acari: Ixodidae) parazitizing mammals of Mexico. *Rev Mex Biodiversidad* 2006;**77**:123−7.
4. Aragão HB. Ixodidas brasileiros e de algunos paízes limitrophes. *Mem Inst Oswaldo Cruz* 1936;**31**:759−843.
5. Jones EK, Clifford CM, Keirans JE, Kohls GM. The ticks of Venezuela (Acarina: Ixodoidea) with a key to the species of *Amblyomma* in the Western Hemisphere. *Brigham Young Univ Sci Bull Biol Ser* 1972;**17**(4):1−40.
6. Labruna MB, Camargo LMA, Terrassini FA, Ferreira F, Schumaker TST, Camargo EP. Ticks (Acari: Ixodidae) from the state of Rondônia , western Amazon, Brazil. *Syst Appl Acarol* 2005;**10**:17−32.
7. Soares HS, Barbieri ARM, Martins TF, Minervino AHH, de Lima JTR, Marcili A, et al. Ticks and rickettsial infection in the wildlife of two regions of the Brazilian Amazon. *Exp Appl Acarol* 2015;**65**:125−40.

Amblyomma parvitarsum
Neumann, 1901

Neumann, L.G. (1901) Révision de la famille des ixodidés (4a mémoire). *Mémoires de la Société Zoologique de France*, 14, 249−372.

DISTRIBUTION

Argentina, Bolivia, Chile, and Peru.[1] The report of one female of *A. parvitarsum* on the penguin *Spheniscus magellanicus* in Brazil[2] is considered an occasional record. Therefore, the distribution of *A. parvitarsum* does not include Brazil.

Biogeographic distribution in the Southern Cone of America: Atacama, Monte, Prepuna, Puna (South American Transition Zone); Central Patagonia, Coquimbo (Andean Region) (see Fig. 1 in the introduction).

HOSTS

The principal hosts for adults of *A. parvitarsum* are South American camelids, while all records of larvae were made on lizards of the family Liolaemidae[1,3] (Table 2.15). There are also records of adults of *A. parvitarsum* on cattle, horse, and the birds *Spheniscus magellanicus* and *Rhea pennata*. Nevertheless, the record on a horse is provisionally valid here because it requires confirmation.

ECOLOGY

A. parvitarsum is principally associated to Andean environments, and it is also present in the Patagonian Plateau,[1,4] where its principal hosts for adult stages prevail. Free-living adults of *A. parvitarsum* are usually found in manure heaps of the camelid species they use as hosts. Adult ticks were found on camelids in all seasons but prevailing in the spring and summer, while larvae were collected on lizards in autumn, spring, and summer.[1] However, the duration of the life cycle of *A. parvitarsum* is unknown.

TABLE 2.15 Hosts of Adults (A) and Larvae (L) of *A. parvitarsum*

MAMMALIA		SPHENISCIFORMES	
ARTIODACTYLA		Sphenicidae	
Bovidae		*Spheniscus magellanicus*	A
Cattle	A	REPTILIA	
Camelidae		SQUAMATA	
Lama glama	A	Liolaemidae	
L. guanicoe	A	*Liolaemus alticolor*	L
Vicugna vicugna	A	*L. andinus*	L
PERISSODACTYLA		*L. copiapensis*	L
Equidae		*L. eleodori*	L
Horse	A	*L. jamesi*	L
AVES		*L. nigripes*	L
STRUTHIONIFORMES		*L. patriciaiturrae*	L
Rhea pennata	A	*L. pleopholis*	L

Main hosts for adult ticks underlined.

SANITARY IMPORTANCE

A. parvitarsum is a common parasite of the domestic llama (*Lama glama*). A high prevalence of infection with a novel spotted fever group *Rickettsia* of unknown pathogenicity was reported in different populations of *A. parvitarsum* by Ogrzewalska et al.[5] There are no records of this tick biting humans.

DIAGNOSIS

Male (Fig. 2.52): large body size, total length 4.4 mm (3.2−4.5), breadth 3.1 mm (2.4−3.7). Body outline oval elongated, narrower in the anterior

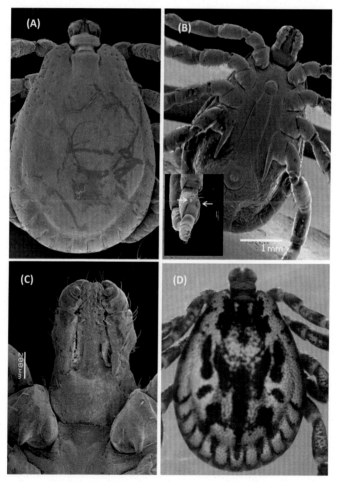

FIGURE 2.52   *A. parvitarsum* male: (A) dorsal view, (B) ventral view and image of the two spines present on the tibia of legs II−IV in both male and female ticks, (C) capitulum ventral view, (D) scutum ornamentation.

part; scapulae rounded; cervical grooves deep and short, comma-shaped; marginal groove incomplete. Eyes orbited. Scutum ornate, postero-median spot narrow, extending to the level of spiracular plates; postero-accessory spots short; lateral spots distinct, the first and second lateral spots conjoined; cervical spots straight, directed posteriorly; punctuations larger in the anterior third of the scutum. Carena absent. Basis capituli dorsally subrectangular, cornua short. Hypostome spatulate, dental formula 3/3. Genital aperture located at level of coxa II, U-shaped. Legs: coxa I with two distinct, triangular, short subequal spurs, the external longer than the internal; coxae II—III each with a triangular, short blunt spur; coxa IV with a narrow, sharp, long spur, reaching the anus level; trochanters without spurs; presence of two spines on the tibia of legs II—IV. Spiracular plates comma-shaped.

Female (Fig. 2.53): large body size, total length 5.0 mm (4.4—5.3), breadth 3.3 mm (2.9—3.5). Body outline oval. Scutum length 1.9 mm (1.7—2.0), breadth 2.2 mm (2.0—2.5); scapulae rounded; cervical grooves deep anteriorly, shallow posteriorly, sigmoid in shape, extending posteriorly to the posterior third of the scutum. Eyes orbited. Chitinous tubercles at posterior body margin absent. Scutum ornate, extensively pale yellowish, central area short and narrow; punctuations small, larger in the antero-lateral fields. Notum glabrous. Basis capituli dorsally rectangular, cornua absent, porose areas oval. Hypostome spatulate, dental formula 3/3. Genital aperture located at level of coxae II, U-shaped. Legs: coxa I with two distinct, triangular, short subequal spurs, the external longer than the internal; coxa II—IV each with a triangular, short blunt spur, spur on coxa IV longer than spur on coxae II—III; trochanters without spurs; presence of two spines on the tibia of legs II—IV. Spiracular plates comma-shaped.

Nymph (Fig. 2.54): the nymph of *A. parvitarsum* was described from just one unfed tick, the only available specimen. Large body size, total length 2.1 mm, breadth 1.8 mm. Body outline oval. Chitinous tubercles at posterior body margin absent. Scutum with surface slightly shagreened and few punctuations. Eyes flat, located on lateral scutal angles at the level of scutal mid-length. Cervical grooves long, reaching the scutal median third, deeper at the anterior third. Basis capituli dorsally rectangular, without cornua. Hypostome spatulate, dental formula 2/2. Legs: coxa I with two subequal triangular spurs, the external slightly stouter than the internal; coxae II—IV with a small triangular spur each.

Taxonomic notes: Santos Dias[6] included *A. parvitarsum* in the subgenus *Anastosiella* with *A. maculatum* Koch, 1844, *A. tigrinum* Koch, 1844, *A. triste* Koch, 1844, *A. neumanni* Ribaga, 1902, *A. pecarium* Dunn, 1933, and *A. brasiliense*, Aragão, 1908. Camicas et al.[7] have also included *A. parvitarsum* within this subgenus along with *A. Neumanni* neumanni not Neumanni, *A. maculatum*, *A. tigrinum*, *A. triste*, *A. parvitarsum*, *A. aureolatum* (Pallas, 1772), and *A. ovale* (Koch, 1884). However, Estrada-Peña et al.[8] have stated that these species do not constitute a natural group, and this assertion is

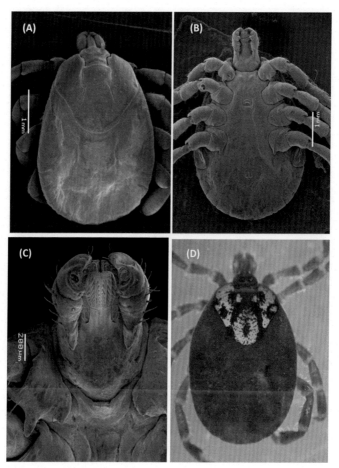

FIGURE 2.53   *A. parvitarsum* female: (A) dorsal view, (B) ventral view, (C) capitulum ventral view, (D) scutum ornamentation.

supported by phylogenetic analyses performed with both mitochondrial (16S rDNA sequences) and nuclear (18S rDNA sequences) markers (Figs. 2.11 and 2.15). In these phylogenetic trees it can be observed that *A. parvitarsum* represents an independent lineage within Neotropical species of the genus *Amblyomma*. Boero[4] described the male of *A. parvitarsum* without the two spines on the tibia of legs II—IV. See also Taxonomic notes of *A. neumanni* for the confusion of *A. parvitarsum* as *A. neumanni* in Voltzit.[9]

DNA sequences with relevance for tick identification and phylogeny: the following DNA sequences of *A. parvitarsum* genes are available in the Gen Bank: 16S rDNA (accession number AY498561), 12S rDNA (accession number AY342272), and 18S rDNA (accession number FJ464423).

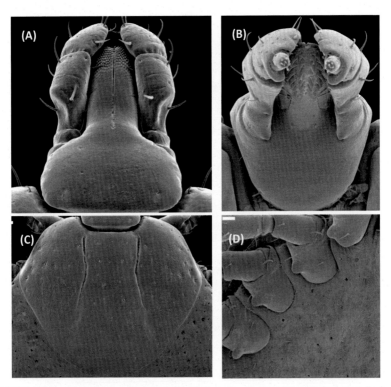

FIGURE 2.54 *A. parvitarsum* nymph: (A) capitulum dorsal view, (B) capitulum ventral view, (C) scutum, (D) coxae. *Figures (A) to (D) are reproductions of figure 1:* A. parvitarsum *nymph. Capitulum, dorsum; Capitulum, venter; Scutum; Coxae I–IV. From Martins TF, Labruna MB, Mangold AJ, Cafrune MM, Guglielmone AA, Nava S. Taxonomic key to nymphs of the genus* Amblyomma *(Acari: Ixodidae) in Argentina, with description and redescription of the nymphal stage of four* Amblyomma *species.* Ticks Tick-borne Dis *2014;5:753–70, with permission of Elsevier.*

REFERENCES

1. Muñoz-Leal S, González-Acuña D, Beltrán-Saavedra F, Limachi JM, Guglielmone AA. *Amblyomma parvitarsum* (Acari: Ixodidae): localities, hosts and host-parasite ecology. *Exp Appl Acarol* 2014;**62**:91–104.

2. Becker GK, Silva RP, Sinkoc AL, Brum JGW. *Amblyomma parvitarsum*, Neumann, 1901 (Acari: Ixodidae) in Magellanic penguins *Spheniscus magellanicus* (Spheniscidae) in Cassinos beach, Rio Grande do Sul, Brazil. *Arq Inst Biol São Paulo* 1997;**64**:81–2.

3. Castillo GN, González-Rivas CJ, Villavicencio HJ, Acosta JC, Nava S. Primer registro de infestación en un reptil por larvas de *Amblyomma parvitarsum* (Acari: Ixodidae) en Argentina. *Cuad Herpetol* 2015;**29**:91–3.

4. Boero JJ. *Las garrapatas de la República Argentina (Acarina: Ixodoidea)*. Buenos Aires: Universidad de Buenos Aires; 1957.

5. Ogrzewalska M, Nieri-Bastos FA, Marcili A, Nava S, Gonzalez-Acuña D, Muñoz-Leal S. A novel spotted fever group Rickettsia infecting the tick *Amblyomma parvitarsum* in highlands of Argentina and Chile. *Ticks Tick-borne Dis* 2016;**7**:439−42.

6. Santos Dias JAT. Nova contribuição para o estudo da sistemática do género *Amblyomma* Koch, 1844 (Acarina-Ixodoidea). *García de Orta Sér Zool* 1993;**19**:11−19.

7. Camicas JL, Hervy JP, Adam F, Morel PC. *Les tiques du monde. Nomenclature, stades décrits, hôtes, répartition (Acarida, Ixodida)*. Paris: ORSTOM; 1998.

8. Estrada-Peña A, Venzal JM, Mangold AJ, Cafrune MM, Guglielmone AA. The *Amblyomma maculatum* Koch, 1844 (Acari: Ixodidae: Amblyomminae) tick group: diagnostic characters, description of the larva of *A. parvitarsum* Neumann, 1901, 16S rDNA sequences, distribution and hosts. *Syst Parasitol* 2005;**60**:99−112.

9. Voltzit OV. A review of Neotropical Amblyomma *Amblyomma* NOT Amblyomma species (Acari: Ixodidae). *Acarina* 2007;**15**:3−134.

Amblyomma parvum
Aragão, 1908

Aragão, H.B. (1908) Algunas novas especies de carrapatos brazileiros. *Brazil Medico*, 22, 111−115.

DISTRIBUTION

Argentina, Bolivia, Brazil, Colombia, Costa Rica, El Salvador, French Guiana, Guatemala, Mexico, Nicaragua, Panama, Paraguay, and Venezuela.[1] *A. parvum* has been also recorded in the Nearctic Region,[2] but there is no evidence that this tick species is established there.

Biogeographic distribution in the Southern Cone of America: Chaco (Neotropical Region) and Monte (South American Transition Zone) (see Fig. 1 in the introduction).

HOSTS

Adults of *A. parvum* present low host specificity. They feed on large wild and domestic mammals of different orders[1,3] (Table 2.16). The principal hosts for immature stages are small and medium-size rodents with several recent records mainly from passeriform birds[1,4−11] (Table 2.16). Almost all larvae and nymphs of *A. parvum* collected in Argentina were recovered from *Galea musteloides* (Rodentia: Caviidae).[4]

ECOLOGY

A. parvum is adapted to environments with low humidity and relatively high temperatures. In South America, this tick is principally distributed in the dry

TABLE 2.16 Hosts of Adults (A), Nymphs (N), and Larvae (L) of *A. parvum*

MAMMALIA		Hominidae	
ARTIODACTYLA		Human	A
Bovidae		**RODENTIA**	
Cattle	AN	**Caviidae**	
Domestic buffalo	A	*Dolichotis salinicola*	N
Goat	ANL	*Galea musteloides*	NL
Sheep	AN	*G. spixii*	N
Cervidae		*Hydrochoerus hydrochaeris*	A
Mazama gouazoubira	AN	*Kerodon rupestris*	A
Odocoileus virginianus	A	**Chinchillidae**	
Suidae		*Lagostomus maximus*	A
Pigs (domestic and feral)	AN	**Cricetidae**	
Tayassuidae		*Graomys centralis*	L
Catagonus wagneri	A	*Sigmodon hispidus*	A
Pecari tajacu	N	**Echimyidae**	
CARNIVORA		*Thrichomys apereoides*	N
Canidae		**Muridae**	
Canis latrans	A	*Mus musculus*	N
Cerodocyon thous	A	**AVES**	
Chrysocyon brachyurus	A	**ACCIPITRIFORMES**	
Domestic dog	A	**Cathartidae**	
Lycalopex gymnocercus	AN	*Coragyps atratus*	A
Urocyon cinereoargenteus	A	**COLUMBIFORMES**	
Felidae		**Columbidae**	
Domestic cat	A	*Leptotila verrauxi*	N
Herpailurus yagouaroundi	A	**PASSERIFORMES**	
Leopardus geoffroyi	A	**Cardinalidae**	
L. pardalis	A	*Cyanocompsa brissonii*	NL
Panthera onca	A	**Emberizidae**	
Puma concolor	A	*Saltator aurantiirostris*	N
Mustelidae		*S. similis*	L
Galictis cuja	A	*Saltatricula multicolor*	NL

(Continued)

TABLE 2.16 (Continued)

Procyonidae		Sicalis flaveola	L
Nasua nasua	A	Sporophila caerulescens	N
Procyon cancrivorus	A	S. nigricollis	N
CHIROPTERA		Zonotrichia capensis	NL
Phyllostomidae		**Parulidae**	
Carollia subrufa	A	M. flaveola	NL
CINGULATA		**Thamnophilidae**	
Dasypodidae		Myrmorchilus strigilatus	N
Chaetophractos vellerosus	A	Sakesphorus cristatus	N
Dasypus keppleri	A	Taraba major	N
D. novemcinctus	A	Thamnophilus capistratus	N
Tolypeutes matacus	A	T. pelzelni	N
DIDELPHIMORPHIA		**Thraupidae**	
Didelphidae		Coryphospyngus pileatus	L
Didelphis albiventris	NL	Dacnis cayana	N
LAGOMORPHA		Rhamphocelus carbo	N
Leporidae		Turdus rufus	N
Sylvilagus floridanus	A	**Troglodytidae**	
PERISSODACTYLA		Cantorchilus longirostris	N
Equidae		Troglodytes musculus	N
Donkey	A	**Turdidae**	
Horse	AN	Turdus amaurochalinus	NL
Tapiridae		T. rufiventris	N
Tapirus terrestris	A	**Tyrannidae**	
PILOSA		Casiornis rufus	N
Myrmecophagidae		Cnemotriccus fuscatus	N
Myrmecophaga tridactyla	AN	Myiarchus tyrannulus	NL
Tamandua tetradactyla	AN	Pitangus sulphuratus	N
PRIMATES		Stigamtura napensis	N
Cebidae		Tyrannus savanna	N
Sapajus apella	A	Xolmis velatus	N

diagonal region in Argentina, Bolivia, Brazil, and Paraguay which contains the biogeographic provinces of Caatinga, Cerrado, and Chaco.[1,12] *A. parvum* had one generation per year, larvae are active from late summer to early winter with the peak in autumn, nymphs are found from early winter to early spring with the peak in middle winter, and adults are found during spring and summer with the peak in early and middle summer.[13]

SANITARY IMPORTANCE

Adults of *A. parvum* are common parasites of goats and cattle, and they are also aggressive to humans[1,3,4,13−15]; they are usually found attached to the periocular region of cattle and goats, where inflammation occurs at the site of tick attachment.[13] Specimens of *A. parvum* have been found naturally infected with the causative agent of the Q fever in humans, *Coxiella burnetii*, but also with *Ehrlichia* cf. *E. chaffeensis* which produces human monocytic ehrlichiosis, and *"Candidatus Rickettsia andeanae,"*[16−19] a bacteria of unknown pathogenicity.

DIAGNOSIS

Male (Fig. 2.55): small body size, total length 2.1 mm (1.9−2.3), breadth 1.6 mm (1.5−1.8). Body outline oval, scapulae rounded, cervical grooves short, comma-shaped; marginal groove complete. Eyes flat. Scutum inornate; punctuations numerous, moderately deep, uniformly distributed. Carena present. Basis capituli subrectangular dorsally, cornua long, article I of palps ventrally with a large blunt spur directed posteriorly. Hypostome spatulate, dental formula 3/3. Genital aperture located at level of coxa II, U-shaped. Legs: coxa I with two distinct, short blunt spurs, subequal in size; coxae II−IV with a triangular spur each; trochanters with spurs. Spiracular plates comma-shaped.

Female (Fig. 2.56): small body size, total length 2.5 mm (2.3−2.6), breadth 1.9 mm (1.8−2.1). Body outline oval. Scutum length 1.4 mm (1.3−1.5), breadth 1.5 mm (1.4−1.6); scapulae rounded, cervical grooves short and comma-shaped. Eyes flat. Chitinous tubercles at the posterior body margin absent. Scutum inornate, punctuations numerous, moderately deep, uniformly distributed. Notum with short setae, evenly distributed. Basis capituli subtriangular dorsally, cornua short, porose areas oval, article I of palps ventrally with a large blunt spur directed posteriorly. Hypostome spatulate, dental formula 3/3. Genital aperture located at level of coxa II, U-shaped. Legs: coxa I with two distinct, short blunt spurs, subequal in size; coxae II−IV with a triangular spur each; trochanters with spurs. Spiracular plates comma-shaped.

Nymph (Fig. 2.57): small body size, total length 1.2 mm (1.1−1.2), breadth 0.9 mm (0.8−0.9). Body outline oval. Chitinous tubercles at the posterior body margin absent. Scutum with few medium punctuations evenly

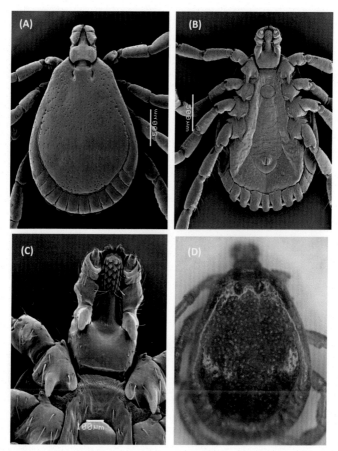

FIGURE 2.55   *A. parvum* male: (A) dorsal view, (B) ventral view, (C) capitulum ventral view, (D) scutum ornamentation.

distributed. Eyes slightly bulging, located on lateral scutal angles at the level of scutal mid-length, Cervical grooves long, deep anteriorly. Basis capituli subrectangular dorsally, with small cornua. Hypostome spatulate, dental formula 2/2. Legs: coxa I with two spurs, the external longer than the internal; coxae II−IV with a small triangular spur each.

Taxonomic notes: the presence of carena is described for the males of *A. parvum*, but in some specimens this morphological feature is almost imperceptible. Nava et al.[1] have suggested that *A. parvum* from Argentina and Brazil could be different species because their 16S rDNA gene sequences show dissimilarities. However, Nava et al.[12] have demonstrated that *A. parvum* populations from Argentina and Brazil belong to the same specific entity. Lado et al.[20] identified molecularly two supported clusters of *A. parvum*, a Central American lineage and a Brazilian-Argentinian lineage, by

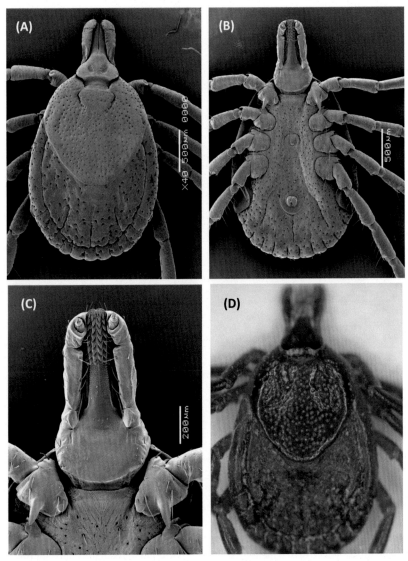

FIGURE 2.56 *A. parvum* female: (A) dorsal view, (B) ventral view, (C) capitulum ventral view, (D) scutum ornamentation.

using mitochondrial (the small and the large ribosomal subunits 12S rDNA and 16S rDNA, the cytochrome oxidase I and II, and the control region or d-loop) and nuclear (internal transcribed spacer 2) markers. These authors have suggested that these two lineages may correspond to different taxonomic entities. See also Taxonomic notes of *A. auricularium* (Conil, 1878).

DNA sequences with relevance for tick identification and phylogeny: the following DNA sequences of *A. parvum* are available in the Gen Bank:

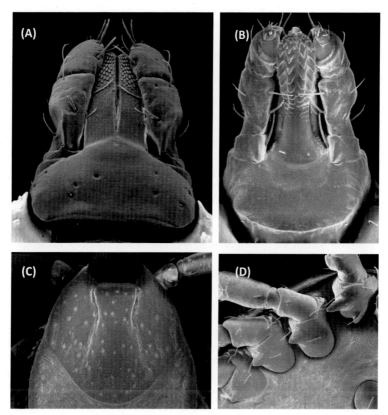

FIGURE 2.57 *A. parvum* nymph: (A) capitulum dorsal view, (B) capitulum ventral view, (C) scutum, (D) coxae. *Figures (A) to (D) are reproductions of figure 11:* A. parvum *nymph. Capitulum, dorsum; Capitulum, venter; Scutum; Coxae I—IV. From Martins TF, Labruna MB, Mangold AJ, Cafrune MM, Guglielmone AA, Nava S. Taxonomic key to nymphs of the genus* Amblyomma *(Acari: Ixodidae) in Argentina, with description and redescription of the nymphal stage of four* Amblyomma *species.* Ticks Tick-borne Dis *2014;5:753—70, with permission of Elsevier.*

16S rDNA gene (accession numbers EU306136—EU306157, KT820252—KT820314), 12S rDNA gene (accession numbers KT820213—KT820251, AY342293), control region or d-loop sequences (accession numbers KT820315—KT820347), cytochrome oxidase I (accession numbers KT820415—KT82016), cytochrome oxidase II (accession numbers KT820400—KT8204414), and ITS2 sequences (accession numbers KT820348—KT820355).

REFERENCES

1. Nava S, Szabó MPJ, Mangold AJ, Guglielmone AA. Distribution, hosts, 16S rDNA sequences and phylogenetic position of the Neotropical tick *Amblyomma parvum* (Acari: Ixodidae). *Ann Trop Med Parasitol* 2008;**102**:409—25.
2. Corn JL, Hanson B, Okraska CR, Muiznieks B, Morgan V, Mertins JW. First at-large record of *Amblyomma parvum* (Acari: Ixodidae) in the United States. *Syst Appl Acarol* 2012;**17**:3—6.

3. Guglielmone AA, Nava S. Las garrapatas argentinas del género *Amblyomma* (Acari: Ixodidae): distribución y hospedadores. *Rev Invest Agropec* 2006;**35**(3):135−55.

4. Nava S, Mangold AJ, Guglielmone AA. The natural hosts for larvae and nymphs of *Amblyomma neumanni* and *Amblyomma parvum* (Acari: Ixodidae). *Exp Appl Acarol* 2006;**40**:123−31.

5. Horta MC, Nascimento GF, Martins TF, Labruna MB, Machado LCP, Nicola PA. Ticks (Acari: Ixodida) parasitizing free-living wild animals in the Caatinga biome in the State of Pernambuco, northeastern Brazil. *Syst Appl Acarol* 2011;**16**:207−11.

6. Saraiva DG, Fournier GFSR, Martins TF, Leal KPG, Vieira FN, Camara EMVC, et al. Ticks (Acari: Ixodidae) associated with terrestrial mammals in the state of Minas Gerais, southeastern Brazil. *Exp Appl Acarol* 2012;**58**:159−66.

7. Ogrzewalska M, Literak I, Martins TF, Labruna MB. Rickettsial infections in ticks from wild birds in Paraguay. *Ticks Tick-borne Dis* 2014;**5**:83−9.

8. Lugarini C, Martins TF, Ogrzewalska M, Vasconcelos NCT, Ellis VA, Oliveira JB. Rickettsial agents in avian ixodid ticks in northeast Brazil. *Ticks Tick-borne Dis* 2015;**6**:364−75.

9. Sponchiado J, Melo GL, Martins TF, Krawczak FS, Labruna MB, Cáceres NC. Association patterns of ticks (Acari: Ixodida: Ixodidae, Argasidae) of small mammals in Cerrado fragments, western Brazil. *Exp Appl Acarol* 2015;**65**:389−91.

10. Rodríguez-Vivas RI, Apanaskevich DA, Ojeda-Chin MM, Trinidad-Martínez I, Reyes-Novelo E, Esteve-Gassent MD, et al. Ticks collected from humans, domestic animals, and wildlife in Yucatán. Mexico. *Vet Parasitol* 2016;**215**:106−13.

11. Witter R, Martins TF, Campos AK, Melo ALT, Correa SHR, Morgado TO, et al. Rickettsial infection in ticks (Acari: Ixodidae) of wild animals in Midwestern Brazil. *Ticks Tick-borne Dis* 2016;**7**:415−23.

12. Nava S, Gerardi M, Szabó MPJ, Mastropaolo M, Martins TF, Labruna MB, et al. Different lines of evidence used to delimit species in ticks: a study of the South American populations of *Amblyomma parvum* (Acari: Ixodidae). *Ticks Tick-borne Dis* 2016;**7**:1168−79.

13. Nava S, Mangold AJ, Guglielmone AA. Aspects of the life cycle of *Amblyomma parvum* (Acari: Ixodidae) under natural conditions. *Vet Parasitol* 2008;**156**:270−6.

14. Guglielmone AA, Mangold AJ, Viñabal AE. Ticks (Ixodidae) parasitizing humans in four provinces of northwestern Argentina. *Ann Trop Med Parasitol* 1991;**85**:539−42.

15. Nava S, Caparrós JA, Mangold AJ, Guglielmone AA. Ticks (Acari: Ixodida: Argasidae, Ixodidae) infesting humans in northwestern Córdoba Province, Argentina. *Medicina (Buenos Aires)* 2006;**66**:225−8.

16. Pacheco RC, Moraes-Filho J, Nava S, Brandao PE, Richtzenhain LJ, Labruna MB. Detection of a novel spotted fever group rickettsia in *Amblyomma parvum* ticks (Acari: Ixodidae) from Argentina. *Exp Appl Acarol* 2007;**43**:63−71.

17. Pacheco RC, Echaide IE, Alves RN, Beletti ME, Nava S, Labruna MB. *Coxiella burnetii* in ticks, Argentina. *Emerg Infect Dis* 2013;**19**:344−6.

18. Tomassone L, Nuñez P, Gurtler R, Ceballos LA, Orozco MA, Kitron UD, et al. Molecular detection of *Ehrlichia chaffeensis* in *Amblyomma parvum* ticks, Argentina. *Emerg Infect Dis* 2008;**14**:1953−5.

19. Labruna MB, Mattar S, Nava S, Bermúdez S, Venzal JM, Dolz G, et al. Rickettsioses in Latin America, Caribbean, Spain and Portugal. *Rev MVZ Córdoba* 2011;**16**:2435−57.

20. Lado P, Nava S, Labruna MB, Szabo MPJ, Durden LA, Bermúdez S, et al. *Amblyomma parvum* Aragão, 1908 (Acari: Ixodidae): phylogeography and systematic considerations. *Ticks Tick-borne Dis* 2016;**7**:817−27.

Amblyomma pseudoconcolor
Aragão, 1908

Aragão, H.B. (1908) Mais um novo carrapato brazileiro. *Brazil Medico*, 22, 431−432.

DISTRIBUTION

Argentina, Bolivia, Brazil, French Guiana, Paraguay, Surinam, and Uruguay.[1,2] Keirans and Durden[3] mentioned the presence of *A. pseudoconcolor* in the Nearctic Region, but there is no evidence that this tick species has become established there,[4] while the record of this species by Romero-Castañón et al.[5] from the neotropics of Mexico requires confirmation.

Biogeographic distribution in the Southern Cone of America: Chaco, Pampa (Neotropical Region), Monte (South American Transition Zone), and Central Patagonia (Andean Region) (see Fig. 1 in the introduction).

HOSTS

The principal hosts for all stages of *A. pseudoconcolor* are mammals of the family Dasypodidae, but adults and nymphs of this tick were also recorded on mammals of different orders, and nymphs also from Aves[1,2,6−10] (Table 2.17).

ECOLOGY

A. pseudoconcolor has been collected in a wide variety of vegetation types, which suggests a high plasticity of this tick to adapt to different environments conditions.[1] The distribution of *A. pseudoconcolor* appears to be determined by the range of hosts rather than environmental conditions, probably due to its nidicolous habits. There is no information on population dynamics of *A. pseudoconcolor*.

SANITARY IMPORTANCE

"*Candidatus* Rickettsia andeanae" was detected in *A. pseudoconcolor*, but this rickettsia is currently considered of unknown pathogenicity.[11,12] The only record of *A. pseudoconcolor* on human corresponds to an adult tick in Argentina.[11]

**TABLE 2.17 Hosts of Adults (A), Nymphs (N), and Larvae (L) of
A. pseudoconcolor**

MAMMALIA		*Tolypeutes matacus*	A
ARTIODACTYLA		*Zaedyus pichiy*	ANL
Bovidae		**DIDELPHIMORPHIA**	
Cattle	A	**Didelphidae**	
Domestic buffalo	A	*Didelphis albiventris*	N
Cervidae		*Philander* sp.	A
Mazama gouazoubira	A	**PERISSODACTYLA**	
CARNIVORA		**Tapiridae**	
Canidae		*Tapirus terrestris*	A
Domestic dog	A	**PILOSA**	
Mustelidae		**Myrmecophagidae**	
Galictis sp.	AN	*Myrmecophaga tridactyla*	A
CINGULATA		*Tamandua tetradactyla*	A
Dasypodidae		**PRIMATES**	
Cabassous unicinctus	A	**Hominidae**	
Calyptophractus retusus	AN	Human	A
Chaetophractus vellerosus	AL	**AVES**	
C. villosus	A	**TINAMIFORMES**	
Dasypus hybridus	A	**Tinamidae**	
D. novemcinctus	M	*Nothura boraquira*	N
Euphractes sexcinctus	AN	*N. maculosa*	N
Priodontes maximus	A	*Rhynchotus rufescens*	N

DIAGNOSIS

Male (Fig. 2.58): medium body size, total length 3.1 mm (3.0−3.2), breadth
2.7 mm (2.6−2.8). Body outline oval rounded, scapulae rounded; cervical
grooves short, comma-shaped; marginal groove complete. Eyes flat. Scutum
ornate, with small pale spots adjacent to the marginal groove; few punctuations, moderately deep. Carena absent. Basis capituli subrectangular dorsally,
cornua short, article I of palps ventrally with a large blunt spur directed posteriorly. Hypostome spatulate, dental formula 3/3. Genital aperture located at

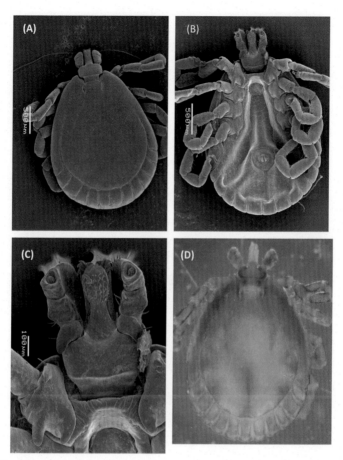

FIGURE 2.58 *A. pseudoconcolor* male: (A) dorsal view, (B) ventral view, (C) capitulum ventral view, (D) scutum ornamentation.

level of coxa II, U-shaped. Legs: coxa I with two distinct, short, triangular spurs, subequal in size; coxae II−IV with a triangular spur each; trochanters with spurs. Spiracular plates comma-shaped.

Female (Fig. 2.59): medium body size, total length 4.0 mm (3.9−4.1), breadth 3.4 mm (3.3−3.5). Body outline oval rounded. Scutum length 1.9 mm (1.6−2.0), breadth 2.3 mm (2.3−2.4); scapulae rounded, cervical grooves short and comma-shaped. Eyes flat. Chitinous tubercles at the posterior body margin absent. Scutum ornate, with small pale spots on a yellowish-brown ground; punctuations small, moderately deep. Notum glabrous. Basis capituli subtriangular dorsally, cornua short, porose areas oval, article I of palps ventrally with a large blunt spur directed posteriorly. Hypostome spatulate, dental formula 3/3. Genital aperture located at level of coxa II, U-shaped. Legs: coxa I with two distinct, short triangular spurs,

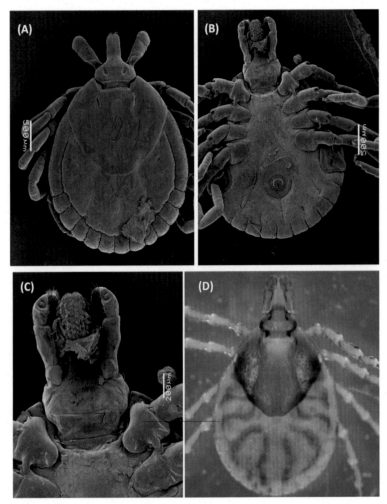

FIGURE 2.59 *A. pseudoconcolor* female: (A) dorsal view, (B) ventral view, (C) capitulum ventral view, (D) scutum ornamentation.

subequal in size; coxae II–IV with a triangular spur each; trochanters with spurs. Spiracular plates comma-shaped.

Nymph (Fig. 2.60): small body size, total length 1.4 mm (1.3–1.5), breadth 1.2 mm (1.0–1.3). Body outline oval. Chitinous tubercles at the posterior body margin absent. Scutum with few medium punctuations evenly distributed. Eyes flat, located on lateral scutal angles at the level of scutal mid-length. Cervical grooves long, reaching the scutal postero-lateral margin, deeper at the anterior third. Basis capituli rectangular dorsally, without cornua. Hypostome spatulate, dental formula 2/2. Legs: coxa I with two spurs, the external longer than the internal; coxae II–IV with a small triangular spur each.

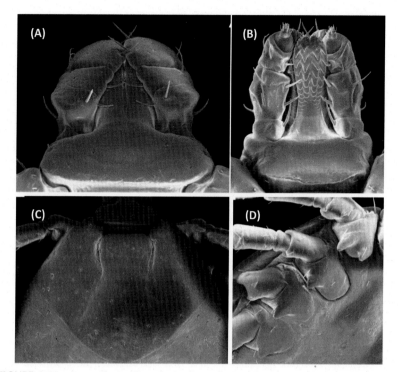

FIGURE 2.60   *A. pseudoconcolor* nymph: (A) capitulum dorsal view, (B) capitulum ventral view, (C) scutum, (D) coxae. *Figures (A) to (D) are reproductions of figure 21:* A. pseudoconcolor *nymph. Capitulum, dorsum; Capitulum, venter; Scutum; Coxae I—IV. From Martins TF, Labruna MB, Mangold AJ, Cafrune MM, Guglielmone AA, Nava S. Taxonomic key to nymphs of the genus* Amblyomma *(Acari: Ixodidae) in Argentina, with description and redescription of the nymphal stage of four* Amblyomma *species.* Ticks Tick-borne Dis *2014;5:753—70, with permission of Elsevier.*

Taxonomic notes: see Taxonomic notes of *A. auricularium* (Conil, 1878). *DNA sequences with relevance for tick identification and phylogeny*: DNA sequences of *A. pseudoconcolor* are available in the Gen Bank as follows: mitochondrial genes 16S rRNA (accession numbers AY628134— AY628137) and 12S rRNA (accession number AY342294), and sequences of the nuclear genome (5.8S ribosomal RNA gene, internal transcribed spacer 2, 28S ribosomal RNA gene) (accession number AY995180).

REFERENCES

1. Guglielmone AA, Estrada-Peña A, Luciani CA, Mangold AJ, Keirans JE. Host and distribution of *Amblyomma auricularium* (Conil 1878) and *Amblyomma pseudoconcolor* Aragão, 1908 (Acari: Ixodidae). *Exp Appl Acarol* 2003;**29**:131—9.

2. Robbins RG, Deem SL, Noss AJ, Greco V. First report of *Amblyomma pseudoconcolor* Aragão (Acari: Ixodida: Ixodidae) from Bolivia, with a new record of this tick from the grey brocket deer, *Mazama gouazoupira* (G. Fischer) (Mammalia: Artiodactyla: Cervidae). *Proc Entomol Soc Wash* 2003;**105**:1053−5.

3. Keirans JE, Durden LA. Invasion: exotic ticks (Acari: Argasidae, Ixodidae) imported into the United States. A review and new records. *J Med Entomol* 2001;**38**:850−61.

4. Guglielmone AA, Robbins RG, Apanaskevich DA, Petney TN, Estrada-Peña A, Horak IG. *The hard ticks of the world (Acari: Ixodida: Ixodidae)*. London: Springer; 2014.

5. Romero-Castañon S, Ferguson BG, Güiris D, González D, López S, Paredes A, et al. Comparative parasitology of wild and domestic ungulates in the Selva Lacandona, Chiapas, México. *Comp Parasitol* 2008;**75**:115−26.

6. Aragão HB. Ixodidas brasileiros e de algunos países limitrophes. *Mem Inst Oswaldo Cruz* 1936;**31**:759−843.

7. Keirans JE. George Henry Falkiner Nuttall and the Nuttall tick catalogue. *US Dep Agric Agric Res Serv Miscel Publ* 1985;**1438**:1785.

8. Superina M, Guglielmone AA, Mangold AJ, Nava S, Lareschi M. New distributional and host records for *Amblyomma pseudoconcolor* Aragão, 1908 (Acari: Ixodidae: Amblyomminae). *Syst Appl Acarol* 2004;**9**:41−3.

9. Guglielmone AA, Nava S. Las garrapatas argentinas del género *Amblyomma* (Acari: Ixodidae): distribución y hospedadores. *Rev Invest Agropec* 2006;**35**(3):135−55.

10. Martins TF, Furtado MM, Jácomo AA, Silveira L, Sollmann R, Torres NM, Labruna MB. Ticks on free-living wild mammals in Emas National Park, Goiás State, central Brazil. *Syst Appl Acarol* 2001;**16**:201−6.

11. Tomassone L, Nuñez P, Ceballos LA, Gurtler RE, Kitron U, Farber M. Detection of "*Candidatus* Rickettsia sp. strain Argentina" and *Rickettsia bellii* in *Amblyomma* ticks (Acari: Ixodidae) from Northern Argentina. *Exp Appl Acarol* 2010;**52**:93−100.

12. Parola P, Paddock CD, Scolovschi C, Labruna MB, Mediannokov O, Kernif T, et al. Update on tick-borne rickettsioses around the world: a geographic approach. *Clin Microbiol Rev* 2013;**26**:657−702.

Amblyomma pseudoparvum Guglielmone, Mangold, and Keirans, 1990

Guglielmone, A.A., Mangold, A.J. & Keirans, J.E. (1990) Redescription of the male and female of *Amblyomma parvum* Aragão, 1908, and description of the nymph and larva, and description of all stages of *Amblyomma pseudoparvum* sp. n. (Acari: Ixodida: Ixodidae). *Acarologia*, 31, 143−159.

DISTRIBUTION

Argentina. Guglielmone et al.[1] mentioned the presence of *A. pseudoparvum* ticks from Brazil in the US National Tick Collection, Georgia Southern University, Statesboro, Georgia, USA, but no specimens of *A. pseudoparvum* from Brazil was found in a revision of that collection.[2] Therefore, Brazil is provisionally excluded from the distributional range of *A. pseudoparvum*.

Biogeographic distribution in the Southern Cone of America: Chaco (Neotropical Region) (see Fig. 1 in the introduction).

HOSTS

The principal host for adults of *A. pseudoparvum* is a rodent, *Dolichotis salinicola*, but this tick was also collected on other species of mammals[3–5] (Table 2.18). The few records of immature stages correspond to nymphs collected on *D. salinicola* and cattle (Table 2.18).

ECOLOGY

The distribution of *A. pseudoparvum* is restricted to areas of the Chaco Biogeographic Province where the climate is hot and dry (annual rainfall between 500 and 800 mm concentrated in summer) with vegetation characterized by xerophilic forest.[3] Most of the records of adults of *A. pseudoparvum* were made in spring and summer.[3]

TABLE 2.18 Hosts of Adults (A) and Nymphs (N) of *A. pseudoparvum*

MAMMALIA		*Tolypeutes matacus*	A
ARTIODACTYLA		PRIMATES	
Bovidae		Hominidae	
Cattle	AN	Human	A
Cervidae		RODENTIA	
Mazama gouazoubira	A	Caviidae	
CARNIVORA		*Dolichotis salinicola*	AN
Canidae		Chinchillidae	
Domestic dog	A	*Lagostomus maximus*	A
CINGULATA			
Dasypodidae			
Chaetophractus villosus	A		

Main host for adult ticks underlined.

SANITARY IMPORTANCE

Although there are few records of *A. pseudoparvum* on man, cattle, and dog, this tick does not appear to be relevant from a medical and veterinary perspective.

DIAGNOSIS

Male (Fig. 2.61): small body size, total length 1.8 mm (1.6−2.0), breadth 1.1 mm (1.1−1.3). Body outline oval elongated, scapulae rounded, cervical grooves short, comma-shaped; marginal groove complete. Eyes flat. Scutum inornate; punctuations sparse, moderately deep, uniformly distributed. Carena absent. Basis capituli subrectangular dorsally, cornua long, article I of palps ventrally with a large blunt spur directed posteriorly. Hypostome

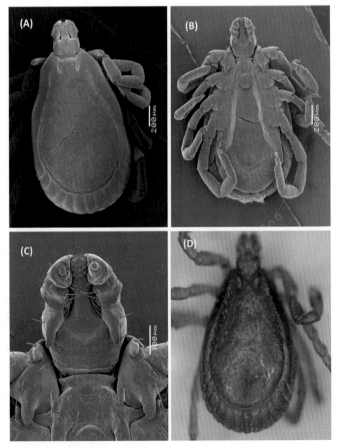

FIGURE 2.61   *A. pseudoparvum* male: (A) dorsal view, (B) ventral view, (C) capitulum ventral view, (D) scutum ornamentation.

spatulate, dental formula 3/3. Genital aperture located at level of coxa II, U-shaped. Legs: coxa I with two distinct spurs, the external longer than the internal; coxae II—IV with a triangular spur each; trochanters with spurs. Spiracular plates comma-shaped.

Female (Fig. 2.62): small body size, total length 2.2 mm (2.0—2.5), breadth 1.5 mm (1.2—1.7). Body outline oval. Scutum length 1.3 mm (1.1—1.4), breadth 1.2 mm (1.1—1.3); scapulae rounded, cervical grooves short and comma-shaped. Eyes flat. Chitinous tubercles at the posterior body margin absent. Scutum inornate, punctuations numerous and deep, forming an evident depression on each lateral field; punctuations scarce around eyes. Notum with numerous, long, white setae, uniformly distributed. Basis capituli subtriangular dorsally, cornua short, porose areas oval, article I of palps

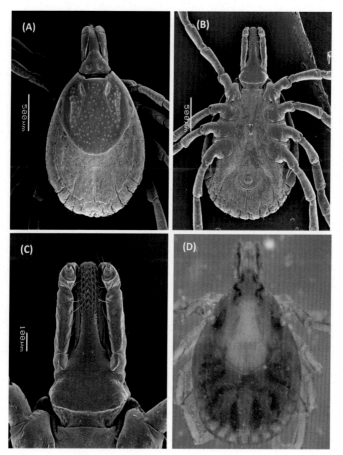

FIGURE 2.62 *A. pseudoparvum* female: (A) dorsal view, (B) ventral view, (C) capitulum ventral view, (D) scutum ornamentation.

ventrally with a large blunt spur directed posteriorly. Hypostome spatulate, dental formula 3/3. Genital aperture located at level of coxa II, U-shaped. Legs: coxa I with two distinct, short blunt spurs, subequal in size; coxae II—IV with a triangular spur each; trochanters with spurs. Spiracular plates comma-shaped.

Nymph (Fig. 2.63): small body size, total length 1.1 mm (1.0—1.2), breadth 0.8 mm (0.7—0.9). Body outline oval. Chitinous tubercles at the posterior body margin absent. Scutum with few medium punctuations, more numerous on lateral fields. Eyes flat, located on lateral scutal angles at the level of scutal mid-length. Cervical grooves long, deep anteriorly. Basis capituli rectangular dorsally, with small cornua. Hypostome spatulate, dental formula 2/2. Legs: coxa I with two spurs, the external longer than the internal; coxae II—IV with a small triangular spur each.

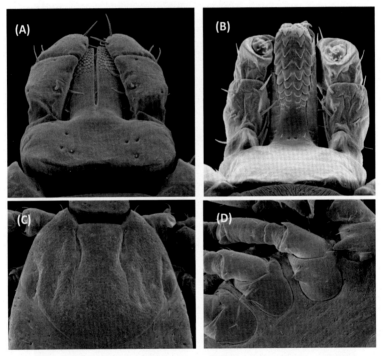

FIGURE 2.63 *A. pseudoparvum* nymph: (A) capitulum dorsal view, (B) capitulum ventral view, (C) scutum, (D) coxae. *Figures (A) to (D) are reproductions of figure 12:* A. pseudoparvum *nymph. Capitulum, dorsum; Capitulum, venter; Scutum; Coxae I—IV. From Martins TF, Labruna MB, Mangold AJ, Cafrune MM, Guglielmone AA, Nava S. Taxonomic key to nymphs of the genus* Amblyomma *(Acari: Ixodidae) in Argentina, with description and redescription of the nymphal stage of four* Amblyomma *species. Ticks Tick-borne Dis 2014;5:753—70, with permission of Elsevier.*

Taxonomic notes: see Taxonomic notes of *A. auricularium* (Conil, 1878). *DNA sequences with relevance for tick identification and phylogeny*: the following DNA sequences of *A. pseudoparvum* are available in the Gen Bank: 16S rDNA gene (accession numbers FJ627952, KT820360), 12S rDNA gene sequences (accession number KT820356), control region or d-loop (accession number KT820357), cytochrome oxidase I (accession number KT820365), and 18S rDNA (accession number FJ464421).

REFERENCES

1. Guglielmone AA, Estrada-Peña A, Keirans JE, Robbins RG. Ticks (Acari: Ixodida) of the neotropical zoogeographic region. Special publication of the integrated consortium on ticks and tick-borne diseases-2. Houten (The Netherlands): Atalanta; 2003.
2. Dantas-Torres F, Onofrio VC, Barros-Battesti DM. The ticks (Acari: Ixodida: Argasidae, Ixodidae) of Brazil. *Syst Appl Acarol* 2009;**14**:30−46.
3. Guglielmone AA, Mangold AJ, Keirans JE. Redescription of the male and female of *Amblyomma parvum* Aragão, 1908, and description of the nymph and larva, and description of all stages of *Amblyomma pseudoparvum* sp. n. (Acari: Ixodida: Ixodidae). *Acarologia* 1990;**31**:143−59.
4. Guglielmone AA, Nava S. Las garrapatas argentinas del género *Amblyomma* (Acari: Ixodidae): distribución y hospedadores. *Rev Invest Agropec* 2006;**35**(3):135−55.
5. Guglielmone AA, Robbins RG, Apanaskevich DA, Petney TN, Estrada-Peña A, Horak IG. *The hard ticks of the world (Acari: Ixodida: Ixodidae)*. London: Springer; 2014.

Amblyomma rotundatum Koch, 1884

Koch, C.L. (1844) Systematische Übersicht über die Ordnung der Zecken. *Archiv für Naturgeschichte*, 10, 217−239.

DISTRIBUTION

Argentina, Bolivia, Brazil, Colombia, Costa Rica, Dominica, French Guiana, Grenada, Guadeloupe, Guatemala, Jamaica, Martinique, Mexico, Montserrat, Panama, Paraguay, Peru, Surinam, Trinidad and Tobago, and Venezuela.[1−3] *A. rotundatum* is also established in the Nearctic Region.[4] Pietzsch et al.[5] have mentioned the introduction of *A. rotundatum* into the Palearctic Region, but there is no evidence of its establishment in this region.[6] Vogelsang[7] allegedly found larvae of *A. rotundatum* (as *A. agamum*) in Uruguay but Venzal et al.[8] considered this record doubtful.

Biogeographic distribution in the Southern Cone of America: Chaco (Neotropical Region) (see Fig. 1 in the introduction).

HOSTS

Amphibians and reptiles are the principal hosts for all parasitic stages of
A. rotundatum; most records on free-ranging hosts correspond to toads and
the reptile *B. constrictor* (Table 2.19),[6] but the range of reptile hosts
for this species is continuously increasing.[9,10] There are also records of

TABLE 2.19 Hosts of Females (F), Nymphs (N), and Larvae (L) of
A. rotundatum, a Parthenogenetic Species

AMPHIBIA		*Eunectes murinus*	FNL
ANURA		**Colubridae**	
Bufonidae		*Spilotes pullatus*	FN
Anaxyrus terrestris	L	*Philodryas olfersii*	N
Rhinella bergi	F	*Waglerophis merremi*	F
R. crucifer	FN	**Dipsadidae**	
R. granulosa	F	*Hydrodynastes gigas*	F
R. icterica	FN	*Oxyrhopus guibei*	F
R. jimi	FN	**Elapidae**	
R. marina	FNL	*Micrurus* sp.	FNL
R. ornate	FN	*M. ibiboboca*	F
R. schneideri	FNL	**Iguanidae**	
Pipidae		*Iguana iguana*	FN
Pipa pipa	F	**Phrynosomatidae**	
Leptodactylidae		*Phrynosoma* sp.	FN
Physalaemus nattereri	F	**Scincidae**	
Scaphiopodidae		*Mabuya mabuya*	NL
Spea bombrifons	FN	**Teiidae**	
MAMMALIA		*Ameiva ameiva*	F
ARTIODACTYLA		*Salvator* sp.	L
Suidae		*S. merianae*	N
Domestic pig	F	**Tropiduridae**	
CARNIVORA		*Plica plica*	F
Procyonidae		*Tropidurus* sp.	FNL[a]
Nasua nasua	F	*T. torquatus*	N

(Continued)

TABLE 2.19 (Continued)

CHIROPTERA		Viperidae	
Phyllostomidae		*Bothrops alternatus*	F
Choeroniscus minor	F	*B. atrox*	FN
CINGULATA		*B. lanceolatus*	F
Dasypodidae		*B. leucurus*	F
Dasypus novemcinctus	F	*B. moojeni*	F
PILOSA		*B. neuwiedi*	F
Bradypodidae		*Crotalus durissus*	F
Bradypus variegatus	F	*Lachesis muta*	FNL
Myrmecophagidae		TESTUDINES	
Tamandua tetradactyla	F	Chelidae	
PRIMATES		*Hydromedusa tectifera*	F
Hominidae		*Mesoclemmys vanderhaegei*	F
Human	F	*Platemys platycephala*	F
REPTILIA		Geoemydidae	
CROCODYLIA		*Heosemys annandalii*	F
Alligatoridae		*Rhinoclemmys areolata*	F
Paleosuchus palpebrosus	F	Kinosternidae	
P. trigonatus	FN	*Kinosternon scorpiodes*	F
SQUAMATA		Testudinidae	
Boidae		*Chelonoidis carbonaria*	F
Boa constrictor	FNL[a,b]	*C. chilensis*	F
Epicrates cenchria	FNL	*C. denticulata*	F

[a]A. rotundatum *is a parthenogenetic species with only two records of males naturally infesting* Tropidurus sp.[17] *in one case, and* B. constrictor *for the second*[b].
[b]*Martins TF, Venzal JM, Terassini FA, Costa FB, Marcili M, Camargo LMA et al. New tick records from the State of Rôndonia, western Amazon, Brazil. Exp Appl Acarol 2014;**62**:121–8.*

A. rotundatum on mammals belonging to different orders (Table 2.19).[6] Scott and Durden[11] found *A. rotundatum* on a migratory bird in Canada.

ECOLOGY

A. rotundatum is a parthenogenetic tick with a three-host life cycle, although a facultative two-host life cycle was also reported.[12–15] Consequently, the records of this species correspond mostly to immature stages and females. Keirans and Oliver[16] described the male of *A. rotundatum* by using two specimens reared from a laboratory colony, but these males were teratological. The first description of a male collected on a free-living host under natural conditions was performed by Labruna et al.[17] These authors described a male with no obvious morphological anomalies which was collected on a lizard (*Tropidurus* sp.). Although the biology of *A. rotundatum* was studied at the laboratory,[12−15,18] the information on the life cycle of this tick under natural conditions is inexistent.

SANITARY IMPORTANCE

Deleterious and lethal effects have been reported in hosts experimentally infested with *A. rotundatum*.[12,15,19] This tick is the vector of *Hemolivia stellata*,[20] a parasite of blood red and endothelial cells of toads, but also a potential vector of *Hepatozoon* spp.[21] *Rickettsia bellii* infection in *A. rotundatum* ticks was also recorded by Horta et al.[22]; the pathogenicity of this bacteria is unknown. There is just one record of an adult of *A. rotundatum* parasitizing humans.

DIAGNOSIS

Male (Fig. 2.64): measures are from one specimen described in Keirans and Oliver.[16] Medium body size, total length 3.8 mm, breadth 3.1 mm. Body outline oval-rounded, scapulae rounded, cervical grooves deep, short, comma-shaped; marginal groove absent. Eyes flat. Scutum ornate, with pale iridescent patches of orange ornamentation on lateral fields, postero-accessory and postero-median spots narrow, lateral spots small but distinct; punctuations numerous, more densely distributed on antero-lateral field. Carena absent. Basis capituli subrectangular dorsally, cornua short. Hypostome spatulate, dental formula 3/3. Genital aperture located at level of coxa II, U-shaped. Legs: coxae I−IV each with two distinct, short, blunt spurs; external spurs on coxa I subequal in size; trochanters without spurs. Spiracular plates comma-shaped.

Female (Fig. 2.65): large body size, total length 5.0 mm (4.9−5.2), breadth 3.8 mm (3.7−4.0) in unfed specimens. Body outline oval. Scutum length 2.2 mm (2.0−2.4), breadth 2.5 mm (2.4−2.6); scapulae rounded; cervical grooves long and deep, sigmoid. Eyes flat. Chitinous tubercles at the postero-body margin absent. Scutum ornate, with an irregular pale spot in each antero-lateral field, and a small pale spot in the posterior field; punctuations few and large, concentrated in the antero-lateral fields. Notum glabrous. Basis capituli dorsally subrectangular, cornua absent, porose areas oval. Hypostome spatulate, dental formula 3/3. Genital aperture located at level of

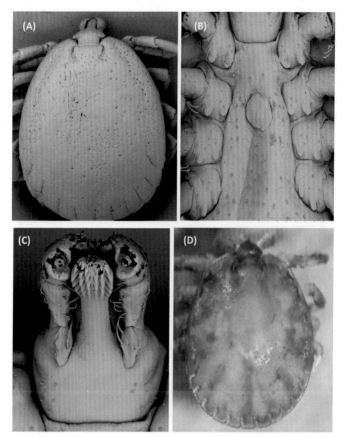

FIGURE 2.64 *A. rotundatum* male: (A) dorsal view, (B) ventral view, (C) capitulum ventral view, (D) scutum ornamentation.

coxa II, U-shaped. Legs: coxae I—IV each with two distinct, short blunt spurs; external spurs on coxae I—IV larger than internal spurs; trochanters without spurs. Spiracular plates comma-shaped.

Nymph (Fig. 2.66): medium body size large, total length 1.6 mm (1.5—1.8), breadth 1.3 mm (1.2—1.4). Body outline oval. Chitinous tubercles at the posterior body margin absent. Scutum with punctuations moderate in number, larger and deeper in the lateral fields. Eyes flat, located on lateral scutal angles at the level of scutal mid-length. Cervical grooves ending at the level of scutal mid-length, deep anteriorly. Basis capituli dorsally subtriangular, without cornua. Hypostome spatulate, dental formula 3/3 apically, 2/2 at the base. Legs: coxa I with two equal triangular spurs; coxa II with two short spurs, the external longer than the internal; coxa III with an external

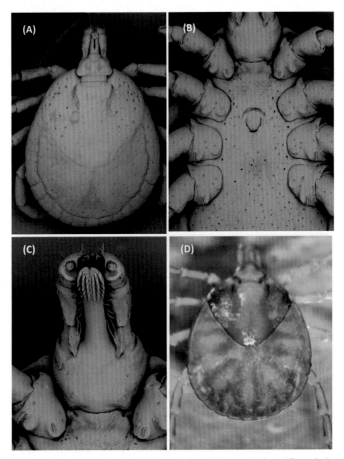

FIGURE 2.65   *A. rotundatum* female: (A) dorsal view, (B) ventral view, (C) capitulum ventral view, (D) scutum ornamentation.

triangular spur, and an internal spur obsolete or absent; coxa IV with a small triangular spur.

Taxonomic notes: see Taxonomic notes of *A. dissimile* Koch, 1844.

DNA sequences with relevance for tick identification and phylogeny: DNA sequences of *A. rotundatum* are available in the Gen Bank as follows: mitochondrial 12S rRNA gene (accession numbers KP987772, AY342250), 16S rRNA gene (accession numbers KP987773, KJ569693, EU805569), cytochrome oxidase II gene (accession numbers KT630577, FJ917664–FJ917668), and cytochrome oxidase I gene (accession number KP987771). Also there are available sequences of 18S rRNA gene (accession numbers

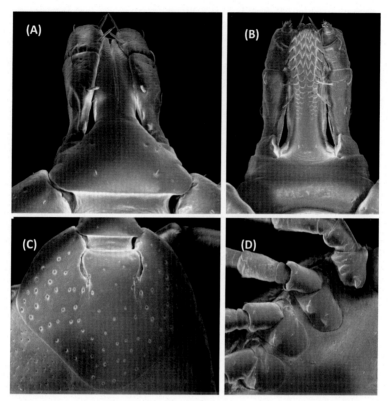

FIGURE 2.66 *A. rotundatum* nymph: (A) capitulum dorsal view, (B) capitulum ventral view, (C) scutum, (D) coxae. *Figures (A) to (D) are reproductions of figure 14:* A. rotundatum *nymph. Capitulum, dorsum; Capitulum, venter; Scutum; Coxae I–IV. From Martins TF, Labruna MB, Mangold AJ, Cafrune MM, Guglielmone AA, Nava S. Taxonomic key to nymphs of the genus* Amblyomma *(Acari: Ixodidae) in Argentina, with description and redescription of the nymphal stage of four* Amblyomma *species.* Ticks Tick-borne Dis *2014;5:753–70, with permission of Elsevier.*

KJ584369), calreticulin gene (accession number AY395251), and 5.8S ribosomal RNA gene, internal transcribed spacer 2, 28S ribosomal RNA gene (accession number AY887112).

REFERENCES

1. Guglielmone AA, Estrada-Peña A, Keirans JE, Robbins RG. Ticks (Acari: Ixodida) of the neotropical zoogeographic region. Special publication of the integrated consortium on ticks and tick-borne diseases-2. Houten (The Netherlands): Atalanta; 2003.
2. Nava S, Lareschi M, Rebollo C, Benítez Usher C, Beati L, Robbins RG, et al. The ticks (Acari: Ixodida: Argasidae, Ixodidae) of Paraguay. *Ann Trop Med Parasitol* 2007; **101**:255–70.

3. Durden LA, Knapp CR, Beati L, Dold S. Reptile-associated ticks from Dominica and the Bahamas with notes on hyperparasitic erythraeid mites. *J Parasitol* 2015;**101**:24–7.

4. Oliver JH, Hayes MP, Keirans JE, Lavender DR. Establishment of the foreign parthenogenetic tick *Amblyomma rotundatum* (Acari: Ixodidae) in Florida. *J Parasitol* 1993;**79**:786–90.

5. Pietzsch M, Quest R, Hillyard PD, Medlock JM, Leach S. Importation of exotic ticks into the United Kingdom via the international trade in reptiles. *Exp Appl Acarol* 2006;**38**:59–65.

6. Guglielmone AA, Nava S. Hosts of *Amblyomma dissimile* Koch, 1844 and *Amblyomma rotundatum* Koch, 1844. *Zootaxa* 2010;**2541**:27–49.

7. Vogelsang E. Garrapatas (Ixodidae) del Uruguay. *Bol Inst Clín Quir Buenos Aires* 1928;**4**:668–70.

8. Venzal JM, Castro O, Cabrera PA, De Souza CG, Guglielmone AA. Las garrapatas de Uruguay: especies, hospedadores, distribución e importancia sanitaria. *Veterinaria (Montevideo)* 2003;**38**:17–28.

9. Martins TF, Teixeira RHF, Labruna MB. Ocorrência de carrapatos em animais silvestres recebidos e atendidos pelo Parque Zoológico Municipal Quinzinho de Barros, Sorocaba, São Paulo, Brasil. *Braz J Vet Res Anim Sci* 2015;**52**:319–24.

10. Witter R, Martins TF, Campos AK, Melo ALT, Correa SHR, Morgado TO, et al. Rickettsial infection in ticks (Acari: Ixodidae) of wild animals in Midwestern Brazil. *Ticks Tick-borne Dis* 2016;**7**:415–23.

11. Scott JD, Durden LA. First record of *Amblyomma rotundatum* tick (Acari: Ixodidae) parasitizing a bird collected in Canada. *Syst Appl Acarol* 2015;**20**:155–61.

12. Aragão HB. Contribução para a sistematica e biolojia dos ixódidas. Partenojeneze em carrapatos. *Mem Inst Oswaldo Cruz* 1912;**4**:96–119.

13. Oba MSP, Schumaker TTS. Estudo da biologia de *Amblyomma rotundatum* (Koch, 1844), em infestações experimentais de *Bufo marinus* (L., 1758) sobre condições variadas de umidade relativa e temperatura do ar. *Mem Inst Butantan* 1993;**47**:195–204.

14. Rodrigues DS, Maciel R, Cunha LM, Leite CR, Oliveira PR. *Amblyomma rotundatum* (Koch, 1844) (Acari: Ixodidae) two host life-cycle on viperidae snakes. *Rev Bras Parasitol Vet* 2010;**19**:174–8.

15. Luz H, Faccini JLH, Pires MS, da Silva HR, Barros-Battesti DM. Life cycle and behavior of *Amblyomma rotundatum* (Acari: Ixodidae) under laboratory conditions and remarks on parasitism of toads in Brazil. *Exp Appl Acarol* 2013;**60**:55–62.

16. Keirans JE, Oliver JH. First description of the male and redescription of the immature stages of *Amblyomma rotundatum* (Acari: Ixodidae), a recently discovered tick in the U.S.A. *J Parasitol* 1993;**79**:860–5.

17. Labruna MB, Terrasini FA, Camargo LMA. First report of the male of *Amblyomma rotundatum* (Acari: Ixodidae) from a field-collected host. *J Med Entomol* 2005;**42**:945–7.

18. Labruna MB, Leite RC, Oliveira PR. Study of the weight of eggs from six ixodid species from Brazil. *Mem Inst Oswaldo Cruz* 1997;**92**:205–7.

19. Hanson BA, Frank PA, Mertins JW, Corn JL. Tick paralysis of a snake caused by *Amblyomma rotundatum* (Acari: Ixodidae). *J Med Entomol* 2007;**44**:155–7.

20. Petit G, Landau I, Baccam D, Lainson R. Description et cycle biologique D'*Hemolivia stellata*, n.g., n.sp., hemogrégarine de crapauds brésiliens. *Ann Parasitol Hum Comp* 1990;**65**:3–15.

21. Smith TG. The genus *Hepatozoon* (Apicomplexa: Adeleina). *J Parasitol* 1996;**82**:565–85.

22. Horta MC, Saraiva DG, Oliveira GMB, Martins TF, Labruna MB. *Rickettsia bellii* in *Amblyomma rotundatum* ticks parasitizing *Rhinella jimi* from northeastern Brazil. *Microb Infect* 2015;**17**:856–8.

Amblyomma sculptum
Berlese, 1888

Berlese, A. (1888) Acari austro-americani. *Bolletino della Società di Entomologia Italiana*, 20, 171−242.

DISTRIBUTION

Argentina, Bolivia, Brazil, and Paraguay[1].
Biogeographic distribution in the Southern Cone of America: Chaco and Yungas (Neotropical Region) (see Fig. 1 in the introduction).

HOSTS

Adults and immature stages of *A. sculptum* present low host specificity. They principally feed on large wild and domestic mammals of different orders, although larvae and nymphs are also parasites of small and medium-sized mammals.[1−5] Aves were recently recorded as hosts for adults, nymphs, and larvae of *A. sculptum* in Brazil[4,6,7] (Table 2.20).

ECOLOGY

A. sculptum has a 1-year life cycle, with larvae concentrated from mid-autumn to mid-winter, nymphs predominating in spring, and adults with the peak of abundance from late spring to late summer[8−11] (as *A. cajennense* in Refs. 8−10). Labruna et al.[12] and Cabrera and Labruna[13] have shown that the life cycle of *A. cajennense* sensu lato (most probably *A. sculptum*) is modulated by behavioral diapause of larvae born during the spring and summer that initiate a simultaneous feeding activity during the autumn. In the Southern Cone of America, *A. sculptum* prevails in the Yungas Biogeographic Province and in the more humid zones of the Chaco Biogeographical Province.[1,14]

SANITARY IMPORTANCE

Immature and adult stages of *A. sculptum* are aggressive common parasites of humans and domestic animals such as cattle and horses.[1,3,11,15] *A. sculptum* is a main vector of *Rickettsia rickettsii*.[16,17] This pathogen is the causative agent of spotted fever in humans, which is the most common *fatal human* tick-borne *disease in South America*. Specimens of *A. sculptum* also have been found naturally infected with "*Candidatus Rickettsia amblyommii*"[18,19] and "*Candidatus Rickettsia andeanae*."[4]

TABLE 2.20 Hosts of Adults (A), Nymphs (N), and Larvae (L) of *A. sculptum*

MAMMALIA		Tapiridae	
ARTIODACTYLA		*Tapirus terrestris*	AN
Bovidae		**PILOSA**	
Cattle	ANL	**Myrmecophagidae**	
Domestic buffalo	AN	*Myrmecophaga tridactyla*	AN
Cervidae		*Tamandua tetradactyla*	AN
Blastoceros dichotomus	A	**CINGULATA**	
Mazama gouazoubira	AN	**Dasypodidae**	
Suidae		*Cabassous unicinctus*	N
Pigs (domestic and feral)	ANL	*Euphractes sexcinctus*	A
Tayassuidae		**PRIMATES**	
Pecari tajacu	AN	**Atelidae**	
CARNIVORA		*Alouatta guariba*	N
Canidae		**Hominidae**	
Cerdocyon thous	N	Human	ANL
Chrysocyon brachyurus	AN	**RODENTIA**	
Domestic dog	AN	**Caviidae**	
Lycalopex vetulus	N	*Hydrochoerus hydrochaeris*	AN
Procyonidae		**Chinchillidae**	
Nasua nasua	A	*Lagostomus maximus*	N
Procyon cancrivorus	AN	**Cricetidae**	
Felidae		*Calomys callosus*	N
Panthera onca	AN	*Rhipidomys macrurus*	N
Puma concolor	A	*Sooretamys agouya*	N
Leopardus pardalis	N	*Thrichomys fosteri*	N
DIDELPHIMORPHIA		**Erethizontidae**	
Didelphidae		*Sphiggurus villosus*	N
Didelphis albiventris	AN	**AVES**	
PERISSODACTYLA		**CARAIAMIFORMES**	
Equidae		**Cariamidae**	
Horse	ANL	*Cariama cristata*	ANL
Mule	A	**GALLIFORMES**	
		Cracidae	
		Crax fasciolata	NL

DIAGNOSIS

Male (Fig. 2.67): medium body size, total length 3.6 mm (3.0–4.0), breadth 2.9 mm (2.4–3.3). Body outline oval rounded, scapulae pointed, cervical grooves short, comma-shaped; marginal groove complete. Eyes flat. Scutum ornate, cervical spots elongated posteriorly, anterior extremity of limiting spots sometimes merging with antero-accessory and ocular spots, lateral spots fused but distinct and clearly delimited by deep punctuations, postero-accessory spots small, postero-median spot narrower than adjacent enameled yellowish stripe; punctuations numerous, uniformly distributed. Carena absent. Basis capituli subrectangular dorsally, cornua short. Hypostome spatulate, dental formula 3/3. Genital aperture located at level of coxa II,

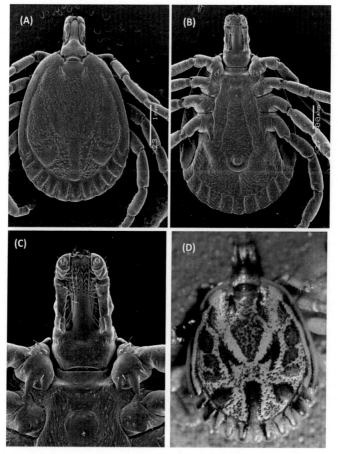

FIGURE 2.67   *A. sculptum* male: (A) dorsal view, (B) ventral view, (C) capitulum ventral view, (D) scutum ornamentation.

U-shaped. Legs: coxa I with two distinct, short blunt spurs, subequal in size; coxae II–III each with a short, triangular spur protruding from ridge-like edge; coxa IV with a pointed spur not reaching level of anus; trochanters without spurs. Spiracular plates comma-shaped.

Female (Fig. 2.68): medium body size, total length 3.9 mm (3.5–4.3), breadth 2.9 mm (2.9–3.5). Body outline oval rounded. Scutum length 1.9 mm (1.2–2.1), breadth 2.2 mm (1.9–2.3); scapulae pointed; cervical grooves sigmoid, deeper in the anterior part. Eyes flat. Chitinous tubercles at the posterior body margin present. Scutum ornate, with an irregular and diffuse central area not reaching the posterior margin of scutum, cervical spots elongated but not merging posteriorly with limiting spots; punctuations numerous, moderately deep, uniformly distributed. Notum with short setae,

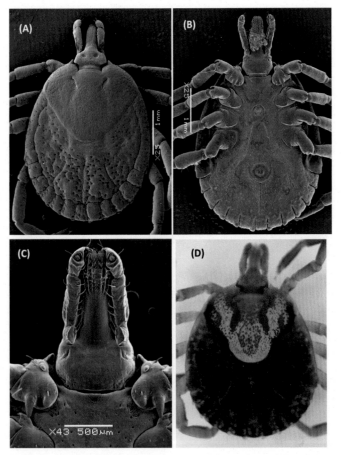

FIGURE 2.68 *A. sculptum* female: (A) dorsal view, (B) ventral view, (C) capitulum ventral view, (D) scutum ornamentation.

more densely distributed on central and posterior fields. Basis capituli subrectangular dorsally, cornua short, porose areas oval. Hypostome spatulate, dental formula 3/3. Genital aperture located at level of coxa II, U-shaped. Legs: coxa I with two distinct, short blunt spurs, the external longer than the internal; coxae II−IV with a triangular spur each; trochanters without spurs. Spiracular plates comma-shaped.

Nymph (Fig. 2.69): medium body size, total length 1.5 mm (1.4−1.6), breadth 1.2 mm (1.1−1.3). Body outline oval. Chitinous tubercles at the posterior body margin absent. Scutum with punctuations large and deep evenly distributed, larger in lateral fields. Eyes bulging, located on lateral scutal angles at the level of scutal mid-length. Cervical grooves deep in the scutal anterior third, followed by a shallow and divergent depression in the scutal median third. Basis capituli rectangular dorsally, without cornua.

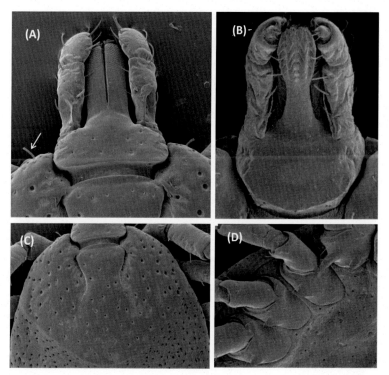

FIGURE 2.69   *A. sculptum* nymph: (A) capitulum dorsal view, (B) capitulum ventral view, (C) scutum, (D) coxae. *White arrow* indicates the antero-lateral seta of coxa I. *Figures (A) to (D) are reproductions of figure 4: A.* sculptum *nymph. Capitulum, dorsum; Capitulum, venter; Scutum; Coxae I−IV. From Martins TF, Labruna MB, Mangold AJ, Cafrune MM, Guglielmone AA, Nava S. Taxonomic key to nymphs of the genus* Amblyomma *(Acari: Ixodidae) in Argentina, with description and redescription of the nymphal stage of four* Amblyomma *species. Ticks Tick-borne Dis 2014;5:753−70, with permission of Elsevier.*

Hypostome spatulate, dental formula 2/2. Legs: coxa I with two spurs, the external longer than the internal and pointed; coxae II−IV with a small triangular spur each. Coxa I with antero-lateral seta shorter than the length of coxa I external spur.

Taxonomic notes: the taxon known until 2013 as *A. cajennense* (Fabricius, 1787) is currently recognized as a complex formed by at least six species, namely *A. cajennense* (Fabricius 1787) (= *A. cajennense* sensu stricto), *A. sculptum* Berlese, 1888, *A. mixtum* Koch, 1844, *A. tonelliae* Nava, Beati, and Labruna, 2014, *A. patinoi* Labruna, Nava, and Beati, 2014, and *A. interandinum* Beati, Nava, and Cáceres, 2014.[1] This new taxonomic scheme was supported by biological, molecular, morphological, and ecological data.[1,7,14,20−23] Neumann[24] synonymized *A. sculptum* with *A. cajennense*, Tonelli-Rondelli[25] resurrected *A. sculptum* as a valid species, and Aragão and Fonseca[26] considered *A. sculptum* as a synonym of *A. cajennense*, but *A. sculptum* is currently a valid species.[1,27]

DNA sequences with relevance for tick identification and phylogeny: sequences of the mitochondrial 16S rRNA gene of *A. sculptum* are available in the Gen Bank (accession numbers KT238826, KP686061, KM519934). In the work of Beati et al.[22] on the phylogeography of *A. cajennense*, the sequences of the 12S rRNA gene, cytochrome oxidase II gene, control region (d-loop), and internal transcribed spacer 2, belonging to the haplotypes YU and AF-MG, are named as belonging to *A. cajennense* but they correspond to sequences of *A. sculptum*.

REFERENCES

1. Nava S, Beati L, Labruna MB, Cáceres AG, Mangold AJ, Guglielmone AA. Reassessment of the taxonomic status of *Amblyomma cajennense* (Fabricius, 1787) with the description of three new species, *Amblyomma tonelliae* n. sp., *Amblyomma interandinum* n. sp. and *Amblyomma patinoi* n. sp., and reinstatement of *Amblyomma mixtum* Koch, 1844 and *Amblyomma sculptum* Berlese, 1888 (Ixodida: Ixodidae). *Ticks Tick-borne Dis* 2014;**5**:252−76.

2. Sponchiado J, Melo GL, Martins TF, Krawczak FS, Labruna MB, Cáceres NC. Association patterns of ticks (Acari: Ixodida: Ixodidae, Argasidae) of small mammals in Cerrado fragments, western Brazil. *Exp Appl Acarol* 2015;**65**:389−91.

3. Ramos VN, Piovezan U, Franco AHA, Rodrigues VS, Nava S, Szabó MPJ. Nellore cattle (*Bos indicus*) and ticks within the Brazilian Pantanal: ecological relationships. *Exp Appl Acarol* 2016;**69**:227−40.

4. Witter R, Martins TF, Campos AK, Melo ALT, Correa SHR, Morgado TO, et al. Rickettsial infection in ticks (Acari: Ixodidae) of wild animals in Midwestern Brazil. *Ticks Tick-borne Dis* 2016;**7**:415−23.

5. Martins TF, Peres MG, Costa FB, Bacchiega TS, Appolinario CM, Antunes JMAP, et al. Ticks infesting wild small rodents in three areas of the state of São Paulo, Brazil. *Ciênc Rural* 2016;**46**:871−5.

6. Luz HR, Faccini JLH, Landulfo GA, Neto SFC, Famadas KM. New records for *Amblyomma sculptum* (Ixodidae) on non-passerine birds in Brazil. *Rev Bras Parasitol Vet* 2016;**25**:124−6.

7. Martins TF, Teixeira RHF, Labruna MB. Ocorrência de carrapatos em animais silvestres recebidos e atendidos pelo Parque Zoológico Municipal Quinzinho de Barros, Sorocaba, São Paulo, Brasil. *Braz J Vet Res Anim Sci* 2015;**52**:319−24.

8. Oliveira PR, Borges LMF, Lopes CML, Leite RC. Population dynamics of the free-living stages of *Amblyomma cajennense* (Fabricius, 1787) (Acari: Ixodidae) on pastures of Pedro Leopoldo, Minas Gerais State, Brazil. *Vet Parasitol* 2000;**92**:295−301.

9. Labruna MB, Kasai N, Ferreira F, Faccini JLH, Gennari SM. Seasonal dynamics of ticks (Acari: Ixodidae) on horses in the state of São Paulo, Brazil. *Vet Parasitol* 2002;**105**:65−77.

10. Oliveira PR, Borges LMF, Leite RC, Freitas CMV. Seasonal dynamics of the Cayenne tick, *Amblyomma cajennense* on horses in Brazil. *Med Vet Entomol* 2003;**17**:412−16.

11. Tarragona EL, Nava S. Unpublished information; 2016.

12. Labruna MB, Amaku M, Metzner JA, Pinter A, Ferreira F. Larval behavioral diapause regulates life cycle of *Amblyomma cajennense* (Acari: Ixodidae) in Southeast Brazil. *J Med Entomol* 2003;**40**:170−8.

13. Cabrera RR, Labruna MB. Influence of photoperiod and temperature on the larval behavioral diapause of *Amblyomma cajennense* (Acari: Ixodidae). *J Med Entomol* 2009;**46**:1303−9.

14. Estrada Peña A, Tarragona EL, Vesco U, De Meneghi D, Mastropaolo M, Mangold AJ, et al. Divergent environmental preferences and areas of sympatry of tick species in the *Amblyomma cajennense* complex (Ixodidae). *Int J Parasitol* 2014;**44**:1081−9.

15. Saracho Bottero MN, Tarragona EL, Mangold AJ, Guglielmone AA, Nava S. Garrapatas (Acari: Ixodidae) parásitas de bovinos en la provincia biogeográfica de las Yungas, Jujuy, Argentina. *Rev FAVE Cienc Vet* 2016; submitted.

16. Szabó MPJ, Pinter A, Labruna MB. Ecology, biology and distribution of spotted-fever tick vectors in Brazil. *Front Cell Infec Microbiol* 2013;**3**:1−9.

17. Labruna MB, Santos FCP, Ogrezewalska M, Nascimento EMM, Colombo S, Marcili A, et al. Genetic identification of rickettsial isolates from fatal cases of Brazilian spotted fever and comparison with *Rickettsia rickettsii* isolates from the American continents. *J Clin Microbiol* 2014;**52**:3788−91.

18. Alves AS, Melo ALT, Amorin MV, Borges AMCM, Silva LGE, Martins TF, et al. Seroprevalence of *Rickettsia* spp. in equids and molecular detection of 'Candidatus Rickettsia amblyommii' in *Amblyomma cajennense* sensu lato ticks from the Pantanal Region of Mato Grosso, Brazil. *J Med Entomol* 2014;**41**:1252−7.

19. Melo ALT, Witter R, Martins TF, Pacheco TA, Alves AS, Chitarra CS, et al. A survey of tick-borne pathogens in dogs and their ticks in the Pantanal biome, Brazil. *Med Vet Entomol* 2016;**30**:112−16.

20. Labruna MB, Soares JF, Martins TF, Soares HS, Cabrera RR. Cross-mating experiments with geographically different populations of *Amblyomma cajennense* (Acari: Ixodidae). *Exp Appl Acarol* 2011;**54**:41−9.

21. Mastropaolo M, Nava S, Guglielmone AA, Mangold AJ. Biological differences between two allopatric populations of *Amblyomma cajennense* (Acari: Ixodidae) in Argentina. *Exp Appl Acarol* 2011;**53**:371−5.

22. Beati L, Nava S, Burkman EJ, Barros-Battesti D, Labruna MB, Guglielmone AA, et al. *Amblyomma cajennense* (Fabricius, 1787) (Acari: Ixodidae), the Cayenne tick: phylogeography and evidence for allopatric speciation. *BMC Evol Biol* 2013;**13**(article267):20.

23. Martins TF, Barbieri AR, Costa FB, Terassini FA, Camargo LM, Peterka CR, et al. Geographical distribution of *Amblyomma cajennense* (sensu lato) ticks (Parasitiformes: Ixodidae) in Brazil, with description of the nymph of *A. cajennense* (sensu stricto). *Parasit Vectors* 2016;**9**(article 186):14 p.

24. Neumann LG. Révision de la famille des ixodidés (3e mémoire). *Mém Soc Zool Fr* 1899;**12**:107−294.

25. Tonelli Rondelli MT, Ixodoidea., Parte I. *Amblyomma ovale* Koch, *Amblyomma cajennense* Fabricius [sic] e le specie a lor affini nuove o poco note. *Riv Parassitol* 1937;**1**:273−8.

26. Aragão HB, Fonseca F. Notas de ixodologia. V. A propósito da validade de algumas espécies do gênero *Amblyomma* do continente Americano (Acari: Ixodidae). *Mem Inst Oswaldo Cruz* 1953;**51**:485−92.

27. Guglielmone AA, Nava S. Names for Ixodidae (Acari: Ixodoidea): valid, synonyms, *incertae sedis, nomina dubia, nomina nuda, lapsus*, incorrect and suppressed names − with notes on confusions and misidentifications. *Zootaxa* 2014;**3767**:1−256.

Amblyomma tigrinum Koch, 1844

Koch, C.L. (1844) Systematische Übersicht über die Ordnung der Zecken. *Archiv für Naturgeschichte*, 10, 217−239.

DISTRIBUTION

Argentina, Bolivia, Brazil, Chile, French Guiana, Paraguay, Peru, Uruguay, and Venezuela.[1] Graham et al.[2] stated that *A. tigrinum* was found in Mexico (Neotropical area) but its presence in the country has not been confirmed.

Biogeographic distribution in the Southern Cone of America: Chaco, Pampa (Neotropical Region), Monte (South American Transition Zone), Santiago, Maule, and Central Patagonia (Andean Region) (see Fig. 1 in the introduction).

HOSTS

The usual hosts for adults of *A. tigrinum* are carnivorous of the family Canidae, while small rodents of the families Caviidae and Cricetidae and birds are the principal hosts for immature stages.[3−9] The complete host range of *A. tigrinum* is presented in Table 2.21.

ECOLOGY

A. tigrinum is a tick with ecological plasticity as shown by its colonization of areas with contrasting climatic conditions.[3,10] This species is distributed in areas belonging to Biogeographic Provinces with marked differences in their ecological characteristics. All parasitic stages of *A. tigrinum* are active throughout the year, although larvae and nymphs reach their peaks of abundance from late spring to middle autumn, and adults in summer.[3,11] It was

TABLE 2.21 Hosts of Adults (A), Nymphs (N), and Larvae (L) of *A. tigrinum*

MAMMALIA		GALLIFORMES	
ARTIODACTYLA		**Odontophoridae**	
Bovidae		*Callipepla californica*	NL
Cattle	A	**PASSERIFORMES**	
Cervidae		**Cardinalidae**	
Blastoceros dichotomus	A	*Saltator aurantiirostris*	NL
Mazama gouazoubira	A	**Cuculidae**	
Suidae		*Guira guira*	NL
Feral pig	A	**Dendrocolaptidae**	
CARNIVORA		*Drymornis bridgesii*	L
Canidae		**Emberizidae**	
Chrysocyon thous	AN	*Coryphospingus cucullatus*	NL
Cerdocyon brachyurus	A	*Lophospingus pusillus*	NL
Domestic dog	AN	*Poospiza torquata*	L
Lycalopex culpaeus	A	*Rhynchospiza strigiceps*	NL
L. gymnocercus	A	*Saltaricula multicolor*	NL
L. griseus	A	*Volatinia jacarina*	L
L. vetulus	A	*Zoonotrichia capensis*	NL
Felidae		**Furnariidae**	
Domestic cat	A	*Asthenes baeri*	L
Panthera onca	AL	*Coryphistera allaudina*	NL
Puma concolor	AL	*Furnarius cristatus*	NL
Procyonidae		*Fu. rufus*	NL
Nasua nasua	A	*Pseudoseisura lophotes*	NL
PRIMATES		*Synallaxis albescens*	NL
Hominidae		*Tarphonomus certhioides*	NL
Human	AN	**Icteridae**	
RODENTIA		*Agelaioides badius*	NL
Caviidae		*Molothrus bonariensis*	NL
Galea musteloides	NL	**Polioptilidae**	
Hydrochoerus hydrochaeris	A	*Polioptila dumicola*	NL

(Continued)

TABLE 2.21 (Continued)

Cricetidae		Rhinocryptidae	
Akodon dolores	NL	Rhinocrypta lanceolata	N
A. molinae	NL	**Thraupidae**	
A. oenos	NL	Paroaria coronata	N
Calomys laucha	N	Sicalis flaveola	L
C. musculinus	L	S. luteola	L
C. venustus	NL	**Troglodytidae**	
Graomys centralis	NL	Troglodytes aedon	NL
Necromys benefatus	NL	**Turdidae**	
Oligoryzomys fulvescens	AN	Turdus amaurochalinus	L
O. longicaudatus	L	**Tyrannidae**	
Phyllotis xanthopygus	L	Pitangus sulphuratus	N
Muridae		**STRIGIFORMES**	
Rattus rattus	L	**Strigidae**	
AVES		Asio clamator	N
COLUMBIFORMES		**TINAMIFORMES**	
Columbidae		**Tinamidae**	
Columbina picui	L	Eudromia elegans	L
Zenaida auriculata	NL	Nothoprocta perdicaria	NL
		Nothura maculosa	N
		Rhynchotus rufescens	N

hypothesized that the life cycle of *A. tigrinum* is principally regulated by temperature with no occurrence of diapause.[11] This fact would permit periods of quiescence and activity modulated by temperature, which explain the distribution of this tick species in habitats with contrasting climatic conditions.[11]

SANITARY IMPORTANCE

Adults of *A. tigrinum* are parasites of dogs in rural and periurban environments,[3,5,12] and they were also recorded parasitizing humans.[13] *A. tigrinum* has been involved as vector of the human pathogen *Rickettsia parkeri*.[14] Adult specimens of *A. tigrinum* were found naturally infected with the bacteria causing Q fever to humans, *Coxiella burnetii*,[15] and *"Candidatus Rickettsia andeanae"*[16,17] a microorganism of unknown pathogenicity.

DIAGNOSIS

Male (Fig. 2.70): medium body size, total length 3.3 mm (3.0−3.6), breadth 2.2 mm (1.8−2.4). Body outline oval elongated, narrower in the anterior part; scapulae rounded; cervical grooves deep anteriorly, shallow posteriorly, sigmoid in shape; marginal groove complete. Eyes flat. Scutum ornate, with reddish brown spots on a pale yellowish ground; postero-median spot extending to the level of spiracular plates; postero-median spot narrower than the area between postero-median and postero-lateral spots; postero-accessory

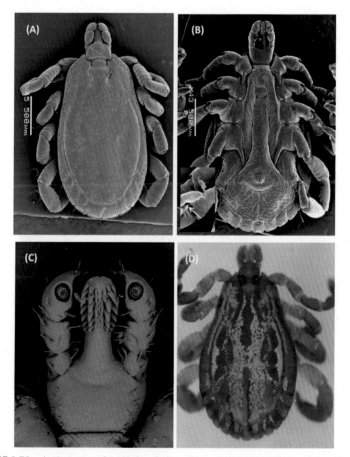

FIGURE 2.70   *A. tigrinum* male: (A) dorsal view, (B) ventral view and image of the spine present on the tibia of legs II−IV in both male and female, (C) capitulum ventral view, (D) scutum ornamentation.

spots parallel to the postero-median spot; lateral spots large, conjoined; cervical spots narrow and divergent posteriorly; central area long, extending to the middle level of the scutum; limiting spots small, not conjoined; punctuations numerous, larger in the anterior and posterior margins of the scutum. Carena absent. Basis capituli dorsally subrectangular, cornua long. Hypostome spatulate, dental formula 3/3. Genital aperture located at level of coxa II, U-shaped. Legs: coxa I with two distinct spurs, the external spur long, narrow, and sharp, the internal spur as a small tubercle; coxae II–III each with a triangular, short blunt spur; coxa IV with a narrow sharp spur, not reaching the anus level; trochanters without spurs; presence of one spine on the tibia of legs II–IV. Spiracular plates comma-shaped.

Female (Fig. 2.71): medium body size, total length 3.4 mm (3.2–3.7), breadth 2.3 mm (2.2–2.4). Body outline oval. Scutum length 1.9 mm

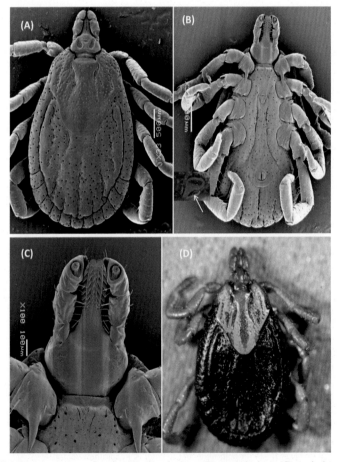

FIGURE 2.71 *A. tigrinum* female: (A) dorsal view, (B) ventral view, (C) capitulum ventral view, (D) scutum ornamentation. *White arrow* in (B) shows the shape of the spiracular plate.

(1.6−2.2), breadth 1.7 mm (1.4−2.0); scapulae rounded, cervical grooves deep anteriorly, shallow posteriorly, sigmoid in shape. Eyes flat. Chitinous tubercles at posterior body margin absent. Scutum ornate, extensively pale yellowish, cervical spots narrow and divergent posteriorly, central area long and narrow, not reaching the posterior margin of the scutum; punctuations numerous, small, larger in the antero-lateral fields. Notum glabrous. Basis capituli dorsally subrectangular, cornua absent, porose areas oval. Hypostome spatulate, dental formula 3/3. Genital aperture located at level of coxa II, U-shaped. Legs: coxa I with two distinct spurs, the external spur long, narrow and sharp, the internal spur as a small tubercle; coxae II−IV each with a triangular, short blunt spur; trochanters without spurs; presence of one spine on the tibia of legs II−IV. Spiracular plates comma-shaped.

Nymph (Fig. 2.72): small body size, total length 1.3 mm (1.1−1.5), breadth 0.8 mm (0.7−0.9). Body outline oval elongated. Chitinous tubercles at posterior body margin absent. Scutum with few punctuations, and postero-lateral margin not sinuous. Eyes flat, located on lateral scutal angles at the

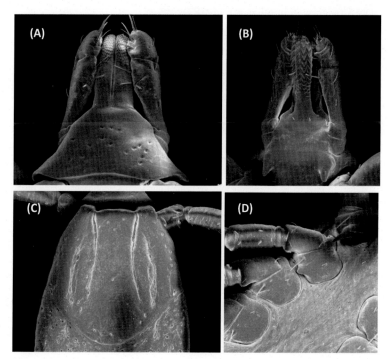

FIGURE 2.72   *A. tigrinum* nymph: (A) capitulum dorsal view, (B) capitulum ventral view, (C) scutum, (D) coxae. *Figures (A) to (D) are reproductions of figure 5:* A. tigrinum *nymph. Capitulum, dorsum; Capitulum, venter; Scutum; Coxae I−IV. From Martins TF, Labruna MB, Mangold AJ, Cafrune MM, Guglielmone AA, Nava S. Taxonomic key to nymphs of the genus* Amblyomma *(Acari: Ixodidae) in Argentina, with description and redescription of the nymphal stage of four* Amblyomma *species. Ticks Tick-borne Dis 2014;5:753−70, with permission of Elsevier.*

level of the beginning of the posterior third of the scutum. Cervical grooves long, reaching the scutal posterior fourth, deeper at the anterior fourth; presence of a rugose depression parallel and external to the posterior half of cervical grooves. Basis capituli dorsally triangular, without cornua, with pseudoauricula and ventral processes. Hypostome spatulate, dental formula 2/2. Legs: coxa I with a long, narrow, and sharp spur; coxa II with a short triangular spur; coxae III−IV with a very short bridge-like spur each.

Taxonomic notes: *A. tigrinum*, *A. triste* Koch, 1844, and *A. maculatum* form the so-called "*A. maculatum* group."[1] These three species are morphologically and phylogenetically closely related. This phylogenetic relationship is based on the concatenated sequences of the small and the large ribosomal subunits 12S rDNA and 16S rDNA (Fig. 2.7). See also Taxonomic notes of *A. neumanni* Ribaga, 1902.

DNA sequences with relevance for tick identification and phylogeny: mitochondrial DNA sequences of *A. tigrinum* genes are available in the Gen Bank as follows: 16S rDNA sequences (accession numbers FJ965339, AY498562, AY836004, AY836005, KF179344, DQ342288−DQ342304) and 12S rDNA (accession number AY342287).

REFERENCES

1. Estrada-Peña A, Venzal JM, Mangold AJ, Cafrune MM, Guglielmone AA. The *Amblyomma maculatum* Koch, 1844 (Acari: Ixodidae: Amblyomminae) tick group: diagnostic characters, description of the larva of *A. parvitarsum* Neumann, 1901, 16S rDNA sequences, distribution and hosts. *Syst Parasitol* 2005;**60**:99−112.

2. Graham OH, Gladney WJ, Beltrán LG. Comparación de la distribución e importanciaeconómica de *Amblyomma maculata* (sic) Koch (Acarina: Ixodidae) en México y los Estados Unidos. *Folia Entomol Mex* 1975;**33**:66−7.

3. Guglielmone AA, Mangold AJ, Luciani CE, Viñabal AE. *Amblyomma tigrinum* (Acari: Ixodidae) in relation to phytogeography of central-northern Argentina with note on hosts and seasonal distribution. *Exp Appl Acarol* 2000;**24**:983−9.

4. Labruna MB, Jorge RSP, Sana DA, Jácomo ATA, Kashivakura CK, Furtado MM, et al. Ticks (Acari: Ixodida) on wild carnivores in Brazil. *Exp Appl Acarol* 2005;**36**:149−63.

5. Guglielmone AA, Nava S. Las garrapatas argentinas del género *Amblyomma* (Acari: Ixodidae): distribución y hospedadores. *Rev Invest Agropec* 2006;**35**(3):135−55.

6. Nava S, Mangold AJ, Guglielmone AA. The natural hosts of larvae and nymphs of *Amblyomma tigrinum* Koch, 1844 (Acari: Ixodidae). *Vet Parasitol* 2006;**140**:124−32.

7. Guglielmone AA, Nava S. Rodents of the subfamily Caviinae as hosts for hard ticks (Acari: Ixodidae). *Mastozool Neotr* 2010;**17**:279−86.

8. Guglielmone AA, Nava S. Rodents of the subfamily Sigmodontinae as hosts for South American hard ticks (Acari: Ixodidae) with hypothesis on life history. *Zootaxa* 2011;**2904**:45−65.

9. Flores FS, Nava S, Batallán G, Tauro LB, Contigiani MS, Diaz LA, et al. Ticks (Acari: Ixodidae) on wild birds in north-central Argentina. *Ticks Tick-borne Dis* 2014;**5**:715−21.

10. González Acuña D, Guglielmone AA. The ticks (Acari: Ixodoidea: Argasidae: Ixodidae) of Chile. *Exp Appl Acarol* 2005;**35**:147−63.

11. Nava S, Mangold AJ, Guglielmone AA. Seasonal distribution of larvae and nymphs of *Amblyomma tigrinum* Koch, 1844 (Acari: Ixodidae). *Vet Parasitol* 2009;**166**:340−2.
12. Debárbora VN, Oscherov EB, Guglielmone AA, Nava S. Garrapatas (Acari: Ixodidae) asociadas a perros en diferentes ambientes de la provincia de Corrientes, Argentina. *InVet* 2011;**13**:45−51.
13. Guglielmone AA, Beati L, Barros-Battesti DM, Labruna MB, Nava S, Venzal JM, et al. Ticks (Ixodidae) on humans in South America. *Exp Appl Acarol* 2006;**40**:83−100.
14. Romer Y, Nava S, Govedic G, Cicuttin G, Denison AM, Singleton J, et al. *Rickettsia parkeri* rickettsiosis in different ecological regions of Argentina and its association with *Amblyomma tigrinum* as a potential vector. *Am J Trop Med Hyg* 2014;**91**:1156−60.
15. Pacheco RC, Echaide IE, Alves RN, Beletti ME, Nava S, Labruna MB. *Coxiella burnetii* in ticks, Argentina. *Emerg Infect Dis* 2013;**19**:344−6.
16. Abarca K, López J, Acosta-Jamett G, Martínez-Valdebenito C. Identificación de *Rickettsia andeanae* en dos regiones de Chile. *Rev Chil Infectol* 2013;**30**:388−94.
17. Saracho Bottero MN, Tarragona EL, Nava S. Spotted fever group rickettsiae in *Amblyomma* ticks likely to infest humans in rural areas from northwestern Argentina. *Medicina (Buenos Aires)* 2015;**74**:391−5.

Amblyomma tonelliae
Nava, Beati, and Labruna, 2014

Nava, S., Beati, L., Labruna, M.B., Cáceres, A.G., Mangold, A.J. & Guglielmone, A.A. (2014) Reassessment of the taxonomic status of *Amblyomma cajennense* (Fabricius, 1787) with the description of three new species, *Amblyomma tonelliae* n. sp., *Amblyomma interandinum* n. sp. and *Amblyomma patinoi* n. sp., and reinstatement of *Amblyomma mixtum* Koch, 1844 and *Amblyomma sculptum* Berlese, 1888 (Ixodida: Ixodidae). *Ticks and Tick-borne Diseases*, 5, 252−276.

DISTRIBUTION

Argentina, Bolivia, and Paraguay.[1]

Biogeographic distribution in the Southern Cone of America: Chaco and Yungas (Neotropical Region) (see Fig. 1 in the introduction).

HOSTS

Adults and immature stages of *A. tonelliae* present low host specificity. They principally feed on large wild and domestic mammals of different orders[1,2] (Table 2.22).

TABLE 2.22 Hosts of Adults (A), Nymphs (N), and Larvae (L) of *A. tonelliae*

MAMMALIA		PERISSODACTYLA	
ARTIODACTYLA		**Equidae**	
Bovidae		Horse	ANL
Cattle	ANL	**PILOSA**	
Cervidae		**Myrmecophagidae**	
Mazama sp.	A	*Myrmecophaga tridactyla*	A
Tayassuidae		*Tamandua tetradactyla*	A
Catagonus wagneri	A	**PRIMATES**	
Pecari tajacu	A	**Hominidae**	
Tayassu pecari	A	Human	ANL
CARNIVORA		**RODENTIA**	
Canidae		**Caviidae**	
Domestic dog	A	*Hydrochoerus hydrochaeris*	A
Felidae			
Puma concolor	A		

ECOLOGY

A. tonelliae has a life cycle with one generation per year.[2] Larvae are active from early autumn to mid-winter with the peak in late autumn, nymphs are found from early winter to spring with the peak in early spring, and the females are found during spring and summer with the peak in early and mid-summer.[2] Although *A. tonelliae* is distributed in both Chaco and Yungas Biogeographic Provinces, it has a marked ecological preference for xeric areas of the Chaco Biogeographic Province.[2,3]

SANITARY IMPORTANCE

Immature and adult stages of *A. tonelliae* are usual parasites of humans and domestic animals such as cattle and horses.[1,2] Tarragona et al.[4] have experimentally demonstrated the vectorial competence of *A. tonelliae* to transmit the human pathogen *Rickettsia rickettsii*. Specimens of *A. tonelliae* were found naturally infected with *"Candidatus Rickettsia amblyommii"* and *Rickettsia* sp. strain El Tunal[5] whose pathogenicities are unknown.

DIAGNOSIS

Male (Fig. 2.73): medium body size, total length 3.7 mm (3.1−3.9), breadth 3.0 mm (2.5−3.3). Body outline oval rounded, scapulae rounded, cervical grooves short, comma-shaped; marginal groove complete. Eyes flat. Scutum ornate, cervical spots elongated posteriorly, posterior branches of limiting spots fused posteriorly, three lateral spots partially fused but clearly recognizable, postero-accessory spots triangular, postero-median spot wider than adjacent enameled yellowish stripe; punctuations numerous, in anterior part of scutum finer and uniformly distributed in median area and larger and deeper laterally, absent on the postero-accessory and postero-median spots. Carena absent. Basis capituli subrectangular dorsally, cornua short. Hypostome spatulate, dental formula 3/3. Genital aperture located at level of

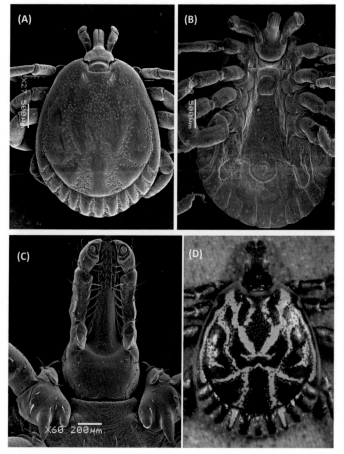

FIGURE 2.73 *A. tonelliae* male: (A) dorsal view, (B) ventral view, (C) capitulum ventral view, (D) scutum ornamentation.

coxa II, U-shaped. Legs: coxa I with two distinct, short blunt spurs, the external longer than the internal; coxae II−III each with a short triangular spur, protruding from ridge-like edge; coxa IV with a pointed spur not reaching level of anus; trochanters without spurs. Spiracular plates comma-shaped.

Female (Fig. 2.74): large body size, total length 4.1 mm (3.7−4.3), breadth 3.3 mm (3.0−3.6). Body outline oval rounded. Scutum length 2.0 mm (1.7−2.1), breadth 2.1 mm (1.7−2.1); scapulae rounded; cervical grooves sigmoid, deeper in the anterior part. Eyes flat. Chitinous tubercles at the posterior body margin present. Scutum ornate, with cervical spots elongated covering cervical grooves and sometimes touching posteriorly limiting spots; punctuations large, numerous, uniformly distributed, deeper in the

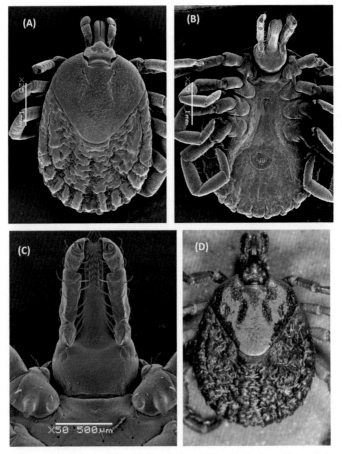

FIGURE 2.74 *A. tonelliae* female: (A) dorsal view, (B) ventral view, (C) capitulum ventral view, (D) scutum ornamentation.

anterior field. Notum corrugated, with long and stout setae, and with three deep grooves. Basis capituli subrectangular dorsally, cornua short, porose areas oval. Hypostome spatulate, dental formula 3/3. Genital aperture located at level of coxa II, V-shaped. Legs: coxa I with two distinct, short blunt spurs, the external longer than the internal; coxae II−IV with a triangular spur each; trochanters without spurs. Spiracular plates comma-shaped.

Nymph (Fig. 2.75): medium body size, total length 1.5 mm (1.4−1.6), breadth 1.2 mm (1.1−1.3). Body outline oval. Chitinous tubercles at the posterior body margin absent. Scutum with surface extensively shagreened (rugose), medium punctuations evenly distributed, larger and deeper laterally. Eyes bulging, located on lateral scutal angles at the level of scutal mid-length. Cervical grooves surpassing eyes level, deeper in the scutal anterior third. Basis capituli rectangular dorsally, without cornua. Hypostome spatulate, dental formula 2/2. Legs: coxa I with two spurs, the external longer than the internal and pointed; coxae II−IV with a small triangular spur each. Coxa I with antero-lateral seta longer than the length of coxa I external spur.

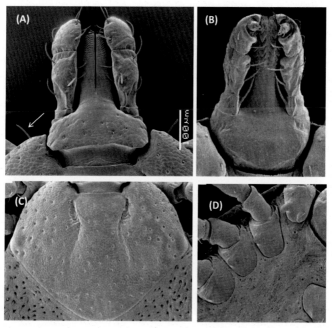

FIGURE 2.75 *A. tonelliae* nymph: (A) capitulum dorsal view, (B) capitulum ventral view, (C) scutum, (D) coxae. *White arrow* indicates the antero-lateral seta of coxa I. *Figures (A) to (D) are reproductions of figure 2:* A. tonelliae *nymph. Capitulum, dorsum; Capitulum, venter; Scutum; Coxae I−IV. From Martins TF, Labruna MB, Mangold AJ, Cafrune MM, Guglielmone AA, Nava S. Taxonomic key to nymphs of the genus* Amblyomma *(Acari: Ixodidae) in Argentina, with description and redescription of the nymphal stage of four* Amblyomma *species. Ticks Tick-borne Dis 2014;5:753−70, with permission of Elsevier.*

Taxonomic notes: see Taxonomic notes of *A. sculptum* Berlese, 1888.
DNA sequences with relevance for tick identification and phylogeny: sequences of the mitochondrial 16S rRNA gene of *A. tonelliae* are available in the Gen Bank (accession numbers KM507359−KM507362). In the work of Beati et al.[6] on the phylogeography of *A. cajennense*, the sequences of the 12S rRNA gene, cytochrome oxidase II gene, control region (d-loop), and internal transcribed spacer 2, belonging to the haplotypes PA and CHS-CHO, are named as belonging to *A. cajennense* sensu lato but they correspond to sequences of *A. tonelliae*.

REFERENCES

1. Nava S, Beati L, Labruna MB, Cáceres AG, Mangold AJ, Guglielmone AA. Reassessment of the taxonomic status of *Amblyomma cajennense* (Fabricius, 1787) with the description of three new species, *Amblyomma tonelliae* n. sp., *Amblyomma interandinum* n. sp. and *Amblyomma patinoi* n. sp., and reinstatement of *Amblyomma mixtum* Koch, 1844 and *Amblyomma sculptum* Berlese, 1888 (Ixodida: Ixodidae). *Ticks Tick-borne Dis* 2014;**5**:252−76.
2. Tarragona EL, Mangold AJ, Mastropaolo M, Guglielmone AA, Nava S. Ecology and genetic variation of *Amblyomma tonelliae* in Argentina. *Med Vet Entomol* 2015;**29**:297−304.
3. Estrada Peña A, Tarragona EL, Vesco U, De Meneghi D, Mastropaolo M, Mangold AJ, et al. Divergent environmental preferences and areas of sympatry of tick species in the *Amblyomma cajennense* complex (Ixodidae). *Int J Parasitol* 2014;**44**:1081−9.
4. Tarragona EL, Soares JF, Borges Costa F, Labruna MB, Nava S. Vectorial competence of *Amblyomma tonelliae* to transmit *Rickettsia rickettsii*. *Med Vet Entomol* 2016;**30**:410−15.
5. Tarragona EL, Cicuttin GL, Mangold AJ, Mastropaolo M, De Salvo MN, Nava S. *Rickettsia* infection in *Amblyomma tonelliae*, a tick species from the *Amblyomma cajennense* complex. *Ticks Tick-borne Dis* 2015;**6**:173−7.
6. Beati L, Nava S, Burkman EJ, Barros-Battesti D, Labruna MB, Guglielmone AA, et al. *Amblyomma cajennense* (Fabricius, 1787) (Acari: Ixodidae), the Cayenne tick: phylogeography and evidence for allopatric speciation. *BMC Evol Biol* 2013;**13**(article 267):20.

Amblyomma triste
Koch, 1844

Koch, C.L. (1844) Systematische Übersicht über die Ordnung der Zecken. *Archiv für Naturgeschichte*, 10, 217−239.

DISTRIBUTION

Argentina, Bolivia, Brazil, Chile, Colombia, Ecuador, Mexico, Paraguay, Peru, Uruguay, and Venezuela.[1−3] See Taxonomic notes regarding the records of *A. triste* from Ecuador and Peru mentioned in Guglielmone et al.[2]

and for Chile in Abarca et al.[3] *A. triste* has been also found established in the Nearctic Region[4] (Mertins et al., 2010).

Biogeographic distribution in the Southern Cone of America: Chaco, Pampa (Neotropical Region), and Atacama (South American Transition Zone) (see Fig. 1 in the introduction).

HOSTS

Although the marsh deer *Blastocerus dichotomus* is the primeval host for adults of *A. triste*, another large wild and domestic mammalian species of different orders are also suitable hosts for adults of this tick.[5-8] Small and medium-size rodents of the families Cricetidae and Caviidae are the principal hosts for immature stages.[7-11] The complete host range of *A. triste* is presented in Table 2.23.

ECOLOGY

A. triste has a 1-year life cycle, where immature stages are active principally in summer, and adults reach their peak of abundance from late winter to mid-spring, gradually declining toward late spring and early summer.[7,8,12] This tick species is strongly associated to wetlands and environment prone to flooding in Argentina, Brazil, and Uruguay,[7,8,13] but the ecological preferences of *A. triste* in other South American countries have not been studied.

SANITARY IMPORTANCE

Adults of *A. triste* are usual parasites of cattle and dogs, and they are aggressive to humans.[7,8,12] *A. triste* is the principal vector of the human pathogen *Rickettsia parkeri* in the Southern Cone of America.[14-20]

DIAGNOSIS

Male (Fig. 2.76): medium body size, total length 3.5 mm (3.0–3.9), breadth 2.2 mm (1.9–2.4). Body outline oval elongated, narrower in the anterior part; scapulae rounded; cervical grooves deep anteriorly, shallow posteriorly, sigmoid in shape; marginal groove complete. Eyes flat. Scutum ornate, with reddish brown spots on a pale yellowish ground; postero-median spot extending to the level of spiracular plates; postero-accessory spots parallel to the postero-median spot; postero-median spot wider than the area between postero-median and postero-lateral spots; lateral spots large, conjoined; cervical spots narrow and divergent posteriorly; central area long, extending to the middle level of the scutum; punctuations numerous, larger in the anterior and posterior margins of the scutum. Carena absent, tubercles present. Basis capituli dorsally subrectangular, cornua long. Hypostome spatulate, dental

TABLE 2.23 Hosts of Adults (A), Nymphs (N), and Larvae (L) of *A. triste*

MAMMALIA		PILOSA	
ARTIODACTYLA		**Myrmecophagidae**	
Bovidae		*Myrmecophaga tridactyla*	N
Cattle	A	**PRIMATES**	
Goat	A	**Hominidae**	
Cervidae		Human	A
Blastoceros dichotomus	A	**RODENTIA**	
Mazama sp.	A	**Caviidae**	
Ozotoceros bezoarticus	A	*Cavia aperea*	NL
Suidae		*Hydrochoerus hydrochaeris*	AN
Domestic pig	A	**Cricetidae**	
CARNIVORA		*Akodon azarae*	NL
Canidae		*Calomys callosus*	NL
Domestic dog	AN	*Holochilus brasiliensis*	NL
Cerdocyon thous	AN	*H. sciureus*	N
Chrisocyon brachyurus	A	*Necromys obscurus*	N
Lycalopex vetulus	A	*Oligoryzomys flavescens*	NL
Felidae		*O. nigripes*	NL
Domestic cat	A	*Oxymycterus nasutus*	NL
Herpailurus yagouaroundi	A	*O. rufus*	NL
Leopardus colocolo	N	*Scapteromys aquaticus*	NL
Panthera onca	AL	*S. tumidus*	NL
Puma concolor	AL	**AVES**	
CHIROPTERA		**PASSERIFORMES**	
Vespertilionidae		**Emberizidae**	
Myotis albescens	N	*Poospiza nigrorufa*	NL
DIDELPHIMORPHIA		*Sporophila caerulescens*	L
Didelphidae		*Zonotrichia capensis*	NL
Didelphis marsupialis	A	**Furnariidae**	
Monodelphis dimidiata	NL	*Furnarius leucopus*	L
PERISSODACTYLA		*Fu. rufus*	NL
Equidae		*Phacellodomus ruber*	NL
Horse	A	**Turdidae**	
Tapiridae		*Turdus rufiventris*	L
Tapirus terrestris	A		

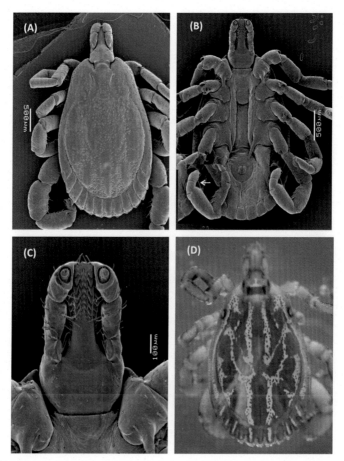

FIGURE 2.76   *A. triste* male: (A) dorsal view, (B) ventral view, (C) capitulum ventral view and image of the spine present on the tibia of legs II−IV in both male and female, (D) scutum ornamentation.

formula 3/3. Genital aperture located at level of coxa II, U-shaped. Legs: coxa I with two distinct spurs, the external spur triangular and sharp, the internal spur as a small tubercle; coxae II−III with a triangular, short blunt spur each; coxa IV with a narrow sharp spur, not reaching the anus level; trochanters without spurs; presence of one spine on the tibia of legs II−IV. Spiracular plates oval.

Female (Fig. 2.77): medium body size, total length 3.6 mm (3.4−3.9), breadth 2.4 mm (2.3−2.5). Body outline oval. Scutum length 1.9 mm (1.7−2.0), breadth 1.7 mm (1.7−1.8); scapulae rounded, cervical grooves deep anteriorly, shallow posteriorly, sigmoid in shape. Eyes flat. Chitinous tubercles at posterior body margin present. Scutum ornate, extensively pale

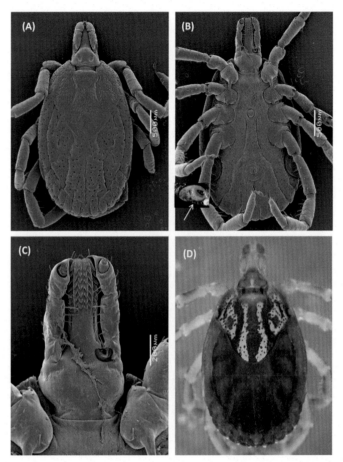

FIGURE 2.77 *A. triste* female: (A) dorsal view, (B) ventral view, (C) capitulum ventral view, (D) scutum ornamentation. *White arrow* in (B) shows the shape of spiracular plate.

yellowish, cervical spots narrow and divergent posteriorly, central area long and narrow, reaching the posterior margin of the scutum; punctuations numerous, small, larger in the antero-lateral fields. Notum glabrous. Basis capituli dorsally subrectangular, cornua absent, porose areas oval. Hypostome spatulate, dental formula 3/3. Genital aperture located at level of coxa II, U-shaped. Legs: coxa I with two distinct spurs, the external spur triangular and sharp, the internal spur as a small tubercle; coxae II—IV each with a triangular, short blunt spur; trochanters without spurs; presence of one spine on the tibia of legs II—IV. Spiracular plates oval.

Nymph (Fig. 2.78): small body size, total length 1.3 mm (1.2—1.3), breadth 0.8 mm (0.7—0.8). Body outline oval elongated. Chitinous tubercles

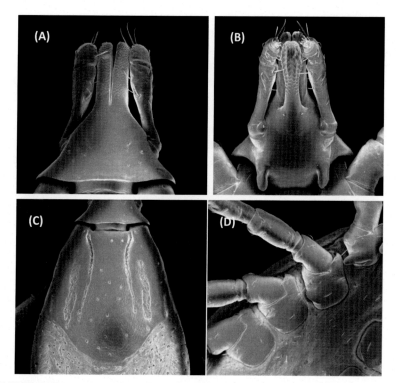

FIGURE 2.78   *A. triste* nymph: (A) capitulum dorsal view, (B) capitulum ventral view, (C) scutum, (D) coxae. *Figures (A) to (D) are reproductions of figure 6:* A. triste *nymph. Capitulum, dorsum; Capitulum, venter; Scutum; Coxae I–IV. From Martins TF, Labruna MB, Mangold AJ, Cafrune MM, Guglielmone AA, Nava S. Taxonomic key to nymphs of the genus* Amblyomma *(Acari: Ixodidae) in Argentina, with description and redescription of the nymphal stage of four* Amblyomma *species.* Ticks Tick-borne Dis *2014;5:753–70, with permission of Elsevier.*

at posterior body margin absent. Scutum with few punctuations, and postero-lateral margin sinuous. Eyes flat, located on lateral scutal angles at the level of the beginning of the posterior third of the scutum. Cervical grooves long, reaching the scutal posterior fourth, deeper at the anterior fourth; presence of a rugose depression parallel and external to the posterior half of cervical grooves. Basis capituli dorsally triangular, without cornua, with pseudoauricula and ventral processes. Hypostome spatulate, dental formula 2/2. Legs: coxa I with a triangular external spur, its base reaching the internal coxal margin; coxa II with a short triangular spur; coxae III–IV with a very short bridge-like spur each.

Taxonomic notes: Guglielmone et al.[2] stated that *A. triste* is established in Argentina, Brazil, Colombia, Ecuador, Peru, Uruguay, and Venezuela;

recently Mastropaolo et al.[1] and Abarca et al.[3] reported its presence in Bolivia and Chile, respectively. The type locality for *A. triste* is Montevideo, Uruguay,[23] and the populations of this tick from Argentina, Brazil, and Uruguay are morphologically and molecularly similar representing bona fide *A. triste*. These bona fide specimens have oval spiracular plates and tibia II–IV with a stout spine and a long seta, while *A. triste* of Chile in Abarca et al.[3] and Ecuador (deposited in the US National Tick Collection and examined by Santiago Nava) have comma-shaped spiracular plates and tibia II–IV with two spines (one stout, the other slender). Mendoza-Uribe and Chávez-Chorocco[22] found specimens morphologically compatible with the specimens from Chile and Ecuador and identified them as *A. maculatum*. Abarca et al.[3] found only 0.5% differences in sequences of 16S rDNA gene from tick populations of Chile and Uruguay; this small difference supported their decision to nominate the Chilean ticks as *A. triste*. Nevertheless, the same study found a low nucleotide difference with *A. maculatum* (1%). The scenario depicted shows that the separation of *A. triste* and *A. maculatum* is not resolved. We have provisionally treated those tick populations from Chile, Ecuador, and Peru as *A. triste*, but a morphological and molecular comparative study of Nearctic and Neotropical populations of *A. triste* and *A. maculatun* is needed to determine whether these ticks are conspecific.

See also Taxonomic notes of *A. neumanni* Ribaga, 1902, and *A. tigrinum* Koch, 1844.

DNA sequences with relevance for tick identification and phylogeny: mitochondrial DNA sequences of *A. triste* are available in the Gen Bank as follows: 16S rDNA gene (accession numbers KP739869, AY498563, JN180848), 12S rDNA gene (accession number AY342286), and sequences of the nuclear genome (the fragment 5.8S ribosomal RNA gene-internal transcribed spacer 2-28S ribosomal RNA gene) (accession number AY887114).

REFERENCES

1. Mastropaolo M, Beltrán-Saavedra LF, Guglielmone AA. The Ticks (Acari: Ixodida: Argasidae, Ixodidae) of Bolivia. *Ticks Tick-borne Dis* 2014;**5**:186–94.
2. Guglielmone AA, Estrada-Peña A, Keirans JE, Robbins RG. Ticks (Acari: Ixodida) of the neotropical zoogeographic region. Special publication of the integrated consortium on ticks and tick-borne diseases-2. Houten (The Netherlands): Atalanta; 2003.
3. Abarca K, López J, Acosta-Jamett G, Lepe P, Soares JF, Labruna MB. A third *Amblyomma* species and the first tick-borne rickettsia in Chile. *J Med Entomol* 2012;**49**:219–22.
4. Mertins JW, Moorhouse AS, Alfred JT, Hutcheson HJ. *Amblyomma triste* (Acari: Ixodidae): new North American collection records, including the first from the United States. *J Med Entomol* 2010;**47**:536–42.

5. Labruna MB, Jorge RSP, Sana DA, Jácomo ATA, Kashivakura CK, Furtado MM, et al. Ticks (Acari: Ixodida) on wild carnivores in Brazil. *Exp Appl Acarol* 2005;**36**:149–63.

6. Guglielmone AA, Nava S. Las garrapatas argentinas del género *Amblyomma* (Acari: Ixodidae): distribución y hospedadores. *Rev Invest Agropec* 2006;**35**(3):135–55.

7. Venzal JM, Estrada-Peña A, Castro O, De Souza CG, Félix ML, Nava S, et al. *Amblyomma triste* Koch, 1844 (Acari: Ixodidae): hosts and seasonality of the vector of *Rickettsia parkeri* in Uruguay. *Vet Parasitol* 2008;**155**:104–9.

8. Nava S, Mangold AJ, Mastropaolo M, Venzal JM, Fracassi N, Guglielmone AA. Seasonal dynamics and hosts of *Amblyomma triste* (Acari: Ixodidae) in Argentina. *Vet Parasitol* 2011;**181**:301–8.

9. Guglielmone AA, Nava S. Rodents of the subfamily Caviinae as hosts for hard ticks (Acari: Ixodidae). *Mastozool Neotr* 2010;**17**:279–86.

10. Guglielmone AA, Nava S. Rodents of the subfamily Sigmodontinae as hosts for South American hard ticks (Acari: Ixodidae) with hypothesis on life history. *Zootaxa* 2011;**2904**:45–65.

11. Wolf RW, Aragona M, Muñoz-Leal S, Pinto LB, Melo ALT, Braga IA, et al. Novel *Babesia* and *Hepatozoon* agents in the Brazilian Pantanal, with the first record of the tick *Ornithodoros guaporensis* in Brazil. *Ticks Tick-borne Dis* 2016;**7**:449–56.

12. Monje LD, Costa FB, Colombo VC, Labruna MB, Antoniazzi LR, Gamietea I, et al. Dynamics of exposure to *Rickettsia parkeri* in cattle in the Paraná River Delta, Argentina. *J Med Entomol* 2016;**53**:660–5.

13. Guglielmone AA, Nava S, Mastropaolo M, Mangold AJ. Distribution and genetic variation of *Amblyomma triste* (Acari: Ixodidae) in Argentina. *Ticks Tick-borne Dis* 2013;**4**:386–90.

14. Venzal JM, Portillo A, Estrada-Peña A, Castro O, Cabrera PA, Oteo JA. *Rickettsia parkeri* in *Amblyomma triste* from Uruguay. *Emerg Infect Dis* 2004;**10**:1493–5.

15. Pacheco RC, Venzal JM, Richtzenhain LJ, Labruna MB. *Rickettsia parkeri* in Uruguay. *Emerg Infect Dis* 2006;**12**:1804–5.

16. Seijo A, Picollo M, Nicholson W, Paddock CD. Fiebre manchada por rickettsias en el Delta del Paraná. Una enfermedad emergente. *Medicina (Buenos Aires)* 2007;**67**:723–6.

17. Nava S, Elshenawy Y, Eremeeva ME, Sumner JW, Mastropaolo M, Paddock CD. *Rickettsia parkeri* in Argentina. *Emerg Infect Dis* 2008;**14**:1894–7.

18. Conti-Díaz IA, Moraes-Filho J, Pacheco RC, Labruna MB. Serological evidence of *Rickettsia parkeri* as the etiological agent of rickettsiosis in Uruguay. *Rev Inst Med Trop São Paulo* 2009;**51**:337–9.

19. Romer Y, Seijo AC, Crudo F, Nicholson WL, Varela-Stokes A, Lash RR, et al. *Rickettsia parkeri* rickettsiosis, Argentina. *Emerg Infect Dis* 2011;**17**:1169–73.

20. Cicuttin G, Nava S. Molecular identification of *Rickettsia parkeri* infecting *Amblyomma triste* ticks in an area of Argentina where cases of rickettsiosis were diagnosed. *Mem Inst Oswaldo Cruz* 2013;**108**:123–5.

21. Keirans JE. George Henry Falkiner Nuttall and the Nuttall tick catalogue. *US Dep Agric Agric Res Serv Miscel Publ* 1985;**1438**:1785.

22. Mendoza Uribe L, Chávez Chorocco J. Ampliación geográfica de siete especies de *Amblyomma* (Acari: Ixodidae) y primer reporte de *A. oblongoguttatum* Koch, 1844 para Perú. *Rev Peru Entomol* 2004;**44**:69–72.

23. Koch CL. Systematische Übersicht über die Ordnung der Zecken. *Arch Natur* 1844;**10**:217–19.

Genus *Dermacentor*

The genus *Dermacentor* comprises actually 40 species (6% of the total of Ixodidae in the world) widespread among continents but being more prevalent in lands of Laurasian origin (Nearctic and Palearctic) than in ancient Gondwanian lands. The most common parasitic cycle is three-host cycle but a few species have one-host parasitic cycle, and almost all of them feed usually on Mammalia. The genus *Dermacentor* has only one nonexclusive representative in the Southern Cone of America, *D. nitens*, a one-host tick of veterinary importance that drew the attention of few regional tick workers.

Dermacentor nitens
Neumann, 1897

Neumann, L.G. (1897) Révision de la famille des ixodidés (2e mémoire). *Mémoires de la Société Zoologique de France*, 10, 324−420.

DISTRIBUTION

D. nitens is widely distributed in the Neotropical Region from northern Argentina to southern Mexico (Chile and Uruguay excluded) and the Caribbean and Galapagos Islands, and it is also present in the Nearctic Region.[1,2]

    Biogeographic distribution in the Southern Cone of America: Chaco and Yungas (Neotropical Region) (see Fig. 1 in the introduction).

HOSTS

Horse is the principal host for *D. nitens*,[3,4] but it has been also found on a variety of mammals as presented in Table 2.24.

ECOLOGY

D. nitens is a one-host tick species usually found in the ears of its hosts, but the nasal diverticulum, the perineal region and tail, the area of the mane, and the ventral midline of the body are also sites of attachment for this tick species.[3,5−7] Studies performed in Brazil by Borges and Oliveira[3] have shown that *D. nitens* can develop three or four generations per year. The seasonal dynamics of *D. nitens* was also studied in Brazil, where the high temperatures and relative humidity occurring during spring and summer lead to a shortening of the nonparasitic phase, high egg production, and hatching rates.[3,8] Seasonality of *D. nitens* on horses in Brazil was characterized by three major peaks of activity, the first two peaks were detected during the

TABLE 2.24 Hosts of *D. nitens,* a One-Host Tick Species (Parasitic Stages not Shown)

MAMMALIA	Leopardus tigrinus
ARTIODACTYLA	Panthera onca
Bovidae	Puma concolor
Cattle	Herpestidae
Domestic buffalo	Herpestes javanicus
Goat	PERISSODACTYLA
Cervidae	Equidae
Blastoceros dichotomus	Donkey
Mazama gouazoubira	Horse
Odocoileus virginianus	Mule
Suidae	Tapiridae
Domestic pig	Tapirus bairdii
CARNIVORA	PILOSA
Canidae	PRIMATES
Domestic dog	Hominidae
Cerdocyon thous	Human
Lycalopex vetulus	RODENTIA
Felidae	Caviidae
Domestic cat	Hydrochoerus hydrochaeris

Main host underlined.

wet and warm period of the year, while a third peak was found in the drier and colder part of the year.[5]

SANITARY IMPORTANCE

The infestation with *D. nitens* in the ears and nasal cavity of horses produces inflammatory reactions which may favor the occurrence of myiasis, and deformation of the ears.[3] This tick species has been involved as vector of diseases of equids caused by the protozoan *Babesia caballi.*[1,2] *D. nitens* has been also found parasitizing other domestic animals as cattle, dog, cat, domestic pig, mule, donkey, and goat. Although the *Rickettsia rickettsii* and *Ehrlichia chaffensis* were detected in *D. nitens* from Panama,[9] the role of this tick as vector of these human pathogens is probably insignificant because human infestation with this tick is unusual.[1]

DIAGNOSIS

Male (Fig. 2.79): total length 2.5 mm, breadth 1.8 mm. Body outline oval elongated, with moderately convex lateral margins; scapulae pointed; cervical grooves shallow, converging anteriorly *then diverging posteriorly*. Eyes small. Scutum inornate, surface smooth with a few punctuations concentrated on the antero-lateral scutal fields, setae present, festoons seven in number. Basis capituli dorsally rectangular, with short cornua. Hypostome spatulate, dental formula 4/4, palpal segment IV terminal. Legs: coxa I with two distinct, blunt short spurs, subequal in size; coxae II−III each with two triangular, short blunt spurs, the external larger than the internal; coxa IV well developed, with a triangular, short blunt external spur, the internal spur as a small tubercle. Spiracular plates subcircular, with few, very large goblet cells, usually in numbers of six or seven.

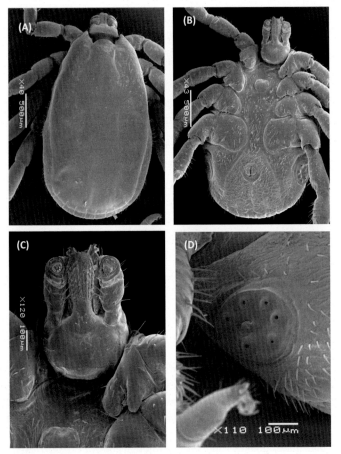

FIGURE 2.79 *D. nitens* male: (A) dorsal view, (B) ventral view, (C) capitulum ventral view, (D) spiracular plate.

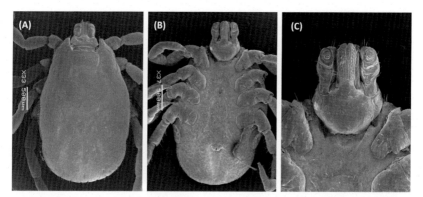

FIGURE 2.80   *D. nitens* female: (A) dorsal view, (B) ventral view, (C) capitulum ventral view. Spiracular plate as in the male, see Fig. 2.79.

Female (Fig. 2.80): total length 2.8 mm, breadth 1.9 mm. Scutum length 1.6 mm, breadth 1.6 mm. Body outline oval elongated, with moderately convex lateral margins; scapulae pointed; cervical grooves shallow, narrow, and linear. Eyes small. Scutum inornate, surface smooth, punctuation more apparent in the anterior field, setae present, festoons seven in number, notum with setae. Basis capituli dorsally rectangular, without cornua, porose areas oval. Hypostome spatulate, dental formula 4/4, palpal segment IV terminal. Legs: coxa I with two distinct, blunt short spurs subequal in size; coxae II–III each with two triangular, short blunt spurs, the external larger than the internal; coxa IV with a triangular, short, blunt external spur, the internal spur as a small tubercle. Spiracular plates subcircular, with few, very large goblet cells, usually in numbers of six or seven.

Nymph (Fig. 2.81): total length 1.1 mm, breadth 0.8 mm. Scutum wider than long, cervical grooves short, postero-lateral margin convex sinuous. Eyes small at lateral scutal angles at the level of the beginning of the posterior third of the scutum. Basis capituli dorsally subrectangular, without cornua. Hypostome spatulate, dental formula 3/3. Legs: coxa I with two small, triangular spurs, the external longer than the internal; coxae II–IV each with a triangular, small external spur. Spiracular plate subcircular, with few, very large goblet cells, usually in numbers of three to six.

Taxonomic notes: the names *Anocentor nitens* and *Otocentor nitens* were largely used in the tick literature, but they are synonyms of *D. nitens*.[10]

DNA sequences with relevance for tick identification and phylogeny: DNA sequences of *D. nitens* are available in the Gen Bank as follows: mitochondrial 16S rRNA gene (accession number AY375436), 12S rRNA gene (accession numbers JF523336, AY008688), and cytochrome oxidase II gene (accession numbers KF200162, KF200169, KF200121, KF200099, KF200094, AY008679). There also are available sequences of 18S rRNA gene (accession number KC769621), 28S rRNA gene (accession number

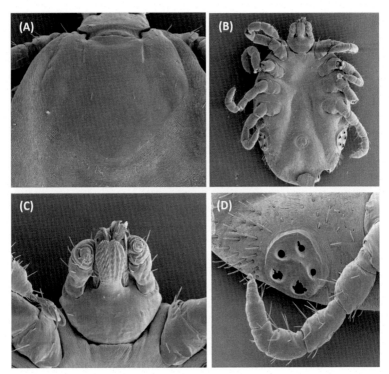

FIGURE 2.81 *D. nitens* nymph: (A) scutum, (B) ventral view, (C) capitulum ventral view, (D) spiracular plate.

KC769642), internal transcribed spacer 2 (accession number AF271274), and of the fragment 5.8S ribosomal RNA gene-internal transcribed spacer 2-28S ribosomal RNA gene (accession number KC503275).

REFERENCES

1. Yunker CE, Keirans JE, Clifford CM, Easton ER. *Dermacentor* ticks (Acari: Ixodoidea: Ixodidae) of the New World: a scanning electron microscopy atlas. *Proc Entomol Soc Wash* 1986;**88**:609−27.
2. Guglielmone AA, Estrada-Peña A, Keirans JE, Robbins RG. Ticks (Acari: Ixodida) of the neotropical zoogeographic region. Special publication of the integrated consortium on ticks and tick-borne diseases-2. Houten (The Netherlands): Atalanta; 2003.
3. Borges LMF, Oliveira PR. The tropical horse tick, *Anocentor nitens* (Neumann, 1897) (Acari: Ixodidae). *Trends Entomol* 2003;**3**:165−8.
4. Guglielmone AA, Robbins RG, Apanaskevich DA, Petney TN, Estrada-Peña A, Horak IG. *The hard ticks of the world (Acari: Ixodida: Ixodidae)*. London: Springer; 2014.
5. Borges LMF, Oliveira PR, Ribeiro MFB. Seasonal dynamics of *Anocentor nitens* on horses in Brazil. *Vet Parasitol* 2000;**89**:165−71.

6.   Labruna MB, Kerber CE, Ferreira F, Faccini JLH, De Waal DT, Gennari SM. Risk factors to tick infestations and their occurrence on horses in the state of São Paulo, Brazil. *Vet Parasitol* 2001;**97**:1−14.

7.   Labruna MB, Amaku M. Rhythm of engorgement and detachment of *Anocentor nitens* females feeding on horses. *Vet Parasitol* 2006;**137**:316−32.

8.   Borges LMF, Oliveira PR, Ribeiro MFB. Seasonal dynamics of the free-living phase of *Anocentor nitens* at Pedro Leopoldo, Minas Gerais, Brazil. *Vet Parasitol* 1999;**87**:73−81.

9.   Bermúdez SE, Ereemeva M, Karpathy SE, Samudio F, Zambrano ML, Zaldivar Y, et al. Detection and identification of rickettsial agents in ticks from domestic animals in eastern Panama. *J Med Entomol* 2009;**46**:856−61.

10.   Guglielmone AA, Nava S. Names for Iodidae (Acari: Ixodoidea): valid, synonyms, *incertae sedis*, *nomina dubia*, *nomina nuda*, *lapsus*, incorrect and suppressed names − with notes on confusions and misidentifications. *Zootaxa* 2014;**3767**:1−256.

Genus *Haemaphysalis*

The genus *Haemaphysalis* is the second largest tick genus of Ixodidae with 166 species (23% of the total worldwide) characterized by a three-host parasitic cycle. Species of *Haemaphysalis* are abundant in the Oriental Region and in a lesser extent in the Afrotropical (Ethiopian) Region, both of Gondwanian origin, with only four species for the Nearctic and Neotropical Regions. They feed largely on mammals or mammals and birds, with five species feeding exclusively on Aves. Only two species are established not exclusively in the Southern Cone of America and known from regional records dispersed through time.

Haemaphysalis juxtakochi Cooley, 1946

Cooley, R.A. (1946) The genera *Boophilus*, *Rhipicephalus*, and *Haemaphysalis* (Ixodoidea) of the New World. *National Institute of Health Bulletin*, 187, 54 pp.

DISTRIBUTION

Argentina, Belize, Bolivia, Brazil, Colombia, Costa Rica, Ecuador, French Guiana, Guyana, Mexico, Panama, Paraguay, Peru, Surinam, Trinidad and Tobago, Uruguay, and Venezuela.[1−4] This species is also established in the Nearctic Region.[3]

Biogeographic distribution in the Southern Cone of America: Chaco, Pampa, Parana Forest, and Yungas (Neotropical Region) (see Fig. 1 in the introduction).

HOSTS

Species of the family Cervidae are the most common hosts for *H. juxtakochi* adults[5,6] but they were also found on a variety of mammals and occasionally on birds (Table 2.25). Large wild and domestic mammals are also hosts for immature stages of *H. juxtakochi*, but larvae and nymphs were found as well on small rodents and birds[5,7−15] (Table 2.25).

ECOLOGY

Ecological information about *H. juxtakochi* refers just to distribution and host association.

SANITARY IMPORTANCE

Adults and immature stages of *H. juxtakochi* are occasional parasites of cattle and humans.[16,17] This tick species was found infected with *Rickettsia rhipicephali* in Brazil,[18] but the pathogenicity of this bacteria is unknown.

DIAGNOSIS

Male (Fig. 2.82): total length 2.0 mm, breadth 1.3 mm (Argentinean specimens; see Taxonomic notes). Body outline oval, scapulae rounded, cervical grooves short and shallow; marginal groove absent. Eyes absent. Scutum inornate; punctuations numerous, fine, uniformly distributed. Basis capituli rectangular dorsally, cornua long; article II of palps extending laterally; article III of palps with a long, retrograde, ventral spur. Hypostome spatulate, dental formula 4/4. Genital aperture located at level of coxa II, U-shaped. Legs: coxae I−IV each with an internal, triangular blunt spur, spur on coxa I long, spurs on coxa II−IV short; trochanter of leg I with a conspicuous spur. Spiracular plates comma-shaped.

Female (Fig. 2.83): total length 2.7 mm, breadth 1.6 mm (Argentinean specimens; see Taxonomic notes). Body outline oval. Scutum length 1.1 mm, breadth 1.1 mm, inornate; punctuations numerous, fine, uniformly distributed; scapulae rounded; cervical grooves long, deep in the anterior part, shallow posteriorly. Eyes absent. Basis capituli rectangular dorsally, cornua long, porose areas oval; article II of palps extending laterally; article III of palps with a long, retrograde, ventral spur. Hypostome spatulate, dental formula 4/4. Genital aperture located at level of coxae II, U-shaped. Legs: coxae I−IV

TABLE 2.25 Hosts of Adults (A), Nymphs (N), and Larvae (L) of
H. juxtakochi

MAMMALIA		Horse	A
ARTIODACTYLA		Mule	A
Bovidae		**Tapiridae**	
Cattle	ANL	*Tapirus bairdii*	A
Goat	A	*T. terrestris*	A
Sheep	N	**PRIMATES**	
Camelidae		**Cebidae**	
Vicugna vicugna	A	*Cebus kaapori*	A
Cervidae		**Hominidae**	
Axis axis	A	Human	AN
Mazama americana	ANL	**RODENTIA**	
M. bororo	ANL	**Cricetidae**	
M. gouazoubira	ANL	*Akodon azarae*	L
M. nana	A	*Oxymycterus rufus*	L
M. temama	A	**Dasyproctidae**	
Odocoileus virginianus	A	*Dasyprocta azarae*	N
Ozotoceros bezoarticus	A	*D. fuliginosa*	NL
Suidae		*D. leporina*	NL
Pigs (domestic and feral)	AN	*D. punctata*	N
Tayassuidae		**Erethizontidae**	
Tayassu pecari	A	*Coendou rothschildi*	N
Pecari tajacu	NL	**Sciuridae**	
CARNIVORA		*Sciurus aestuans*	N
Canidae		*S. igniventris*	L
Canis latrans	N	**AVES**	
Cerdocyon thous	AN	**FALCONIFORMES**	
Chrysocyon brachyurus	N	**Accipitridae**	
Domestic dog	AN	*Harpia harpyja*	N
Felidae		**GALLIFORMES**	
Puma concolor	A	**Cracidae**	
Procyonidae		*Penelope obscura*	L

(*Continued*)

TABLE 2.25 (Continued)

Nasua nasua	NL	**PASSERIFORMES**	
CINGULATA		**Corvidae**	
Dasypodidae		*Cyanocorax chrysops*	NL
Tolypeutes matacus	A	**Emberizidae**	
DIDELPHIMORPHIA		*Arremon flavirostris*	L
Didelphidae		**Thamnophilidae**	
Didelphis aurita	N	*Hylophylax naevius*	NL
LAGOMORPHA		*Willisornis poecilinotus*	L
Leporidae		**Thraupidae**	
Sylvilagus brasiliensis	NL	*Pyrrochoma ruficeps*	A
S. floridanus	A	**Turdidae**	
PERISSODACTYLA		*Turdus albicollis*	NL
Equidae	AN	*T. rufiventris*	NL

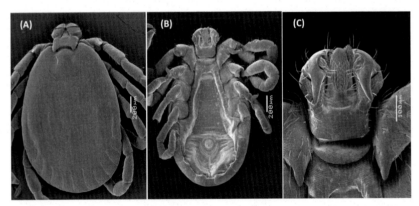

FIGURE 2.82 *H. juxtakochi* male: (A) dorsal view, (B) ventral view, (C) capitulum ventral view.

each with an internal rounded small spur; trochanter of legs I with a conspicuous spur. Spiracular plates comma-shaped.

Nymph (Fig. 2.84): total length 1.3 mm, breadth 1.1 mm. Body outline oval, scapulae rounded, cervical grooves short and shallow. Eyes absent. Scutum impunctate. Basis capituli quadrangular dorsally, cornua long; article II of palps extending laterally; article III of palps with a long, retrograde, ventral spur. Ventral cornua present. Hypostome spatulate, dental formula 2/2.

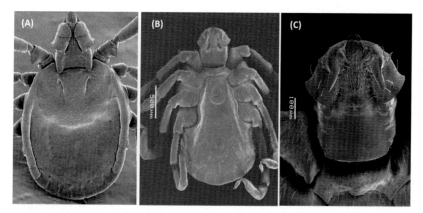

FIGURE 2.83   *H. juxtakochi* female: (A) dorsal view, (B) ventral view, (C) capitulum ventral view.

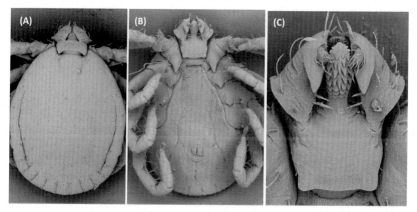

FIGURE 2.84   *H. juxtakochi* nymph: (A) dorsal view, (B) ventral view, (C) capitulum ventral view.

Legs: coxae I–IV each with an internal, triangular blunt spur, spur on coxa I long, spurs on coxae II–IV short; trochanter of leg I with a conspicuous spur.

Taxonomic notes: great discrepancies were observed in the sizes of males and in a lesser extent of females of *H. juxtakochi* by different authors and no sizes for the nymph of this species were found. The measures above correspond to unfed Argentinean specimens obtained from vegetation.

DNA sequences with relevance for tick identification and phylogeny: the sequences of *H. juxtakochi* available in the Gen Bank correspond to the mitochondrial genes 16S rRNA (accession numbers AY762323, AY762324, GQ891953), 12S rRNA (accession number KF195929), and cytochrome oxidase I (accession numbers KF200120, KF200092, KF200091, KF200081, KF200077).

REFERENCES

1. Jones EK, Clifford CM, Keirans JE, Kohls GM. The ticks of Venezuela (Acarina: Ixodoidea) with a key to the species of *Amblyomma* in the Western Hemisphere. *Brigham Young Univ Sci Bull Biol Ser* 1972;**17**(4):1−40.
2. Varma MGR. Ticks (Ixodidae) of British Honduras. *Trans R Soc Trop Med Hyg* 1973;**67**:92−102.
3. Guglielmone AA, Estrada-Peña A, Keirans JE, Robbins RG. Ticks (Acari: Ixodida) of the neotropical zoogeographic region. Special publication of the integrated consortium on ticks and tick-borne diseases-2. Houten (The Netherlands): Atalanta; 2003.
4. Guglielmone AA, Romero J, Venzal JM, Nava S, Mangold AJ, Villavicencio J. First record of *Haemaphysalis juxtakochi* Cooley, 1946 (Acari: Ixodidae) from Peru. *Syst Appl Acarol* 2005;**10**:33−5.
5. Labruna MB, Camargo LMA, Terrassini FA, Ferreira F, Schumaker TST, Camargo EP. Ticks (Acari: Ixodidae) from the state of Rondônia, western Amazon, Brazil. *Syst Appl Acarol* 2005;**10**:17−32.
6. Guglielmone AA, Robbins RG, Apanaskevich DA, Petney TN, Estrada-Peña A, Horak IG. *The hard ticks of the world (Acari: Ixodida: Ixodidae)*. London: Springer; 2014.
7. Arzua M, Onofrio VC, Barros-Battesti DM. Catalogue of the tick collection (Acari, Ixodida) of the Museu de História Natural Capão da Imbuia, Curitiba, Paraná, Brazil. *Rev Bras Zool* 2005;**22**:623−32.
8. Ogrzewalska M, Uezu A, Labruna MB. Ticks (Acari: Ixodidae) infesting wild birds in the eastern Amazon, northern Brazil, with notes on rickettsial infections in ticks. *Parasitol Res* 2010;**106**:809−16.
9. Nava S, Mangold AJ, Mastropaolo M, Venzal JM, Fracassi N, Guglielmone AA. Seasonal dynamics and hosts of *Amblyomma triste* (Acari: Ixodidae) in Argentina. *Vet Parasitol* 2011;**181**:301−8.
10. Debárbora VN, Nava S, Cirignoli S, Guglielmone AA, Poi ASG. Ticks (Acari: Ixodidae) parasitizing endemic and exotic wild mammals in the Esteros del Iberá wetlands, Argentina. *Syst Appl Acarol* 2012;**17**:243−50.
11. Ogrzewalska M, Saraiva DG, Moraes-Filho J, Martins TF, Costa FB, Pinter A, et al. Epidemiology of Brazilian spotted fever in the Atlantic Forest, state of São Paulo, Brazil. *Parasitology* 2012;**139**:1283−300.
12. Bermúdez SE, González D, García G. Ticks (Acari: Ixodidae, Argasidae) of coyotes in Panama. *Syst Appl Acarol* 2013;**18**:112−15.
13. Nava S, Guglielmone AA. A meta-analysis of host specificity in Neotropical hard ticks (Acari: Ixodidae). *Bull Entomol Res* 2013;**103**:216−24.
14. Flores FS, Nava S, Batallán G, Tauro LB, Contigiani MS, Diaz LA, et al. Ticks (Acari: Ixodidae) on wild birds in north-central Argentina. *Ticks Tick-borne Dis* 2014;**5**:715−21.
15. Lamattina D, Tarragona EL, Costa SA, Guglielmone AA, Nava S. Ticks (Acari: Ixodidae) of northern Misiones Province, Argentina. *Syst Appl Acarol* 2014;**19**:393−8.
16. Guglielmone AA, Beati L, Barros-Battesti DM, Labruna MB, Nava S, Venzal JM, et al. Ticks (Ixodidae) on humans in South America. *Exp Appl Acarol* 2006;**40**:83−100.
17. Guglielmone AA, Nava S. Epidemiología y control de las garrapatas de los bovinos en la Argentina. In: Nari A, Fiel C, editors. *Enfermedades parasitarias con importancia clínica y productiva en rumiantes: fundamentos epidemiológicos para su diagnóstico y control*. Buenos Aires: Editorial Hemisferio Sur; 2013; pp. 441−56.
18. Labruna MB, Camargo LMA, Camargo EP, Walker DH. Detection of a spotted fever group *Rickettsia* in the tick *Haemaphysalis juxtakochi* in Rondônia, Brazil. *Vet Parasitol* 2005;**127**:169−74.

Haemaphysalis leporispalustris (Packard, 1869)

Ixodes leporispalustris. Packard, A.S. (1869) Report of the curator of Articulata. *First Annual Report of the Trustees of the Peabody Academy of Sciences*, pp. 56–69.

DISTRIBUTION

Argentina, Bolivia, Brazil, Colombia, Costa Rica, Guatemala, Mexico, Panama, Paraguay, Peru, Venezuela, and a few islands in the Lesser Antilles.[1,2] This species is also established in the Nearctic Region.[2]

Biogeographic distribution in the Southern Cone of America: Chaco, Parana Forest, and Yungas (Neotropical Region) (see Fig. 1 in the introduction).

HOSTS

In the Neotropical Region, all parasitic stages of *H. leporispalustris* are usually found on wild rabbits, but grown-dwelling birds are also frequently recorded as hosts for immature stages.[3–9] The complete host range of *H. leporispalustris* is presented in Table 2.26.

TABLE 2.26 Hosts of Adults (A), Nymphs (N), and Larvae (L) of *H. leporispalustris*

MAMMALIA		PASSERIFORMES	
DIDELPHIMORPHIA		Corvidae	
Didelphidae		*Cyanocorax chrysops*	L
Metachirus nudicaudatus	N	Emberizidae	
LAGOMORPHA		*Arremon flavirostris*	L
Leporidae		*A. semitorquatus*	NL
Lepus europaeus	A	*Arremonops conirostris*	L
Sylvilagus brasiliensis	ANL	Parulidae	
S. floridanus	ANL	*Myiothlypis bivittata*	L
RODENTIA		Thamnophilidae	
Cricetidae		*Myrmeciza exsul*	NL
Peromyscus sp.	L	Thraupidae	

(Continued)

TABLE 2.26 (Continued)

Dasyproctidae		Tricothraupis melanops	
Dasyprocta azarae	A	Volatinia jacarina	L
D. punctata	AL	**Troglodytidae**	L
Echimyidae		Thryothorus genibarbis	
Proechimys guyannensis	N	T. nigricapillus	L
PRIMATES		Troglodytes musculus	L
Hominidae		**Turdidae**	L
Human	N	Catharus dryas	
AVES		Turdus rufiventris	L
GALLIFORMES		**Tyrannidae**	NL
Tinamidae		Corythopis delalandi	
Crypturellus tataupa	N	Fluvicola nengeta	L
		Myiarchus tyrannulus	L

Main hosts for adult ticks and immature stages underlined.

ECOLOGY

Labruna et al.[10] investigated the life cycle of this species under laboratory conditions in Brazil, but there are no studies on ecological aspects of *H. leporispalustris* under field conditions in the Neotropical Region, at least to the knowledge of the authors of this book. Information about population dynamics of *H. leporispalustris* performed in the Nearctic Region (United States and Canada) demonstrated that adults are active principally in spring and early summer while the peak of activity of larvae and nymphs occur in late summer and autumn.[11-13] The climbing behavior and the "drop-off" rhythm of *H. leporispalustris* are determinants of its nidicolous habits, which lead to a high host specificity.[14,15]

SANITARY IMPORTANCE

Spotted fever group *Rickettsia rickettsii* and a member of the typhus group *R. canadensis* (originally named *Rickettsia canada*) were isolated from *H. leporispalustris*.[16-18] Because the parasitism of *H. leporispalustris* in humans is a rare event,[17] the sanitary importance of this tick is associated to its participation in the enzootic cycle of those *Rickettsia* species. *H. leporispalustris* is also involved in the enzootic cycle of *Francisella tularensis*, the causal agent of tularemia, in North America.[20,21]

DIAGNOSIS

Male (Fig. 2.85): total length 1.7 mm, breadth 1.2 mm. Body outline oval, scapulae rounded, cervical grooves short and deep; marginal groove absent. Eyes absent. Scutum inornate; punctuations numerous, fine, uniformly distributed. Basis capituli rectangular dorsally, cornua long; ventral process conspicuous; article II of palps extending laterally; article III of palps with a short, retrograde, ventral spur. Hypostome spatulate, dental formula 3/3. Genital aperture located at level of coxa II, U-shaped. Legs: coxae I–IV with an internal rounded blunt spur each; trochanter of leg I with a conspicuous spur. Spiracular plates comma-shaped.

Female (Fig. 2.86): total length 2.0 mm, breadth 1.3 mm. Body outline oval. Scutum length 0.8 mm, breadth 0.7 mm, inornate; punctuations numerous, fine, uniformly distributed; scapulae rounded; cervical grooves long,

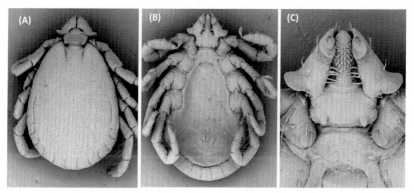

FIGURE 2.85   *H. leporispalustris* male: (A) dorsal view, (B) ventral view, (C) capitulum ventral view.

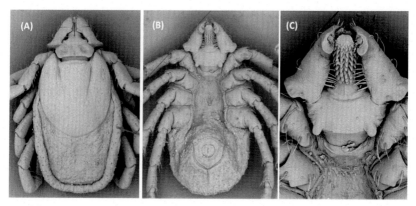

FIGURE 2.86   *H. leporispalustris* female: (A) dorsal view, (B) ventral view, (C) capitulum ventral view.

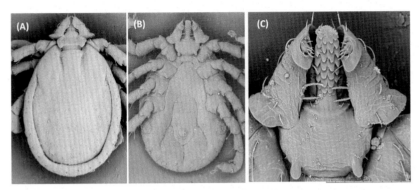

FIGURE 2.87 *H. leporispalustris* nymph: (A) dorsal view, (B) ventral view, (C) capitulum ventral view.

deep in the anterior part, shallow posteriorly. Eyes absent. Basis capituli rectangular dorsally, cornua long, porose areas oval; ventral process conspicuous; article II of palps extending laterally; article III of palps with a short, retrograde, ventral spur. Hypostome spatulate, dental formula 3/3. Genital aperture located at level of coxa II, V-shaped. Legs: coxae I–IV with an internal rounded small spur each; trochanter of legs I with a conspicuous spur. Spiracular plates oval.

Nymph (Fig. 2.87): total length 1.2 mm, breadth 0.8 mm. Body outline oval, scapulae rounded, cervical grooves long, deep in the anterior part, shallow posteriorly. Eyes absent. Scutum impunctate. Basis capituli rectangular dorsally, cornua long; ventral process conspicuous; article II of palps extending laterally; article III of palps with a short, retrograde, ventral spur. Hypostome spatulate, dental formula 2/2. Legs: coxae I–IV with an internal rounded small spur each.

DNA sequences with relevance for tick identification and phylogeny: DNA sequences of *H. leporispalustris* are available in the Gen Bank as follows: mitochondrial genes 16S rRNA (accession number L34309) and 12S rRNA (accession numbers AM410574, U95873, JX192832), and sequences of the nuclear markers 18S rRNA (accession number JX573124), 28S rRNA (accession numbers JX573132, AF120310), and 5.8S ribosomal RNA gene and internal transcribed spacer 2 fragment (accession numbers JQ868582, JQ868580, JQ868574, JQ868568, JQ868565, GU993306, GU993307).

REFERENCES

1. Kohls GM. New records of ticks from the Lesser Antilles. *Stud Fauna Curaçao Caribb Isl* 1969;**28**:126–34.
2. Guglielmone AA, Estrada-Peña A, Keirans JE, Robbins RG. Ticks (Acari: Ixodida) of the neotropical zoogeographic region. Special publication of the integrated consortium on ticks and tick-borne diseases-2. Houten (The Netherlands): Atalanta; 2003.

3. Hoffmann A. Monografía de los Ixodoidea de México. I parte. *Rev Soc Mex Hist Nat* 1962;**23**:191−307.

4. Fairchild GB, Kohls GM, Tipton VJ. The ticks of Panama (Acarina: Ixodoidea). In: Wenzel WR, Tipton VJ, editors. *Ectoparasites of Panama*. Chicago (IL): Field Museum of Natural History; 1966. p. 167−219.

5. Jones EK, Clifford CM, Keirans JE, Kohls GM. The ticks of Venezuela (Acarina: Ixodoidea) with a key to the species of *Amblyomma* in the Western Hemisphere. *Brigham Young Univ Sci Bull Biol Ser* 1972;**17**(4):1−40.

6. Ogrzewalska M, Saraiva DG, Moraes-Filho J, Martins TF, Costa FB, Pinter A, et al. Epidemiology of Brazilian spotted fever in the Atlantic Forest, state of São Paulo, Brazil. *Parasitology* 2012;**139**:1283−300.

7. Flores FS, Nava S, Batallán G, Tauro LB, Contigiani MS, Díaz LA, et al. Ticks (Acari: Ixodidae) on wild birds in north-central Argentina. *Ticks and Tick-borne Diseases* 2014;**5**:715−21.

8. Guglielmone AA, Robbins RG, Apanaskevich DA, Petney TN, Estrada-Peña A, Horak IG. *The hard ticks of the world (Acari: Ixodida: Ixodidae)*. London: Springer; 2014.

9. Zeringóta V, Maturano R, Santolin ÍD, McIntosh D, Famadas KM, Daemon E, et al. New host records of *Haemaphysalis leporispalustris* (Acari: Ixodidae) on birds in Brazil. *Parasitol Res* 2016;**115**:2107−10.

10. Labruna MB, Leite RC, Faccini JL, Ferreira F. Life cycle of the tick *Haemaphysalis leporispalustris* (Acari: Ixodidae) under laboratory conditions. *Exp Appl Acarol* 2000;**24**:683−94.

11. Campbell A, Ward RM, Garvie MB. Seasonal activity and frequency distributions of ticks (Acari: Ixodidae) infesting snowshoe hares in Nova Scotia, Canada. *J Med Entomol* 1980;**17**:22−9.

12. Kollars TM, Oliver JH. Host associations and seasonal occurrence of *Haemaphysalis leporispalustris*, *Ixodes brunneus*, *I. cookie*, *I. dentatus*, and *I. texanus* (Acari: Ixodidae) in southeastern Missouri. *J Med Entomol* 2003;**40**:103−7.

13. Gabriele-Rivet V, Arsenault J, Badcock J, Cheng A, Edsall J, Goltz J, et al. Different ecological niches for ticks of public health significance in Canada. *Plos One* 2015;**10**(article e0131282):19 p.

14. George JE. Drop-off rhythms of engorged rabbit ticks, *Haemaphysalis leporispalustris* (Packard, 1869) (Acari: Ixodidae). *J Med Entomol* 1971;**8**:461−79.

15. Camin JH, Drenner RW. Climbing behavior and host-finding of larval rabbit ticks (*Haemaphysalis leporispalustris*). *J Parasitol* 1978;**64**:905−9.

16. McKiel JA, Bell EJ, Lackman DB. *Rickettsia canadensis*: a new member of the typhus group of ricketssiae isolated from *Haemaphysalis leporispalustris* Packard. *Can J Microbiol* 1967;**13**:503−10.

17. Fuentes L, Calderón A, Hun L. Isolation and identification of *Rickettsia rickettsii* from the rabbit tick *Haemophysalis leporispalustris* in the Atlantic zone of Costa Rica. *Am J Trop Med Hyg* 1985;**34**:564−7.

18. Hun L, Cortés X, Taylor L. Molecular characterization of *Rickettsia rickettsii* isolated from human clinical samples and from the rabbit tick *Haemaphysalis leporispalustris* collected at different geographic zones in Costa Rica. *Am J Trop Med Hyg* 2008;**79**:899−902.

19. Guglielmone AA, Beati L, Barros-Battesti DM, Labruna MB, Nava S, Venzal JM, et al. Ticks (Ixodidae) on humans in South America. *Exp Appl Acarol* 2006;**40**:83−100.

20. Philip CB, Parker R. Occurrence of tularaemia in the rabbit tick (*Haemaphysalis leporispalustris*) in Alaska. *Publ Health Rep* 1938;**53**:574−5.

21. Black DM, Thompson J. Tularaemia in British Columbia. *Can Med Assoc J* 1958;**78**:16−18.

Genus *Ixodes*

Ixodes is the largest and most probably the oldest genus of Ixodidae with 247 species, which represents 34% of the total forming the family, and all these species appeared to be three-host ticks. *Ixodes* are well distributed around the world but prevails in lands of Gondwanian origins (Afrotropical, Australasian, Neotropical, and Oriental). Host usage of *Ixodes* is characterized by 25 species feeding exclusively on Aves for a total of 31 species of Ixodidae with the same feeding habit. However, most of *Ixodes* feed on Mammalia and in a lesser extent on mammals and birds with one species not established in the Southern Cone of America bounds to Squamata. A total of 16 species of *Ixodes* are found in the region, 10 of them are established exclusively here. Several studies have been done locally due to sanitary and phylogenetic importance of *Ixodes*.

Ixodes abrocomae
Lahille, 1916

Lahille, F. (1916) Descripción de un nuevo ixódido chileno. *Revista Chilena de Historia Natural*, 20, 107−108.

DISTRIBUTION

Chile.[1]

  Biogeographic distribution in the Southern Cone of America: Coquimbo (Andean Region) (see Fig. 1 in the introduction).

HOSTS

Small rodents of the families Abrocomidae and Cricetidae are the only hosts recorded for adults of *I. abrocomae*[1] (Table 2.27).

TABLE 2.27 Hosts of Adults (A) of *I. abrocomae*

MAMALIA		Cricetidae	
RODENTIA		*Abrothrix longipilis*	A
Abrocomidae		*A. olivaceous*	A
Abrocoma bennetti	A	*Phyllotis xanthopygus*	A

ECOLOGY

Besides the few data on host association and geographical distribution, there is no information on the ecology of *I. abrocomae*.

SANITARY IMPORTANCE

The capacity of *I. abrocomae* to transmit pathogens has not been investigated to date, and no records on humans and domestic animals have been found.

DIAGNOSIS

Male (Fig. 2.88): total length 1.8 mm, breadth 0.8 mm. Body outline oval, scapulae rounded; cervical grooves shallow, converging anteriorly then diverging as shallow depressions. Scutum inornate; punctuations numerous, fine, uniformly distributed; setae scattered all over scutum but absent on postero-central field. Ventral surface covered by scattered setae, very few setae on anal plate; ventral plates as shown in Fig. 2.88. Basis capituli pentagonal dorsally, without cornua; presence of triangular and small auriculae; articles II and III of palps subequal in length, suture inconspicuous. Hypostome short, notched, dental formula 2/2 with an additional row of three small denticles in the anterior portion. Genital aperture located between coxae II and III. Legs: coxa I with two subequal spurs reaching coxa II; coxae II–III with a short triangular external spur each; coxa IV with a short

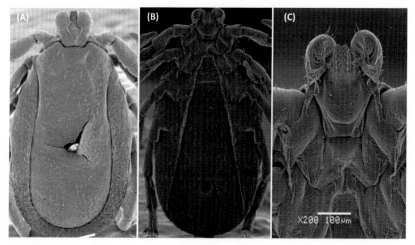

FIGURE 2.88 *I. abrocomae* male: (A) dorsal view, (B) ventral view, (C) capitulum ventral view. *Figure (B) is a reproduction of figure 2 of male ventral view from Guglielmone AA, Nava S, Bazán-León EA, Vásquez R, Mangold AJ. Redescription of the male and description of the female of* I. abrocomae *Lahille, 1916 (Acari: Ixodidae). Syst Parasitol 2010;77:153–60, with permission of Springer.*

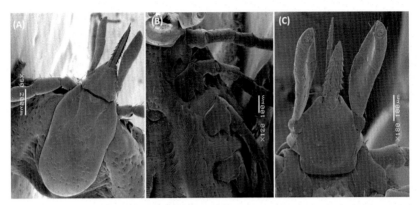

FIGURE 2.89   *I. abrocomae* female: (A) dorsal view, (B) ventral view, (C) capitulum ventral view. *Figures (B) and (C) are reproductions of figures 8 (basis capituli, ventral view of female) and 11 (coxae of female) from Guglielmone AA, Nava S, Bazán-León EA, Vásquez R, Mangold AJ. Redescription of the male and description of the female of* Ixodes abrocomae *Lahille, 1916 (Acari: Ixodidae). Syst Parasitol 2010;77:153–60, with permission of Springer.*

triangular central spur; coxae II—IV with an internal spur as a small tubercle. Spiracular plates rounded.

Female (Fig. 2.89): total body length for slightly engorged specimen 2.5 mm, breadth 1.5 mm. Body outline oval, scapulae pointed; cervical grooves shallow, converging anteriorly then diverging as shallow depressions. Scutum length 1.1 mm, breadth 0.6 mm, inornate, with sinuous outline and presence of few and scattered setae; punctuations evident, more numerous in the posterior field. Notum with numerous, whitish setae. Ventral surface covered by scattered setae. Basis capituli triangular dorsally, cornua small, porose area oval, auriculae rounded; article II of the palp longer than article III, suture conspicuous. Hypostome pointed, dental formula 3/3 (not shown in Fig. 2.89 because the hypostome of the female is broken). Genital aperture located between coxae II and III. Legs: coxae I—III with a single short external spur each; coxa IV with a single short central spur. Spiracular plates rounded.

Taxonomic notes: *I. abrocomae* form a phylogenetic group with *I. stilesi* Neumann, 1911, *I. sigelos* Keirans, Clifford, and Corwin, 1976, and *I. neuquenensis* Ringuelet, 1947 (Fig. 2.90). The distribution of these species is restricted to areas belonging to the Andean Region and South American Transition Zone sensu Morrone.[2] Larva and nymph of *I. abrocomae* are unknown. See Keirans et al.[2] for a figure of the female of *I. sigelos* with a complete hypostome.

DNA sequences with relevance for tick identification and phylogeny: there are just two sequences of the mitochondrial 16S rRNA gene of *I. abrocomae* in the Gen Bank (accession numbers GU188043, GU188044).

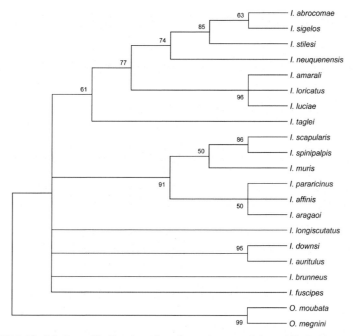

FIGURE 2.90   Maximum-likelihood condensed tree constructed from 16S rDNA sequences. The phylogenetic tree was generated with GTR model and a discrete Gamma-distribution.

REFERENCES

1. Guglielmone AA, Nava S, Bazán-León EA, Vásquez RA, Mangold AJ. Redescription of the male and description of the female of *Ixodes abrocomae* Lahille, 1916 (Acari: Ixodidae). *Syst Parasitol* 2010;**77**:153−60.
2. Morrone JJ. Biogeographic areas and transition zones of Latin America and the Caribbean Islands based on panbiogeographic and cladistic analyses of the entomofauna. *Ann Rev Entomol* 2006;**51**:467−94.

Ixodes aragaoi
Fonseca, 1935

Fonseca, F. (1935) Notas de acareologia. XV. Ocorrência de subspecies de *Ixodes ricinus* (L., 1758) no estado de S. Paulo (Acarina, Ixodidae). *Memórias do Instituto Butantan*, 9, 131−135.

DISTRIBUTION

Brazil and Uruguay.[1]

   Biogeographic distribution in the Southern Cone of America: Pampa (Neotropical Region) (see Fig. 1 in the introduction).

TABLE 2.28 Hosts of Adults (A), Nymphs (N), and Larvae (L) of *I. aragaoi*

MAMMALIA		PERISSODACTYLA	
ARTIODACTYLA		Equidae	
Bovidae		Horse	A
Cattle	A	RODENTIA	
Cervidae		Cricetidae	
Mazama gouazoubira	A	*Akodon* sp.	L
M. rufina	A	*A. azarae*	N
CARNIVORA		*Oligoryzomys nigripes*	N
Canidae		AVES	
Domestic dog	A	PASSERIFORMES	
Felidae		Furnariidae	
Puma concolor	A	*Phacellodomus striaticollis*	NL
DIDELPHIMORPHIA		*Syndactyla rufosuperciliata*	NL
Didelphidae		Parulidae	L
Monodelphis sp.	N	*Basileuterus leucoblepharus*	L
		Turdidae	
		Turdus albicollis	NL
		T. rufiventris	NL

HOSTS

Large wild and domestic mammals are listed as hosts for adults of *I. aragaoi*, but deer of the genus *Mazama* appear to be the principal hosts for this tick stage[1,2] (Table 2.28). Small rodents and birds are the hosts for larvae and nymphs[1,3,4] (Table 2.28). The records of immature stages of *I. pararicinus* on birds and small rodents mentioned in Venzal et al.[3,4] correspond, in fact, to *I. aragaoi* (see Taxonomic notes).

ECOLOGY

Besides the available information on host association and geographical distribution, there is no information on the ecology of *I. aragaoi*.

SANITARY IMPORTANCE

Adults of *I. aragaoi* are frequently found as parasites of cattle. Different genospecies of the *Borrelia burgdorferi* sensu lato group were found infecting *I. aragaoi* in Uruguay[5] but to date, there is no epidemiological evidence to involve *I. aragaoi* as vector of human pathogenic *Borrelia* genospecies. In fact, cases of human parasitism by *I. aragaoi* ticks were not recorded so far.

DIAGNOSIS

Male (Fig. 2.91): total length 1.9 mm, breadth 1.1 mm. Body outline oval, scapulae rounded; cervical grooves shallow, diverging posteriorly. Scutum inornate; punctuations numerous, larger and more densely distributed in the central field; presence of long setae irregularly distributed. Ventral surface with numerous punctuations, larger on the median plate; presence of numerous setae, larger on median plate; ventral plates as shown in Fig. 2.91; demarcation between median plate and anal and adanal plates broadly curved. Basis capituli rectangular dorsally, without cornua; ventral outline with trilobed transverse ridge, two lateral lobes and one median lobe triangular and displaced posteriorly; articles II and III of palps subequal in length, suture conspicuous. Hypostome short, with lateral teeth pointed and longer than median teeth. Genital aperture located between coxa III. Legs: coxa I with two spurs, internal spur long, narrow, and sharp, reaching the coxa II, external spur short and rounded; coxae II−IV each with a short rounded external spur. Spiracular plates oval.

Female (Fig. 2.92): total length 1.8 mm, breadth 1.4 mm. Body outline oval, scapulae pointed; cervical grooves shallow, converging anteriorly then diverging as shallow depressions. Scutum length 1.4 mm, breadth 1.3 mm, inornate; outline rounded; presence of few setae, irregularly distributed; punctuations evident, more numerous and larger on the posterior field. Notum with numerous whitish setae. Ventral surface covered by numerous whitish setae. Basis capituli subrectangular dorsally, cornua absent, porose area oval, auriculae rounded; article II of the palp longer than article III,

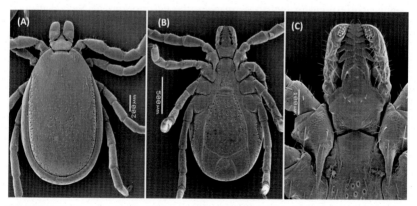

FIGURE 2.91 *I. aragaoi* male: (A) dorsal view, (B) ventral view, (C) capitulum ventral view.

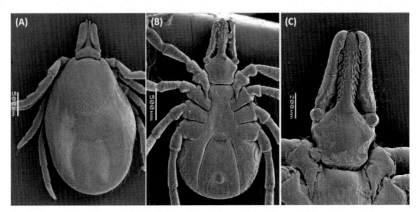

FIGURE 2.92 *I. aragaoi* female: (A) dorsal view, (B) ventral view, (C) capitulum ventral view.

suture conspicuous. Hypostome pointed, dental formula 4/4 for most of length, then 3/3 and 2/2 near base. Genital aperture located between coxa IV. Legs: coxa I with two spurs, the internal spur long, narrow, and sharp, reaching the coxa II, the external spur short and rounded; coxae II–IV each with a short, rounded, external spur. Spiracular plates oval.

Taxonomic notes: all records of *I. pararicinus* from Uruguay that were published until 2014 correspond to *I. aragaoi* as stated in Onofrio et al.[1] Therefore, Uruguay is excluded from the distributional range of *I. pararicinus*, whose current distribution includes Argentina, Colombia, and Peru.[1,6,7] Both *I. aragaoi* and *I. pararicinus* belong to the *I. ricinus* group. These two tick species are phylogenetically closely related (Fig. 2.90) and they are almost indistinguishable morphologically. Larva and nymph of *I. aragaoi* are not formally described. Venzal et al.[3] described larvae and nymphs of *I. pararicinus* using specimens from Argentina and Uruguay. Because it is currently recognized that the taxon previously named as *I. pararicinus* in Uruguay correspond in fact to *I. aragaoi*, the descriptions of Venzal et al.[3] cannot be assigned to any of the two tick species.

DNA sequences with relevance for tick identification and phylogeny: the only DNA sequence named as belonging to *I. aragaoi* in Gen Bank corresponds to a fragment of the mitochondrial 16S rRNA gene (accession number KJ650032). However, the 16s rDNA sequences of *I. pararicinus* from Uruguay that were published in Barbieri et al.[5] and Onofrio et al.[1] (accession numbers JX082322, KJ650033) correspond to sequences of *I. aragaoi* (see above).

REFERENCES

1. Onofrio VC, Ramírez DG, Giovanni DNS, Marcili A, Mangold AJ, Venzal JM, et al. Validation of the taxon *Ixodes aragaoi* Fonseca (Acari: Ixodidae) based on morphological and molecular data. *Zootaxa* 2014;**3860**:361−70.

2. Labruna MB, Jorge RSP, Sana DA, Jácomo ATA, Kashivakura CK, Furtado MM, et al. Ticks (Acari : Ixodida) on wild carnivores in Brazil. *Exp Appl Acarol* 2005;**36**:149−63.

3. Venzal JM, Estrada-Peña A, Barros-Battesti DM, Onofrio VC, Beldoménico PM. *Ixodes (Ixodes) pararicinus* Keirans & Clifford, 1985 (Acari: Ixodidae): description of the immature stages, distribution, hosts and medical/veterinary importance. *Syst Parasitol* 2005;**60**:225−34.

4. Venzal JM, Félix ML, Olmos A, Mangold AJ, Guglielmone AA. A collection of ticks (Ixodidae) from wild birds in Uruguay. *Exp Appl Acarol* 2005;**36**:325−31.

5. Barbieri AM, Venzal JM, Marcili A, Almeida AP, González EM, Labruna MB. *Borrelia burgdorferi* sensu lato infecting ticks of the *Ixodes ricinus* complex in Uruguay: first report for the southern hemisphere. *Vector-borne Zoon Dis* 2013;**13**:147−53.

6. Guglielmone AA, Estrada-Peña A, Keirans JE, Robbins RG. Ticks (Acari: Ixodida) of the neotropical zoogeographic region. Special publication of the integrated consortium on ticks and tick-borne diseases-2. Houten (The Netherlands): Atalanta; 2003.

7. Díaz MM, Nava S, Venzal JM, Sánchez N, Guglielmone AA. Tick collections from the Peruvian Amazon, with new host records for species of *Ixodes* Latreille, 1795 (Acari: Ixodidae) and *Ornithodoros* Koch, 1844 (Acari: Argasidae). *Syst Appl Acarol* 2007; **12**:127−33.

Ixodes auritulus Neumann, 1904

Neumann, L.G. (1904) Notes sur les ixodidés. II. *Archives de Parasitologie*, 8, 444−464.

DISTRIBUTION

I. auritulus is a cosmopolitan tick species whose distribution encompasses the Afrotropical, Australasian, Nearctic, and Neotropical Zoogeographic Regions.[1] In Latin America, *I. auritulus* was recorded in Argentina, Brazil, Chile, Costa Rica, Ecuador, Guatemala, Peru, Venezuela, and Uruguay.[2]

Biogeographic distribution in the Southern Cone of America: Atacama (South American Transition Zone), Santiago, Maule, Magellanic Paramo, Central Patagonia (Andean Region), and Pampa (Neotropical Region) (see Fig. 1 in the introduction).

HOSTS

Aves of different orders are the principal hosts for females, nymphs, and larvae of *I. auritulus*[1−3]; the male of *I. auritulus* is just known for very few nonparasitic specimens. Table 2.29 contains only parasitic records of *I. auritulus* from South and Central America. Rodents are considered exceptional hosts for this tick.[1]

TABLE 2.29 Hosts of Females (F), Nymphs (N), and Larvae (L) of
I. auritulus

MAMMALIA		Parulidae	
RODENTIA		*Myiothlypis leucoblephara*	L
Cricetidae		*Seiurus aurocapillus*	L
Aepeomys lugens	L	**Rhinocryptidae**	
Microryzomys minutus	L	*Scytalopus* sp.	L
Nephelomys albigularis	L	**Thamnophilidae**	
AVES		*Thamnophilus caerulescens*	FN
COLUMBIFORMES		*T. ruficapillus*	F
Columbidae		Thraupidae	
Columbina talpacoti	F	*Anisognathus igniventris*	F
FALCONIFORMES		*A. lacrymosus*	L
Falconidae		*Diglossa albilatera*	L
Phalcoboenus australis	FNL	*Nephelornis oneilli*	F
GALLIFORMES		*Tangara vassori*	L
Cracidae		*Thraupis bonariensis*	FN
Oreophasis derbianus	F	*T. cyanocephala*	F
Penelope sp.	NL	*Trichothraupis melanops*	N
P. superciliaris	N	**Troglodytidae**	
PASSERIFORMES		*Troglodytes* sp.	FN
Conopophagidae		*Thryorchilus browni*	FNL
Conopophaga lineata	NL	*Thryothorus genibarbis*	L
Emberizidae		*T. nigricapillus*	L
Atlapetes pallidinucha	NL	**Turdidae**	
A. schistaceus	L	*Catharus gracilirostris*	L
Haplospiza unicolor	L	*Turdus albicollis*	FN
Junco vulcani	N	*T. amaurochalinus*	N
Poospiza lateralis	N	*T. falcklandii*	FNL
P. nigrorufa	NL	*T. nigrescens*	FL
Phrygilus fruticeli	FN	*T. nigriceps*	FL
Sicalis olivascens	FN	*T. plebejus*	L

(Continued)

TABLE 2.29 (Continued)

Sporophila caerulescens	L	*T. rufiventris*	FNL
Xenospingus concolor	L	**Tyrannidae**	
Zonothrichia capensis	NL	*Elaenia albiceps*	L
Furnariidae		*Knipolegus nigerrimus*	F
Aphrastura spinicauda	NL	**Picidae**	
Clibanornis dendrocolaptoides	FNL	*Colaptes rupicola*	F
Cinclodes antarcticus	F	**PELECANIFORMES**	
C. fuscus	FN	**Ardeidae**	
C. patagonicus	N	*Nycticorax nycticorax*	FN
Pygarrhichas albogularis	L	**PROCELLARIFORMES**	
Synallaxis ruficapilla	FL	**Pelecanoididae**	
Syndactyla rufosuperciliata	N	*Pelecanoides magellani*	FNL
Icteridae		**TINAMIFORMES**	
Curaeus curaeus	FN	**Tinamidae**	
Sturnella loyca	A	*Notophrocta pentlandii*	FN

Records from the Afrotropical, Australasian, and Nearctic Zoogeographic Regions excluded.

ECOLOGY

I. auritulus is distributed in regions with diverse ecological characteristics, which shows the plasticity of this tick to adapt to different environmental conditions. There is substantial information on host association and geographic distribution of *I. auritulus*, but the structure of its life cycle under natural conditions is unknown.

SANITARY IMPORTANCE

There are no records of *I. auritulus* on domestic animals and humans. *I. auritulus* is thought to be involved in the enzootic cycle of *Borrelia burgdorferi* sensu lato.[4,5]

DIAGNOSIS

Male (see Taxonomical notes): total length 2.4 mm, breadth 1.8 mm (see Taxonomical notes). Body outline oval, scapulae rounded; cervical grooves faint and shallow. Scutum inornate, impunctate, and hairless; venter impunctate, with very few setae. Basis capituli rectangular dorsally, without cornua,

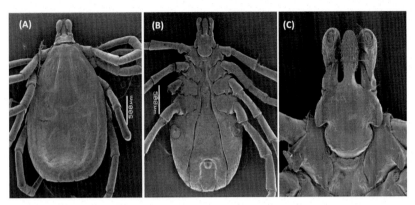

FIGURE 2.93 *I. auritulus* female: (A) dorsal view, (B) ventral view, (C) capitulum ventral view.

auriculae absent, article II and III of palps subequal in length with suture inconspicuous. Hypostome short, notched, dental formula 3/3 in the anterior third, 6/6 in the median portion. Genital aperture located between coxae III and IV. Legs: coxae I–IV with a short external spur each, increasing in size from coxa I to coxa IV; internal spurs absent. Spiracular plates oval.

Female (Fig. 2.93): total length slightly engorged specimen 3.3 mm, breadth 1.8 mm (Argentinean specimen, see taxonomic note). Body outline oval, scapulae rounded; cervical grooves long, shallow, converging anteriorly then diverging as shallow depressions. Scutum length 1.3 mm, breadth 1.0 mm, inornate; outline oval elongated, flattened antero-laterally; presence of few and shallow punctuations; setae short, concentrated anteriorly, and antero-laterally. Notum with scattered, whitish setae. Ventral surface covered by scattered, whitish setae. Basis capituli subtriangular dorsally; cornua pointed; porose area oval; auriculae large, horn-like; article I of the palp with a long internal projection. Hypostome spatulate, dental formula 5/5 apically, then 4/4 for most of the length, then 3/3, and 2/2 near base. Genital aperture located between coxa III. Legs: coxa I with two triangular short spurs, subequal in size; coxae II–III with two short triangular spurs each, the external spur longer than the internal spur; coxa IV with a single short external spur; legs with spurs on trochanters I–IV. Spiracular plates rounded.

Nymph (Fig. 2.94): total length 1.3 mm, breadth 0.7 mm. Body outline oval; cervical grooves long, shallow, converging anteriorly then diverging as shallow depressions. Scutum outline oval elongated, flattened antero-laterally; setae short, concentrated anteriorly, and antero-laterally. Basis capituli subtriangular dorsally, cornua pointed; auriculae large, horn-like; article I of the palp with an internal projection. Hypostome spatulate, dental formula 3/3 for most of the length, then 2/2 near base. Coxae I–III with two short

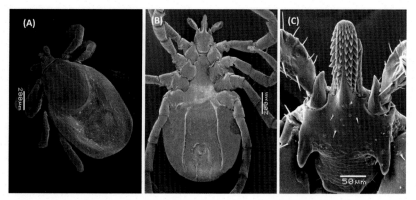

FIGURE 2.94   *I. auritulus* nymph: (A) dorsal view, (B) ventral view, (C) capitulum ventral view.

triangular spurs each, the external spur longer than the internal spur; coxa IV with a single short external spur; legs with spurs on trochanters I–IV.

Taxonomic notes: Arthur[6] has observed morphological variation among ticks assigned to the taxon *I. auritulus*. In the same way, preliminary evidence from mitochondrial 16S rDNA sequences suggests that *I. auritulus* could be a species complex.[3] Because the authors of this book have no available males of *I. auritulus* to obtain original pictures, readers should consult the figures of the male in Arthur,[6] Kohls and Clifford,[7] or Keirans and Clifford.[8] Great discrepancies were observed in the sizes of females of *I. auritulus* and no sizes for nymphs were found; the measures for these stages above are from Argentinean specimens collected from the environment. The measures for the male tick were obtained from Kohls and Clifford.[7]

DNA sequences with relevance for tick identification and phylogeny: the following DNA sequences of *I. auritulus* genes are available in the Gen Bank: 16S rDNA (accession numbers AF113928, AF549845, FJ392274, FJ392275), cytochrome oxidase I sequences (accession number AY059275), 18S rDNA (accession number AF018649), and 28S rDNA (accession number AF120290).

REFERENCES

1. Guglielmone AA, Robbins RG, Apanaskevich DA, Petney TN, Estrada-Peña A, Horak IG. *The hard ticks of the world (Acari: Ixodida: Ixodidae)*. London: Springer; 2014.
2. González-Acuña D, Venzal JM, Keirans JE, Robbins RG, Ippi S, Guglielmone AA. New host and locality records for the *Ixodes auritulus* (Acari: Ixodidae) species group, with a review of host relationship and distribution in the Neotropical Zoogeographic Region. *Exp Appl Acarol* 2005;**37**:147–56.
3. González-Acuña D, Mangold AJ, Robbins RG, Guglielmone AA. New host and locality records for the *Ixodes auritulus* Neumann, 1904 (Acari: Ixodidae) species group in Northern Chile. *Syst Appl Acarol* 2009;**14**:47–50.

4. Morshed MG, Scott JD, Fernando K, Beati L, Mazerolle DF, Geddes G, et al. Migratory songbirds disperse ticks across Canada, and first isolation of the Lyme disease spirochete, *Borrelia burgdorferi*, from the avian tick, *Ixodes auritulus*. *J Parasitol* 2005;**91**:780–90.

5. Scott JD, Anderson JF, Durden LA. Widespread dispersal of *Borrelia burgdorferi*-infected ticks collected from songbirds across Canada. *J Parasitol* 2012;**98**:49–59.

6. Arthur DR. A review of some ticks (Acarina: Ixodidae) of sea birds. Part II. The taxonomic problems associated with the *Ixodes auritulus* – *percavatus* group of species. *Parasitology* 1960;**50**:199–226.

7. Kohls GM, Clifford CM. Three new species of *Ixodes* from Mexico and description of the male of *I. auritulus auritulus* Neumann, *I. conepati* Cooley and Kohls, and *I. lasallei* Méndez and Ortiz (Acarina: Ixodidae). *J Parasitol* 1966;**52**:810–20.

8. Keirans JE, Clifford CM. The genus Ixodes in the United States: a scanning electron microscope study and key to adults. *J Med Entomol* 1978;(Suppl. 2):149.

Ixodes chilensis
Kohls, 1956

Kohls, G.M. (1956) Eight new species of *Ixodes* from Central and South America (Acarina: Ixodidae). *Journal of Parasitology*, 42, 636–649.

DISTRIBUTION

Chile.[1]

Biogeographic distribution in the Southern Cone of America: Maule (Andean Region) (see Fig. 1 in the introduction).

HOSTS

The host of *I. chilensis* is unknown.[1]

ECOLOGY

There is no information on the ecology of *I. chilensis*.

SANITARY IMPORTANCE

The medical and veterinary importance of *I. chilensis* is unknown. No records of *I. chilensis* on humans and domestic animals have been reported.

DIAGNOSIS

Female (see Taxonomic notes): total length of an engorged specimen 7.6 mm, breadth 4.8 mm. Scapulae pointed; cervical grooves shallow, converging anteriorly then diverging as shallow depressions. Scutum length 1.4 mm, breadth 1.3 mm, inornate; surface of anterolateral field rugose; punctuations conspicuous, larger in the posterior filed; posterior margin straight; presence of setae on the anterior; and lateral margins of the scutum. Notum with numerous whitish setae. Ventral surface covered by scattered setae. Basis capituli subrectangular dorsally, cornua absent, porose area oval, auriculae absent. Hypostome: shape and dentition of distal portion is unknown, in the remaining portion the dental portion is 3/3 and then 2/2 at the base. Genital aperture located between coxae II and III. Legs: all coxae with an external spur as small tubercles. Spiracular plates rounded.

Taxonomic notes: the only record of *I. chilensis* corresponds to the holotype specimen, an engorged female which has broken the distal portion of the hypostome. Male, nymph, and larva of *I. chilensis* are unknown. One of the authors of this book (DGA) has recently collected a female tick on a bat in Chile, whose morphology may be compatible with *I. chilensis*. Molecular and morphological analyses of this specimen are currently in progress. Because the authors of this book have no available specimens of *I. chilensis* to obtain original pictures; readers should consult figures of *I. chilensis* in the original description of Kohls.[2]

DNA sequences with relevance for tick identification and phylogeny: there are no sequences of *I. chilensis* in the Gen Bank.

REFERENCES

1. Guglielmone AA, Estrada-Peña A, Keirans JE, Robbins RG. Ticks (Acari: Ixodida) of the neotropical zoogeographic region. Special publication of the international consortium on ticks and tick-borne diseases-2. Houten (The Netherlands): Atalanta; 2003.
2. Kohls GM. Eight new species of *Ixodes* from Central and South America (Acarina: Ixodidae). *J Parasitol* 1956;42:636–49.

Ixodes cornuae
Arthur, 1960

Arthur, D.R. (1960) A review of some ticks (Acarina: Ixodidae) of sea birds. Part II. The taxonomic problems associated with the *Ixodes auritulus – percavatus* group of species. *Parasitology*, 50, 199–226.

DISTRIBUTION

Chile and Ecuador.[1]

Biogeographic distribution in the Southern Cone of America: Magellanic Paramo (Andean Region) (see Fig. 1 in the introduction).

HOSTS

A bird of the family Phasianidae (order Galliformes) is the only host recorded for *I. cornuae*, but its specific scientific name was not provided.[1,2]

ECOLOGY

There is no information on the ecology of *I. cornuae*.

SANITARY IMPORTANCE

There are no records of *I. cornuae* on domestic animals and humans. The capacity of *I. cornuae* to transmit pathogens is unknown.

DIAGNOSIS

Female (see Taxonomic notes): total length of partially engorged specimens 6.0 mm, breadth 7.0 mm. Body outline oval elongated, scapulae rounded; cervical grooves long, shallow, converging anteriorly then diverging as shallow depressions. Scutum length 1.5 mm, breath 1.1 mm, inornate; outline oval elongated, flattened antero-laterally; presence of few and shallow punctuations; setae short, concentrated anteriorly and antero-laterally. Basis capituli subtriangular dorsally; cornua long, horn-like; porose area oval; auriculae large, horn-like; article I of the palp with a long internal projection, and presence of a strong meso-dorsal spur. Hypostome spatulate, dental formula 4/4 for most of the length, then 3/3, and 2/2 near base. Genital aperture located between coxae III. Legs: coxa I with two triangular short spurs, subequal in size; coxae II–III with two short triangular spurs each, the external spur longer than the internal spur; coxa IV with a single short external spur; legs with spurs on trochanters I–IV. Spiracular plates rounded.

Nymph (see Taxonomic notes): no sizes of total length and breadth are available. Scutum oval; cervical grooves shallow, converging anteriorly then diverging as shallow depressions but not reaching postero-lateral edge. Basis capituli subtriangular dorsally; cornua long, horn-like; auriculae large, horn-like; article I of the palp with an internal projection, and presence of a conspicuous meso-dorsal spur. Hypostome spatulate, dental formula 3/3 for most of the length, then 2/2 near base. Coxae I–III with two short triangular spurs, the external spur longer than the internal spur; coxa IV with a single short external spur; legs with spurs on trochanters I–IV.

Taxonomic notes: Arthur[1] considers that *I. cornuae* is a different species from *I. auritulus* due to the presence of a meso-dorsal spur on article I of the palp in *I. cornuae* and the shape of the cornua. No specimens of *I. cornuae* were available to the authors of this book; therefore, the diagnosis presented here is based on the original description and drawings of Arthur[1] and no figures of this tick are shown here. Readers should consult figures for *I. cornuae* in the original description of Arthur.[1] Male and larva of *I. cornuae* are unknown.

DNA sequences with relevance for tick identification and phylogeny: DNA sequences of *I. cornuae* are not available.

REFERENCES

1. Arthur DR. A review of some ticks (Acarina: Ixodidae) of sea birds. Part II. The taxonomic problems associated with the *Ixodes auritulus* – *percavatus* group of species. *Parasitology* 1960;**50**:199–226.
2. Guglielmone AA, Robbins RG, Apanaskevich DA, Petney TN, Estrada-Peña A, Horak IG. *The hard ticks of the world (Acari: Ixodida: Ixodidae)*. London: Springer; 2014.

Ixodes longiscutatus
Boero, 1944

Ixodes longiscutatum. Boero, J.J. (1944) Notas ixodológicas. I) *Ixodes longiscutatum*, nueva especie. *Revista de la Asociación Médica Argentina*, 58, 353–354.

DISTRIBUTION

Argentina and Uruguay.[1]

   Biogeographic distribution in the Southern Cone of America: Chaco and Pampa (Neotropical Region) (see Fig. 1 in the introduction).

HOSTS

Cattle, horse, and rodents of the genus *Cavia* are the hosts recorded for the female of *I. longiscutatus*, while rodents of the families Caviidae and Cricetidae are the hosts for larvae and nymphs[1] (Table 2.30).

ECOLOGY

Besides data of hosts and distribution, there is no information on ecological aspects of *I. longiscutatus*.

SANITARY IMPORTANCE

Females of *I. longiscutatus* were recorded on cattle and horses, but the role of this tick as vector of pathogens is unknown.

TABLE 2.30 Hosts of Females (F), Nymphs (N), and Larvae (L) of
I. longiscutatus

MAMMALIA		C. aperea	NL
ARTIODACTYLA		*C. tschudii*	FNL
Bovidae		**Cricetidae**	
Cattle	F	*Akodon lutescens*	L
PERISSODACTYLA		*Necromys obscurus*	L
Equidae		*Oligoryzomys flavescens*	L
Horse	F	*Oxymycterus nasutus*	NL
RODENTIA		*Scapteromys tumidus*	NL
Caviidae			
Cavia sp.	F		

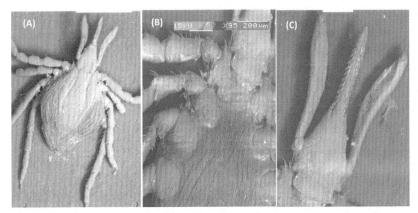

FIGURE 2.95   *I. longiscutatus* female: (A) dorsal view, (B) ventral view, (C) capitulum ventral view.

DIAGNOSIS

Female (Fig. 2.95): total length 2.6 mm, breadth 1.2 mm. Body outline oval, scapulae rounded; cervical grooves long, shallow, converging anteriorly then diverging as shallow depressions reaching the posterior third of the scutum. Scutum length 1.5 mm, breadth 0.7 mm, inornate; outline oval elongated; lateral margins convex; presence of numerous and shallow punctuations, more densely distributed in the posterior field of the scutum; setae numerous and long. Notum with numerous long whitish setae, uniformly distributed. Ventral surface covered by numerous, long, whitish setae. Basis capituli triangular dorsally; cornua triangular; porose area oval; auriculae small. Hypostome pointed, dental formula 3/3 apically and 2/2 at the base. Genital aperture located between coxae III and IV. Legs: coxa I with two triangular short spurs, the external longer than the internal; coxae II—IV with a single short external spur each. Spiracular plates rounded.

Nymph (Fig. 2.96): total length 1.2 mm, breadth 0.7 mm. Body outline oval elongated; cervical grooves long, shallow, converging anteriorly then diverging as shallow depressions. Scutum outline oval elongated; setae few, short; punctuations few, scattered. Notum with numerous long, stout, blunt setae. Ventral surface covered by numerous whitish setae, uniformly distributed. Basis capituli subtriangular dorsally, with a short cornua-like processes basolaterally, and a pair of short, broad projections on the posterior margin; auriculae long, slightly directed laterally; article II of the palps with a strong, bifurcate, basolateral salient, and a dorsal process rounded; article III of the palp with a strong, pointed, retrograde spur; suture between articles II and III

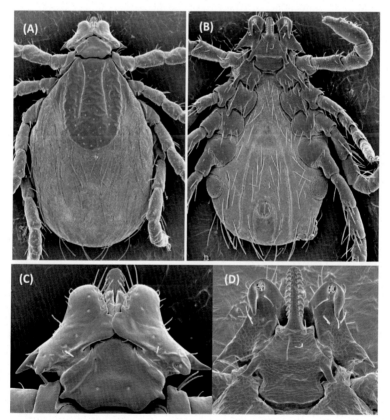

FIGURE 2.96   *I. longiscutatus* nymph: (A) dorsal view, (B) ventral view, (C) capitulum dorsal view, (D) capitulum ventral view.

of the palp barely perceptible. Hypostome blunt, dental formula 2/2. Legs: coxa I with two long, narrow spurs, the external spur longer than the internal spur; coxae II–III with two short triangular spurs each, the external spur longer than the internal; coxa IV with a single short external spur.

Taxonomic notes: at least with the phylogenetic pattern obtained with 16S rDNA sequences, the evolutionary relationship of *I. longiscutatus* with the other Neotropical species of *Ixodes* remains unresolved (Fig. 2.90). Boero[2] described the female of *I. longiscutatus* (spelled *I. longiscutatum*) with a dental formula 2/2, but Venzal et al.[3] (also spelled *I. longiscutatum*) stated that the hypostome described by Boero[2] is broken apically and that the dental formula of the female of *I. longiscutatus* is 3/3 apically and 2/2 at the base. Kohls and Clifford[4] described the larva and nymph of *I. longiscutatus* under the name of *I. uruguayensis* Kohls and Clifford, 1967, as explained in Venzal et al.[3] The male of *I. longiscutatus* remains unknown.

DNA sequences with relevance for tick identification and phylogeny: the only DNA sequence of *I. longiscutatus* available in Gen Bank corresponds to a fragment of the mitochondrial 16S rRNA gene (accession number DQ061294).

REFERENCES

1. Venzal JM, Nava S, Beldoménico PM, Barros-Battesti DM, Estrada-Peña A, Guglielmone AA. Hosts and distribution of *Ixodes longiscutatus* Boero, 1944 (Acari: Ixodidae). *Syst Appl Acarol* 2008;**13**:102−8.
2. Boero JJ. Notas ixodológicas. I) *Ixodes longiscutatum*, nueva especie. *Rev Asoc Méd Argent* 1944;**58**:353−4.
3. Venzal JM, Castro O, Cabrera P, Souza C, Fregueiro G, Barros-Battesti DM, et al. *Ixodes* (*Haemixodes*) *longiscutatum* [sic] Boero (new status) and *I.* (*H.*) *uruguayensis* Kohls & Cilfford, a new synonym of *I.* (*H.*) *longiscutatum* [sic] (Acari: Ixodidae). *Mem Inst Oswaldo Cruz* 2001;**96**:1121−2.
4. Kohls GM, Clifford CM. *Ixodes* (*Haemixodes*) *uruguayensis*, new subgenus, new species (Acarina: Ixodidae) from small rodents in Uruguay. *Ann Entomol Soc Am* 1967;**60**:391−4.

Ixodes loricatus Neumann, 1899

Neumann, L.G. (1899) Révision de la famille des ixodidés (3e mémoire). *Mémoires de la Société Zoologique de France*, 12, 107−294.

DISTRIBUTION

Argentina, Brazil, Paraguay, and Uruguay.[1] *I. loricatus* has been historically considered to range from Argentina to Mexico but it is actually known as a South American species.[1]

Biogeographic distribution in the Southern Cone of America: Chaco, Pampa, and Parana Forest (Neotropical Region) (see Fig. 1 in the introduction).

HOSTS

Marsupials are the principal hosts for adults of *I. loricatus*, while the immature stages feed mostly on small rodents and marsupials.[1-3] The complete list of hosts of *I. loricatus* is given in Table 2.31.

TABLE 2.31 Hosts of Adults (A), Nymphs (N), and Larvae (L) of *I. loricatus*

MAMMALIA		RODENTIA	
CARNIVORA		Caviidae	
Procyonidae		*Cavia* sp.	A
Nasua nasua	A	Cricetidae	
DIDELPHIMORPHIA		*Akodon azarae*	NL
Didelphidae		*Calomys* sp.	NL
Chironectes minimus	A	*Cerradomys subflavus*	A
Didelphis albiventris	ANL	*Euryoryzomys russatus*	N
D. aurita	ANL	*Holochilus brasiliensis*	NL
D. marsupialis	AL	*Hylaeamys megacephalus*	NL
Lutreolina crassicaudata	ANL	*Oligoryzomys flavescens*	NL
Marmosa sp.	AN	*O. nigripes*	N
M. murina	L	*Oxymycterus nasutus*	L
Monodelphis americana	L	*O. rufus*	NL
M. dimidiata	NL	*Rhipidomys* sp.	NL
M. sorex	NL	*Scapteromys aquaticus*	NL
Philander frenatus	ANL	*Zygodontomys* sp.	A
P. opossum	A	Cuniculidae	
		Cuniculus paca	A
		Muridae	
		Rattus norvegicus	L

ECOLOGY

After a study on the population dynamic of *I. loricatus*, Colombo et al.[4] have not found a marked pattern of seasonality for larvae and nymphs of this tick species, a phenomenon probably related to the nidicolous habit of *I. loricatus*. Unpublished information about the seasonality of adults of this species also found no obvious seasonal pattern.[5]

SANITARY IMPORTANCE

There are no records of *I. loricatus* on humans and domestic animals. This tick species has not been involved as vector of pathogenic microorganisms so far. Minoprio et al.[6] described a case of human babesiosis due to bites of *I. loricatus* that is treated here as a diagnostic error.

DIAGNOSIS

Male (Fig. 2.97): total length 3.9 mm, breadth 1.8 mm. Body outline oval, scapulae rounded; cervical grooves shallow, diverging posteriorly; presence of two dorsolateral depressions extending from the level of coxae III to the posterior field of the scutum. Scutum inornate; punctuations numerous, larger and more densely distributed in the postero-lateral field; setae absent. Ventral surface with numerous punctuations, larger on the adanal plates, and presence of numerous setae; ventral plates as shown in Fig. 2.97. Basis capituli subtriangular dorsally, without cornua; auriculae rounded; article II of the palp longer than article III, suture conspicuous. Hypostome short, blunt, dental formula 2/2 for most of the length, 3/3 at the apex. Genital aperture located between coxae II and III. Legs: coxa I with two short spurs, the internal slightly narrower than the external spur; coxae II−IV each with a short, blunt external spur. Spiracular plates oval.

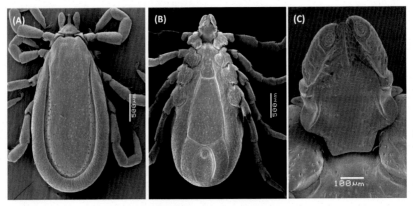

FIGURE 2.97 *I. loricatus* male: (A) dorsal view, (B) ventral view, (C) capitulum ventral view.

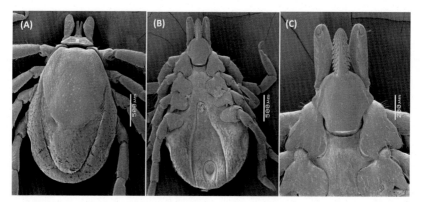

FIGURE 2.98   *I. loricatus* female: (A) dorsal view, (B) ventral view, (C) capitulum ventral view.

Female (Fig. 2.98): total length 3.1 mm, breadth 1.8 mm. Body outline oval, scapulae rounded; cervical grooves shallow, converging anteriorly then diverging as shallow depressions. Scutum length 1.5 mm, breadth 1.2 mm, inornate; outline oval rounded; presence of few and short setae; numerous punctuations, larger on the posterior field. Notum with numerous whitish setae. Ventral surface covered by numerous whitish setae. Basis capituli subtriangular dorsally, cornua absent, porose area oval and transversally elongated, auriculae small and rounded; article II of the palp longer than article III, suture conspicuous. Hypostome short, blunt, dental formula 2/2 for most of the length, 3/3 at the apex. Genital aperture located between coxa III. Legs: coxa I with two short spurs, the internal slightly narrower than the external; coxae II–IV each with a short, blunt external spur. Spiracular plates rounded.

Nymph (Fig. 2.99): total length 1.8 mm, breadth 0. 7 mm. Body outline oval elongated, scapulae rounded; cervical grooves shallow, converging anteriorly then diverging as shallow depressions. Scutum inornate; outline oval; setae few and short, irregularly distributed. Notum with numerous whitish setae. Ventral surface covered by numerous whitish setae. Basis capituli triangular dorsally, cornua absent, auriculae short, triangular. Hypostome short, blunt, dental formula 2/2 for most of the length, then 3/3, and 4/4 near the apex. Legs: coxa I with two short spurs, the internal spur slightly narrower than external spur; coxae II–IV each with a short, blunt external spur.

Taxonomic notes: *I. loricatus*, *I. luciae* Sénevet, 1940, and *I. amarali* Fonseca, 1935, are three species morphologically similar and they share the same pattern of host usage (Didelphidae and Cricetidae).[1,2,7,8] The evidence obtained with the analysis of the mitochondrial 16S rDNA gene sequences of the Neotropical species of *Ixodes* showed that *I. loricatus*, *I. luciae*, and *I. amarali* are phylogenetically closely related (Fig. 2.90).

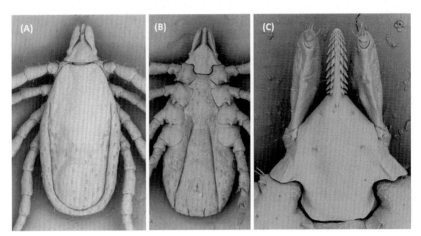

FIGURE 2.99 *I. loricatus* nymph: (A) dorsal view, (B) ventral view, (C) capitulum ventral view.

DNA sequences with relevance for tick identification and phylogeny: DNA sequences of *I. loricatus* are available in the Gen Bank as follows: mitochondrial 16S rRNA gene (accession numbers U95892, AY510269, AY510268, AF549840) and 12S rRNA gene (accession number U95891). There also are available sequences of 28S rRNA gene (accession number AF120291), and 5.8S ribosomal RNA gene, internal transcribed spacer 2, 28S ribosomal RNA gene (accession numbers AF327339-AF327344).

REFERENCES

1. Guglielmone AA, Nava S, Díaz MM. Relationships of South American marsupials (Didelphimorphia, Microbiotheria and Paucituberculata) and hard ticks (Acari: Ixodidae), with distribution of four species of *Ixodes. Zootaxa* 2011;**3086**:1−30.

2. Guglielmone AA, Nava S. Rodents of the subfamily Sigmodontinae (Myomorpha: Cricetidae) as hosts for South American hard ticks (Acari: Ixodidae) with hypothesis on life history. *Zootaxa* 2011;**2904**:45−65.

3. Coelho MG, Ramos VN, Limongi JE, Lemos ERS, Guterres A, Neto SFC, et al. Serologic evidence of the exposure of small mammals to spotted-fever *Rickettsia* and *Ricekttsia bellii* in Minas Gerais, Brazil. *J Infect Dev Countries* 2016;**10**:275−82.

4. Colombo VC, Guglielmone AA, Monje LD, Nava S, Beldoménico PM. Seasonality of immature stages of *Ixodes loricatus* (Acari: Ixodidae) in the Paraná Delta, Argentina. *Ticks Tick-borne Dis* 2014;**5**:701−5.

5. Tarragona EL. Unpublished information; 2016.

6. Minoprio JL, Minoprio JE, Jörg ME. Babesiosis, caso humano con síndrome hemocitofágico observado en Argentina. *Prensa Méd Argent* 1994;**84**:43−8.

7. Onofrio VC, Barros-Battesti DM, Labruna MB, Faccini JLH. Diagnoses and illustrated key to the species of *Ixodes* Latreille, 1795 (Acari: Ixodidae) from Brazil. *Syst Parasitol* 2009;**72**:143−57.

8. Onofrio VC, Labruna MB, Faccini JLH, Barros-Battesti DM. Description of immature stages and redescription of adults of *Ixodes luciae* Sénevet (Acari: Ixodidae). *Zootaxa* 2010;**2495**:53−64.

Ixodes luciae
Sénevet, 1940

Sénevet, G. (1940) Quelques ixodidés de la Guyane française. Espèces nouvelles d'*Ixodes* et d'*Amblyomma*. *6° Congreso Internacional de Entomología*, Madrid, España, setiembre 1935, pp. 891−898.

DISTRIBUTION

Argentina, Belize, Bolivia, Brazil, Colombia, Costa Rica, Ecuador, French Guiana, Guatemala, Honduras, Mexico, Nicaragua, Panama, Peru, Surinam, Trinidad and Tobago, and Venezuela.[1] The presence of *I. luciae* in Ecuador, Costa Rica, Honduras, and Nicaragua requires confirmation.[1]

Biogeographic distribution in the Southern Cone of America: Yungas (Neotropical Region) (see Fig. 1 in the introduction).

HOSTS

The principal hosts for adults of *I. luciae* are marsupials, while small rodents of the family Cricetidae and marsupials are the main hosts for the immature stages.[1−3] The complete host range of *I. luciae* is showed Table 2.32.

ECOLOGY

I. luciae appears to have a continuous life cycle, without a marked seasonal trend of parasitic activity.[2] This pattern is probably related to the nidicolous habits of this tick. *I. luciae* is principally distributed in tropical and subtropical moist broadleaf forests.[4]

SANITARY IMPORTANCE

I. luciae lacks medical and veterinary relevance, at least with the currently available evidence. Its record by Méndez-Arocha and Ortiz[5] on a dog in Venezuela is an unusual host-parasite relationship.

DIAGNOSIS

Male (Fig. 2.100): total length 4.7 mm, breadth 2.0 mm. Body outline oval, scapulae rounded; cervical grooves shallow, diverging posteriorly; presence

TABLE 2.32 Hosts of Adults (A), Nymphs (N), and Larvae (L) of *I. luciae*

MAMMALIA		*M. brevicaudata*	NL
ARTIODACTYLA		*M. sorex*	NL
Cervidae		*Philander andersoni*	A
Mazama gouazoubira	N	*P. opossum*	A
CARNIVORA		*Thylamys cinderella*	N
Canidae		*T. venustus*	N
Domestic dog	A	*Tlacuatzin canescens*	N
DIDELPHIMORPHIA		**RODENTIA**	
Didelphidae		**Cricetidae**	
Caluromys lanatus	A	*Calomys* sp.	N
C. philander	L	*Hylaeamys perenensis*	N
Didelphis albiventris	A	*H. yunganus*	N
D. marsupialis	A	*Ichthyomys pittieri*	A
Lutreolina crassicaudata	A	*Nyctomys sumichrasti*	N
Marmosa constantinae	N	*Oecomys bicolor*	N
M. mexicana	N	*O. concolor*	NL
M. murina	L	*Oligoryzomys microtis*	N
M. demererae	N	*Oryzomys* sp.	N
M. robinsoni	ANL	*Sigmodon* sp.	NL
Marmosops impavidus	A	*Zygodontomys brevicauda*	N
Metachirus nudicaudatus	A	**Cuniculidae**	
Monodelphis sp.	ANL	*Cuniculus paca*	A

of two dorsolateral depressions extending from the level of coxa III to the posterior field of the scutum. Scutum inornate; punctuations numerous, larger, and more densely distributed in the postero-lateral field; setae absent. Ventral surface with numerous punctuations, larger on the adanal plates, and presence of numerous setae; ventral plates as shown in Fig. 2.100. Basis capituli subtriangular dorsally, without cornua; auriculae rounded; article II of the palp longer than article III, suture conspicuous. Hypostome short, blunt, dental formula 2/2 for most of the length, 3/3 at the apex. Genital aperture located between coxae II and III. Legs: coxa I with two spurs, the external spur long, narrow, and sharp, reaching the coxa II, and the internal

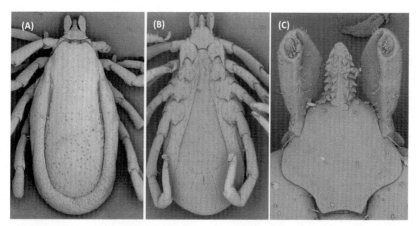

FIGURE 2.100 *I. luciae* male: (A) dorsal view, (B) ventral view, (C) capitulum ventral view.

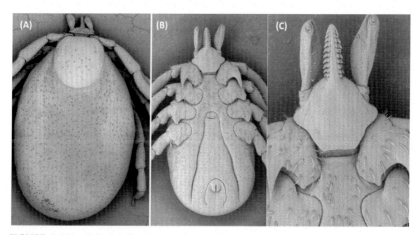

FIGURE 2.101 *I. luciae* female: (A) dorsal view, (B) ventral view, (C) capitulum ventral view.

spur triangular and short; coxae II–IV each with a short blunt external spur. Spiracular plates oval.

Female (Fig. 2.101): total length 4.4 mm, breadth 2.2 mm. Body outline oval, scapulae rounded; cervical grooves shallow, converging anteriorly then diverging as shallow depressions. Scutum length 1.7 mm, breadth 1.3 mm, inornate; outline oval rounded; presence of few and short setae; numerous punctuations, larger on the central and posterior field. Notum with numerous whitish setae. Ventral surface covered by numerous whitish setae. Basis capituli subtriangular dorsally, cornua absent, porose area oval and transversally elongated, auriculae small and rounded; article II of the palp longer than article III, suture conspicuous. Hypostome short, blunt, dental formula 2/2 for

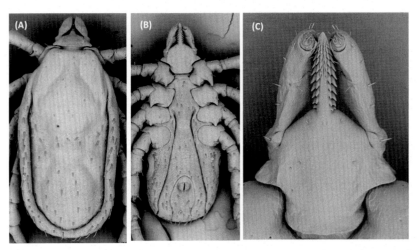

FIGURE 2.102 *I. luciae* nymph: (A) dorsal view, (B) ventral view, (C) capitulum ventral view.

most of the length, 3/3 at the apex. Genital aperture located between coxa III. Legs: coxa I with two spurs, the external long, narrow, and sharp, reaching the coxa II, and the internal triangular and short; coxae II–IV each with a short blunt external spur. Spiracular plates rounded.

Nymph (Fig. 2.102): total length 1.8 mm, breadth 0.7 mm. Body outline oval elongated, scapulae rounded; cervical grooves shallow, converging anteriorly then diverging as shallow depressions. Scutum inornate; outline oval; setae few and short, irregularly distributed. Notum with numerous whitish setae. Ventral surface covered by numerous whitish setae. Basis capituli triangular dorsally, cornua absent, auriculae short, triangular. Hypostome pointed, dental formula 2/2 from base to middle, 3/3 from middle to apical third, and 4/4 near the apex. Legs: coxa I with two spurs, the external longer than the internal; coxae II–IV each with a short, blunt, external spur.

Taxonomic notes: see Taxonomic notes of *I. loricatus* Neumann, 1899.

DNA sequences with relevance for tick identification and phylogeny: DNA sequences of *I. luciae* are available in the Gen Bank as follows: mitochondrial 16S rRNA gene (accession numbers U95894, AF549851) and 12S rRNA gene (accession number U95893), and nuclear 18S rRNA gene (accession number AF115367).

REFERENCES

1. Guglielmone AA, Nava S, Díaz MM. Relationships of South American marsupials (Didelphimorphia, Microbiotheria and Paucituberculata) and hard ticks (Acari: Ixodidae), with distribution of four species of *Ixodes*. *Zootaxa* 2011;**3086**:1–30.
2. Díaz MM, Nava S, Guglielmone AA. The parasitism of *Ixodes luciae* (Acari: Ixodidae) on marsupials and rodents in Peruvian Amazon. *Acta Amazonica* 2009;**39**:997–1002.

3. Guglielmone AA, Nava S. Rodents of the subfamily Sigmodontinae (Myomorpha: Cricetidae) as hosts for South American hard ticks (Acari: Ixodidae) with hypothesis on life history. *Zootaxa* 2011;**2904**:45–65.

4. Guglielmone AA, Robbins RG, Apanaskevich DA, Petney TN, Estrada-Peña A, Horak IG. *The hard ticks of the world (Acari: Ixodida: Ixodidae).* London: Springer; 2014.

5. Méndez Arocha M, Ortiz I. Revisión de las garrapatas venezolanas del género *Ixodes* Latreille, 1795 y estudio de un nuevo *Amblyomma* (Acarina: Ixodidae). *Mem Soc Cienc Nat La Salle* 1958;**18**:196–208.

Ixodes neuquenensis Ringuelet, 1947

Ringuelet, R. (1947) La supuesta presencia de *Ixodes brunneus* Koch en la Argentina y descripción de una nueva garrapata *Ixodes neuquenensis* nov. sp. *Notas del Museo de La Plata*, 12, 207–216.

DISTRIBUTION

Argentina and Chile.[1,2]

    Biogeographic distribution in the Southern Cone of America: Valdivian Forest and Subandean Patagonia (Andean Region) (see Fig. 1 in the introduction).

HOSTS

Dromiciops gliroides (Microbiotheria: Microbiotheriidae) has been considered the only hosts of females, nymphs, and larvae of *I. neuquenensis*.[1–3] This peculiar host has been historically the only member of Microbiotheria, but recently D'Elía et al.[4] demonstrated that three species were contained under the name *D. gliroides*. Nevertheless the published records of *I. neuquenensis* coincided with the distribution of *D. gliroides* although one of the authors (DGA) recently found this tick on *D. bozinovici*.

ECOLOGY

Besides the available data on host association and geographical distribution, there is no information on the ecology of *I. neuquenensis*.

SANITARY IMPORTANCE

There are no records of *I. neuquenensis* on humans and domestic animals. The potential role of *I. neuquenensis* as vector of pathogenic microorganism has not been evaluated to date.

DIAGNOSIS

Female (Fig. 2.103): total body length from a slightly engorged specimen 3.7 mm, breadth 1.9 mm. Body outline oval, scapulae pointed; cervical grooves shallow, converging anteriorly then diverging as shallow depressions. Scutum length 1.4 mm, breadth 1.1 mm, inornate, oval in outline, with large punctuations on lateral and central fields. Notum with numerous, short, whitish setae. Ventral surface covered by scattered setae, and with ventral chitinous plaques medial to coxa I. Basis capituli rectangular dorsally; cornua small; porose area oval; auriculae small and triangular; article II of the palp longer than article III, with suture between article II and III conspicuous. Hypostome blunt, dental formula 3/3 for most of the length, then 2/2 in the posterior third. Genital aperture located between coxa III. Legs: coxa I with two sharp, long, narrow spurs, the external spur longer than the internal spur; coxae II–IV each with two triangular spurs decreasing in size from II to IV, the internal spur smaller than the external. Spiracular plates rounded.

Nymph (Fig. 2.104): total length from slightly engorged specimens 2.0 mm, breadth 1.4 mm. Body outline oval, scapulae pointed; cervical grooves shallow, converging anteriorly then diverging as shallow depressions. Scutum inornate, oval in outline, with large punctuations on lateral and central fields. Notum with scattered whitish setae. Ventral surface covered by scattered setae, and with ventral chitinous plaques medial to coxa I. Basis capituli rectangular dorsally, cornua triangular, auriculae small and triangular, article I of the palp with large ventral anterior and posterior processes, suture between articles II and III of the palp indistinct. Hypostome blunt, dental formula 2/2. Legs: coxa I with two sharp, long, narrow spurs, the external spur longer than the internal; coxae II–IV each with two triangular spurs decreasing in size from II to IV, the internal spur smaller than the external. Spiracular plates rounded.

Taxonomic notes: see Taxonomic notes of *I. abrocomae* Lahille, 1916. The male of *I. neuquenensis* is unknown.

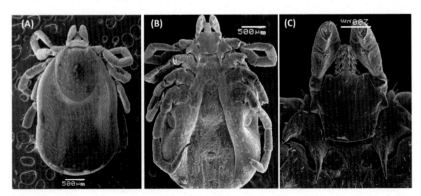

FIGURE 2.103 *I. neuquenensis* female: (A) dorsal view, (B) ventral view, (C) capitulum ventral view.

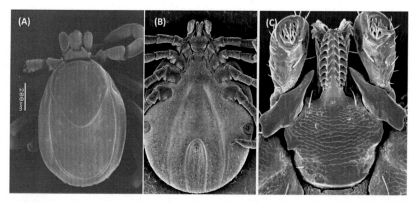

FIGURE 2.104 *I. neuquenensis* nymph: (A) dorsal view, (B) ventral view, (C) capitulum ventral view.

DNA sequences with relevance for tick identification and phylogeny: there are just three sequences of the mitochondrial 16S rRNA gene of *I. neuquenensis* in the Gen Bank (accession number: AY254393−AY254395).

REFERENCES

1. Guglielmone AA, Venzal JM, Amico G, Mangold AJ, Keirans JE. Description of the nymph and larva and redescriptions of the female of *Ixodes neuquenensis* Ringuelet, 1947 (Acari: Ixodidae), a parasite of the endangered Neotropical marsupial *Dromiciops gliroides* Thomas (Microbiotheria: Microbiotheriidae). *Syst Parasitol* 2004;**57**:211−19.
2. Marin-Vial P, González-Acuña D, Celis-Diez L, Cattan PE, Guglielmone AA. Presence of *Ixodes neuquenensis* Ringuelet, 1947 (Acari: Ixodidae) on the endangered Neotropical marsupial Monito del monte (*Dromiciops gliroides* Thomas, 1894, Microbiotheria: Microbiotheriidae) at Chiloé Island, Chile. *Eur J Wildl Res* 2007;**53**:73−5.
3. Guglielmone AA, Nava S, Díaz MM. Relationships of South American marsupials (Didelphimorphia, Microbiotheria and Paucituberculata) and hard ticks (Acari: Ixodidae) with distribution of four species of *Ixodes*. *Zootaxa* 2011;**3086**:1−30.
4. D'Elía G, Hurtado N, D'Anatro A. Alpha taxonomy of *Dromiciops* (Microbiotheriidae) with the description of 2 new species. *J Mammal* 2016;**97**:1136−52.

Ixodes nuttalli
Lahille, 1913

Lahille, F. (1913) Sobre dos *Ixodes* de la República Argentina y la medición de las garrapatas. *Boletín del Ministerio de Agricultura (Argentina)*, **16**, 278−289.

DISTRIBUTION

Argentina and Peru.[1]

Biogeographic distribution in the Southern Cone of America: Monte (South American Transition Zone) (see Fig. 1 in the introduction).

HOSTS

Adults of *I. nuttalli* were collected on *Lagidium viscacia* and *L. peruanum* (Rodentia: Chinchillidae), while the records of larvae and nymphs have been made on *L. peruanum.*[2,3]

ECOLOGY

There is no information on the ecology of *I. nuttalli* apart from the data about hosts and geographic distribution.

SANITARY IMPORTANCE

There are no records of *I. nuttalli* on domestic animals and humans. The capacity of *I. nuttalli* to transmit pathogens has not been investigated to date.

DIAGNOSIS

Male (see Taxonomic notes): total length 3.7 mm, breadth 2.1 mm. Body outline oval, scapulae rounded; cervical grooves shallow, converging anteriorly then diverging as shallow depressions. Scutum inornate; presence of a noticeable pseudoscutum; punctuations on the pseudoscutum numerous, fine, more densely distributed in the central field; the rest of the scutum with few fine punctuations and few short setae. Ventral surface with pre-genital plate broader than long, median plate longer than broad and hexagonal in shape, adanal and anal plates much longer than broad. Basis capituli pentagonal dorsally, without cornua; article II and III of palps subequal in length, suture conspicuous; auriculae short and triangular. Hypostome short, notched, dental formula 3/3. Genital aperture located between coxa II. Legs: coxa I with a short internal spur, and two postero-external spurs, one dorsal and one ventral; coxae II–III with two postero-external spurs each, one dorsal one ventral, and one short internal spur; coxa IV with two short spurs, one internal and one external. Spiracular plates oval.

Female (see Taxonomic notes): total body length 3.7 mm, breadth 2.0 mm. Body outline oval, scapulae rounded; cervical grooves shallow, converging anteriorly then diverging as shallow depressions. Scutum length 1.9 mm, breadth 1.3 mm, inornate, presence of few and scattered setae, punctuations few and moderate in size. Notum with numerous whitish setae. Ventral surface covered by scattered setae. Basis capituli triangular dorsally, cornua small, porose area oval, auriculae rounded; article II of the palp longer than article III, suture conspicuous. Hypostome pointed, dental formula 3/3. Genital aperture located between coxa III. Legs: coxa I with a short internal spur, and two postero-external spurs, one dorsal and one ventral;

coxae II−III with two postero-external spurs each, one dorsal one ventral, and one short internal spur; coxa IV with a short external spur. Spiracular plates rounded.

Nymph (see Taxonomic notes): total body length 1.5 mm, breadth 0.9 mm. Body outline oval; scapulae rounded; cervical grooves shallow, converging anteriorly then diverging as shallow depressions, reaching the posterior margin of the scutum. Basis capituli triangular dorsally, cornua small, auriculae rounded, article I of the palp with ventral anterior and posterior processes. Hypostome blunt, dental formula 2/2. Legs: coxa I with a short internal spur, and two postero-external spurs, one dorsal and one ventral; coxae II−III with two postero-external spurs each, one dorsal one ventral, and one short internal spur; coxa IV with a short external spur.

Taxonomic notes: because the authors of this book were unable to get specimens of *I. nuttalli* for morphological examination, the diagnoses above are based in the descriptions of Lahille[2] and Nuttall.[3] Readers should consult Lahille[2] and Nuttall[3] to view figures of adults, larva, and nymph of *I. nuttalli*.

DNA sequences with relevance for tick identification and phylogeny: no DNA sequences of *I. nuttalli* are available.

REFERENCES

1. Guglielmone AA, Estrada-Peña A, Keirans JE, Robbins RG. Ticks (Acari: Ixodida) of the neotropical zoogeographic region. Special publication of the integrated consortium on ticks and tick-borne diseases-2. Houten (The Netherlands): Atalanta; 2003.
2. Lahille F. Sobre dos *Ixodes* de la República Argentina y la medición de las garrapatas. *Bol Minist Agric (Argentina)* 1913;**16**:278−89.
3. Nuttall GHF. Notes on ticks. IV. Relating to the genus *Ixodes* and including description of three new species and two varieties. *Parasitology* 1916;**8**:294−337.

Ixodes pararicinus
Keirans and Clifford, 1985

Keirans J.E., Clifford, C.M., Guglielmone, A.A. & Mangold, A.J. (1985) *Ixodes (Ixodes) pararicinus* n. sp. (Acari: Ixodoidea: Ixodidae), a South American cattle tick long confused with *Ixodes ricinus*. *Journal of Medical Entomology*, 22, 401−407.

DISTRIBUTION

Argentina, Colombia, and Peru.[1−3]

Biogeographic distribution in the Southern Cone of America: Chaco and Yungas (Neotropical Region) (see Fig. 1 in the introduction).

HOSTS

Large wild and domestic mammals are the principal hosts for adults of *I. pararicinus*, while larvae and nymphs feed mainly on rodents of the family Cricetidae and birds of the order Passeriformes[4–9] (Table 2.33).

TABLE 2.33 Hosts of Adults (A), Nymphs (N), and Larvae (L) of *I. pararicinus*

MAMMALIA		*Calomys fecundus*	NL
ARTIODACTYLA		*Oligoryzomys destructor*	L
Bovidae		**AVES**	
Cattle	A	**PASSERIFORMES**	
Cervidae		**Corvidae**	
Mazama gouazoubira	A	*Cyanacorax chrysops*	NL
Tayassuidae		**Emberizidae**	
Pecari tajacu	A	*Arremon flavirostris*	L
CARNIVORA		*A. torquatus*	L
Canidae		*Atlapetes citrinellus*	L
Domestic dog	N	*Coryphospingus cucullatus*	L
CHIROPTERA		*Zonotrichia capensis*	L
Phyllostomidae		**Furnariidae**	
Desmodus rotundus	A	*Syndactyla rufosuperciliata*	L
DIDELPHIMORPHIA		**Parulidae**	
Didelphidae		*Myioborus brunniceps*	L
Monodelphis adusta	L	*Myiothlypis bivittata*	L
Thylamys venustus	L	**Troglodytidae**	
PERISSODACTYLA		*Troglodytes aedon*	L
Equidae		**Turdidae**	
Horse	A	*Turdus amaurochalinus*	NL
RODENTIA		*T. rufiventris*	ANL
Cricetidae			
C. callosus	N		

ECOLOGY

There is no information on the life cycle and population dynamics of *I. pararicinus* under natural conditions.

SANITARY IMPORTANCE

Adults of *I. pararicinus* are parasites of cattle, horses, and dogs; notably all records on domestic animals correspond to localities in the Yungas Biogeographic Region. Nava et al.[10] found a genospecies of the *Borrelia burgdorferi* sensu lato group infecting *I. pararicinus* from Argentina, but currently there is no epidemiological evidence to involve this tick as vector of human pathogenic *Borrelia* genospecies. There are no records of human infestation by *I. pararicinus*.

DIAGNOSIS

Male (Fig. 2.105): total body length 2.2 mm, breadth 1.1 mm. Body outline oval, scapulae rounded; cervical grooves shallow, diverging posteriorly. Scutum inornate; punctuations numerous, larger and more densely distributed in the central field; presence of long setae irregularly distributed. Ventral surface with numerous punctuations, larger on the median plate; presence of numerous setae, larger on median plate; ventral plates as shown in Fig. 2.105; demarcation between median plate and anal and adanal plates broadly curved. Basis capituli rectangular dorsally, without cornua; ventral outline with trilobed transverse ridge, two laterals lobes and one median lobe triangular and displaced posteriorly; articles II and III of palps subequal in length, suture conspicuous. Hypostome short, with lateral teeth pointed and longer than median teeth. Genital aperture located between coxa III. Legs: coxa I with two spurs, the internal long, narrow, and sharp, reaching the

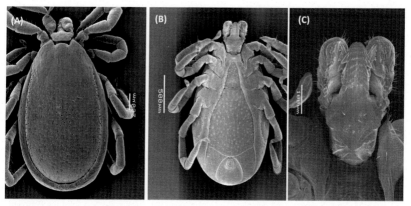

FIGURE 2.105   *I. pararicinus* male: (A) dorsal view, (B) ventral view, (C) capitulum ventral view.

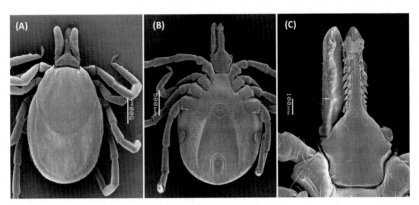

FIGURE 2.106   *I. pararicinus* female: (A) dorsal view, (B) ventral view, (C) capitulum ventral view.

coxa II, the external spur short and rounded; coxae II−IV each with a short rounded external spur. Spiracular plates oval.

Female (Fig. 2.106): total body length from slightly engorged specimens 4.7 mm, breadth 3.0 mm. Body outline oval, scapulae pointed; cervical grooves shallow, converging anteriorly then diverging as shallow depressions. Scutum length 1.4 mm, breadth 1.3 mm, inornate; outline rounded; presence of few setae, irregularly distributed; punctuations evident, more numerous and larger on the posterior field. Notum with numerous whitish setae. Ventral surface covered by numerous whitish setae. Basis capituli subquadrangular dorsally, cornua absent, porose area oval, auriculae rounded; article II of the palp longer than article III, suture conspicuous. Hypostome pointed, dental formula 4/4 for most of length, then 3/3, and 2/2 near base. Genital aperture located between coxa IV. Legs: coxa I with two spurs, the internal spur long, narrow, and sharp, reaching the coxa II, the external spur short and rounded; coxae II−IV each with a short rounded external spur. Spiracular plates oval.

Taxonomic notes: see Taxonomic notes of *I. aragaoi* Fonseca, 1935.

DNA sequences with relevance for tick identification and phylogeny: the only DNA sequence of *I. pararicinus* available in the Gen Bank corresponds to a fragment of the mitochondrial 16S rRNA gene (accession number AF549855). The 16s rDNA sequences of *I. pararicinus* from Uruguay that were published in Onofrio et al. and Barbieri et al.[3,11] (accession numbers JX082322, KJ650033) correspond to sequences of *I. aragaoi* (see also Taxonomic notes of *I. aragaoi*).

REFERENCES

1. Guglielmone AA, Estrada-Peña A, Keirans JE, Robbins RG. Ticks (Acari: Ixodida) of the neotropical zoogeographic region. Special publication of the integrated consortium on ticks and tick-borne diseases-2. Houten (The Netherlands): Atalanta; 2003.
2. Díaz MM, Nava S, Venzal JM, Sánchez N, Guglielmone AA. Tick collection from the Peruvian Amazon, with new host records for species of *Ixodes* Latreille, 1795 (Acari: Ixodidae) and *Ornithodoros* Koch, 1844 (Acari: Argasidae). *Syst Appl Acarol* 2007;**12**:127−33.

3. Onofrio VC, Ramírez DG, Giovanni DNS, Marcili A, Mangold AJ, Venzal JM, et al. Validation of the taxon *Ixodes aragaoi* Fonseca (Acari: Ixodidae) based on morphological and molecular data. *Zootaxa* 2014;**3860**:361−70.

4. Keirans JE, Clifford CM, Guglielmone AA, Mangold AJ. *Ixodes (Ixodes) pararicinus* n. sp. (Acari: Ixodoidea: Ixodidae), a South American cattle tick long confused with *Ixodes ricinus*. *J Med Entomol* 1985;**22**:401−7.

5. Beldoménico PM, Baldi JC, Orcellet V, Peralta JL, Venzal JM, Mangold AJ, et al. Ecological aspects of *Ixodes pararicinus* (Acari: Ixodidae) and other ticks species parasitizing sigmodontin mice (Rodentia: Muridae) in the northwestern Argentina. *Acarologia* 2003;**54**:15−21.

6. Guglielmone AA, Nava S. Las garrapatas de la familia Argasidae y de los géneros *Dermacentor, Haemaphysalis, Ixodes y Rhipicephalus* (Ixodidae) de la Argentina: distribución y hospedadores. *Rev Invest Agropec* 2005;**34**(2):123−41.

7. Autino AG, Nava S, Venzal JM, Mangold AJ, Guglielmone AA. La presencia de *Ixodes luciae* en el noroeste y nuevos huéspedes para *Ixodes pararicinus* y algunas especies de *Amblyomma* (Acari: Ixodidae). *Rev Soc Entomol Argent* 2006;**65**:27−32.

8. Mastropaolo M, Orcellet V, Guglielmone AA, Mangold AJ. *Ixodes pararicinus* Keirans & Clifford 1985 y *Amblyomma tigrinum* Koch, 1884 [sic] (Acari: Ixodidae): nuevos registros para Argentina y Chile. *Rev FAVE Cienc Vet* 2008;**7**:67−70.

9. Flores FS, Nava S, Batallán G, Tauro LB, Contigiani MS, Diaz LA, et al. Ticks (Acari: Ixodidae) on wild birds in north-central Argentina. *Ticks Tick-borne Dis* 2014;**5**:715−21.

10. Nava S, Barbieri AM, Maya L, Colina R, Mangold AJ, Labruna MB, et al. *Borrelia* infection in *Ixodes pararicinus* ticks (Acari: Ixodidae) from northwestern Argentina. *Acta Trop* 2014;**139**:1−4.

11. Barbieri AM, Venzal JM, Marcili A, Almeida AP, González EM, Labruna MB. *Borrelia burgdorferi* sensu lato infecting ticks of the *Ixodes ricinus* complex in Uruguay: first report for the southern hemisphere. *Vector-borne Zoon Dis* 2013;**13**:147−53.

Ixodes schulzei
Aragão and Fonseca, 1951

Aragão, H.B. & Fonseca, F. (1951) Notas de ixodologia. I. Duas novas espécies do gênero *Ixodes* e um novo nomen para *Haemaphysalis kochi* Aragão, 1908 (Acari: Ixodidae). *Memórias do Instituto Oswaldo Cruz*, 49, 567−574.

DISTRIBUTION

Argentina and Brazil.[1,2]

Biogeographic distribution in the Southern Cone of America: Parana Forest (Neotropical Region) (see Fig. 1 in the introduction).

HOSTS

Rodents of the family Cricetidae are the only hosts recorded for both adult and immature stages of *I. schulzei*[2−5] (Table 2.34).

TABLE 2.34 Hosts of Adults (A), Nymphs (N), and Larvae (L) of *I. schulzei*

MAMMALIA		*Euryorzomys russatus*	N
RODENTIA		*Nectomys squamipes*	A
Cricetidae		*Oligoryzomys flavescens*	N
Akodon montensis	NL	*O. nigripes*	N

ECOLOGY

Besides the few data on hosts and geographic distribution, there is no information on the ecology of *I. schulzei*.

SANITARY IMPORTANCE

There are no records of *I. schulzei* on humans and domestic animals, and the potential role of this tick as vector of pathogenic microorganism has not been evaluated to date.

DIAGNOSIS

Female (Fig. 2.107): total body length 4.5 mm, breadth 1.5 mm. Body outline oval elongated, scapulae rounded; cervical grooves shallow, converging anteriorly then diverging as shallow depressions reaching the postero-lateral margin of the scutum. Scutum inornate; outline oval; punctuations shallow, concentrated on anterior and lateral fields. Notum with numerous whitish setae. Ventral surface covered by numerous whitish setae. Basis capituli triangular dorsally, cornua absent, porose area oval and contiguous, auriculae pointed and extended laterally; article II of the palp longer than article III, suture conspicuous. Hypostome pointed, dental formula 2/2. Genital aperture located between coxa III. Legs: coxa I with two short, triangular spurs, the external spur slightly longer than the internal spur; coxae II–IV each with a short, triangular, external spur. Spiracular plates rounded.

Nymph (Fig. 2.108): total body length 2.0 mm, breadth 0.7 mm. Body outline oval elongated, scapulae rounded; cervical grooves short. Scutum inornate; outline oval; with few shallow punctuations. Notum with few whitish setae. Ventral surface covered by few whitish setae. Basis capituli triangular dorsally, cornua absent, auriculae pointed, and extended laterally. Hypostome pointed, dental formula 3/3 in the first and second, then 2/2 in the posterior third. Legs: coxa I with two short triangular spurs, the external spur slightly longer than the internal spur; coxae II–IV each with a short triangular external spur each.

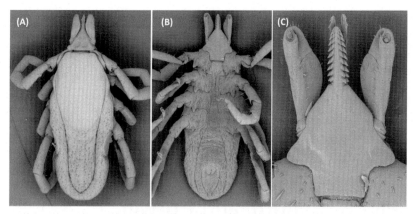

FIGURE 2.107   *I. schulzei* female: (A) dorsal view, (B) ventral view, (C) capitulum ventral view.

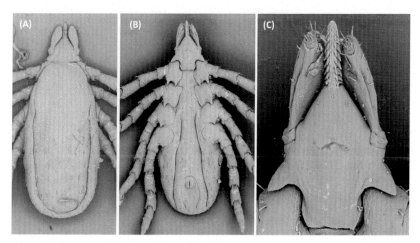

FIGURE 2.108   *I. schulzei* nymph: (A) dorsal view, (B) ventral view, (C) capitulum ventral view.

Taxonomic notes: the male of *I. schulzei* is unknown.

DNA sequences with relevance for tick identification and phylogeny: DNA sequences of *I. schulzei* are not available.

REFERENCES

1. Barros-Battesti DM, Onofrio VC, Faccini JLH, Labruna MB, Arruda-Santos AD, Giacomin FG. Description of the immature stages and redescription of the female of *Ixodes schulzei* Aragão & Fonseca, 1951 (Acari: Ixodidae), an endemic tick species of Brazil. *Syst Parasitol* 2007;**68**:157−66.
2. Lamattina D, Venzal JM, Guglielmone AA, Nava S. *Ixodes schulzei* Aragão & Fonseca, 1951 in Argentina. *Syst Appl Acarol* 2016;**21**:865−7.

3. Labruna MB, Da Silva MJN, De Oliveira MF, Barros-Battesti DM, Keirans JE. New records and laboratory-rearing data for *Ixodes schulzei* (Acari: Ixodidae) in Brazil. *J Med Entomol* 2003;**40**:116−18.
4. Onofrio VC, Nieri Bastos FA, Sampaio JS, Soares JF, Silva MJJ, Barros-Battesti DM. Noteworthy records of *Ixodes schulzei* (Acari: Ixodidae) on rodents from the state of Parana, southern Brazil. *Rev Bras Parasitol Vet* 2013;**22**:159−61.
5. Martins TF, Peres MG, Costa FB, Bacchiega TS, Appolinario CM, Antunes JMAP, et al. Ticks infesting wild small rodents in three areas of the state of São Paulo, Brazil. *Ciênc Rural* 2016;**46**:871−5.

Ixodes sigelos
Keirans, Clifford, and Corwin, 1976

Keirans J.E., Clifford, C.M. & Corwin, D. (1976) *Ixodes sigelos*, n. sp. (Acarina: Ixodidae), a parasite of rodents in Chile, with a method for preparing ticks for examination by scanning electron microscopy. *Acarologia*, 18, 217−225.

DISTRIBUTION

Argentina and Chile.[1]
   Biogeographic distribution in the Southern Cone of America: Coquimbo, Santiago, Central Patagonia, Valdivian Forest, Magellanic Forest, Subandean Patagonia (Andean Region), and Prepuna (South American Transition Zone) (see Fig. 1 in the introduction).

HOSTS

Rodents of the families Abrocomidae, Cricetidae, and Octodontidae are the hosts for all parasitic stages of *I. sigelos*[1−3] (Table 2.35).

ECOLOGY

Besides the data on host association and geographical distribution, there is no information on the ecology of *I. sigelos*.

SANITARY IMPORTANCE

The capacity of *I. sigelos* to transmit pathogens has not been investigated to date, and no records on humans and domestic animals have been found.

TABLE 2.35 Hosts of Females (F), Nymphs (N), and Larvae (L) of *I. sigelos*

MAMMALIA		*Eligmodontia morgani*	L
RODENTIA		*Euneomys chinchilloides*	L
Abrocomidae		*Loxodontomys micropus*	N
Abrocoma bennetti	N	*Oligoryzomys longicaudatus*	FNL
Cricetidae		*Phyllotis xanthopygus*	FNL
Abrothrix longipilis	NL	*Reithrodon auritus*	FN
A. olivaceus	FNL	**Muridae**	
Akodon spegazzini	N	*Rattus norvegicus*	N
A. xanthorinus	N	**Octodontidae**	
Calomys musculinus	L	*Aconaemys fuscus*	FNL
Chelemys macronyx	L	*Octodon degus*	FL

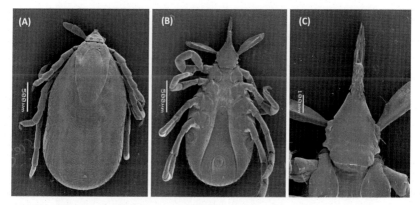

FIGURE 2.109   *I. sigelos* female: (A) dorsal view, (B) ventral view, (C) capitulum ventral view. The hypostome of the specimen showed in this figure is broken. See the image of the complete hypostome of *I. sigelos* female in Keirans et al.[2] whose details are given in the list of references for this species.

DIAGNOSIS

Female (Fig. 2.109): total body length from an slightly engorged specimen 2.5 mm, breadth 1.3 mm. Body outline oval, scapulae pointed; cervical grooves shallow, converging anteriorly then diverging as shallow depressions. Scutum length 1.1 mm, breadth 0.6 mm, inornate, elongate and narrow, punctuations few, setae long and numerous. Notum with numerous

whitish setae. Ventral surface covered by scattered setae. Basis capituli triangular dorsally, cornua triangular, porose area oval, auriculae small, triangular; article II of the palp longer than article III, suture conspicuous.
Hypostome pointed, dental formula 3/3. Genital aperture located between
coxa III. Legs: coxa I with two triangular spurs, the external slightly larger
than the internal spur; coxae II—IV with a single short triangular external
spur. Spiracular plates rounded.

Nymph (Fig. 2.110): total body length 1.1 mm, breadth 0.5 mm. Body
outline oval, scapulae rounded; cervical grooves shallow, converging anteriorly then diverging as shallow depressions. Scutum inornate, oval in outline,
with few setae and punctuations. Notum with scattered whitish setae. Ventral
surface covered by scattered setae. Basis capituli subtriangular dorsally; presence of prominent triangular cornua and distinctive lateral protuberances,
giving it a winged appearance; article I of palp with large anterior and

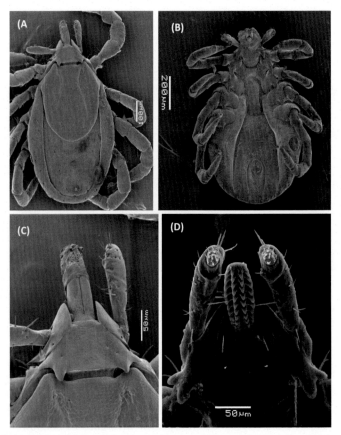

FIGURE 2.110 *I. sigelos* nymph: (A) dorsal view, (B) ventral view, (C) capitulum dorsal
view, (D) capitulum ventral view.

posterior processes, plus a large postero-lateral spur on the posterior process, and a minute lateral protuberance near the point of insertion of article II of palp; auriculae absent. Hypostome blunt, dental formula 2/2. Legs: coxa I with triangular external and internal spurs, the external slightly longer than the internal spur; coxae II−IV each with a single short triangular external spur. Spiracular plates rounded.

Taxonomic notes: *I. sigelos* is treated as a probable synonym of *I. abrocomae* by Camicas et al.[4] but this was not accepted by Guglielmone et al.[5] Sánchez et al.[1] state that more than one species may be included under the name *I. sigelos*. The male of *I. sigelos* is unknown. See also Taxonomic notes of *I. abrocomae* Lahille, 1916.

DNA sequences with relevance for tick identification and phylogeny: there are just three sequences of the mitochondrial 16S rRNA gene of *I. sigelos* in the Gen Bank (accession numbers AF549858, HM014413, HM014414).

REFERENCES

1. Sánchez JP, Nava S, Lareschi M, Udrizar Sauthier DE, Mangold AJ, Guglielmone AA. Host range and geographical distribution of *Ixodes sigelos* (Acari: Ixodidae). *Experimental and Applied Acarology* 2010;**52**:199−205.
2. Keirans JE, Clifford CM, Corwin D. *Ixodes sigelos*, n. sp. (Acarina: Ixodidae), a parasite of rodents in Chile, with a method for preparing ticks for examination by scanning electron microscopy. *Acarologia* 1976;**18**:217−25.
3. Guglielmone AA, González Acuña D, Autino AG, Venzal JM, Nava S, Mangold AJ. *Ixodes sigelos* Keirans, Clifford & Corwin, 1976 (Acari: Ixodidae) in Argentina and southern Chile. *Syst Appl Acarol* 2005;**10**:37−40.
4. Camicas JL, Hervy JP, Adam F, Morel PC. *Les tiques du monde. Nomenclature, stades décrits, hôtes, répartition (Acarida, Ixodida)*. Paris: ORSTOM; 1998.
5. Guglielmone AA, Nava S, Bazán-León EA, Mangold AJ. Redescription of the male and description of the female of *Ixodes abrocomae* Lahille, 1916 (Acari: Ixodidae). *Syst Parasitol* 2010;**77**:153−60.

Ixodes stilesi
Neumann, 1911

Neumann, L.G. (1911) Note rectificative à propos de deux espèces d'Ixodinae. *Archives de Parasitologie*, 14, 415.

DISTRIBUTION

Chile.[1]

Biogeographic distribution in the Southern Cone of America: Maule and Valdivian Forest (Andean Region) (see Fig. 1 in the introduction).

HOSTS

I. stilesi adults have been collected only from *Pudu puda* (Artiodactyla: Cervidae), while nymphs of this tick were collected on *P. puda* and *Oligoryzomys longicaudatus* (Rodentia: Cricetidae).[1−3]

ECOLOGY

Besides the few data on host association and geographical distribution, there is no information on the ecology of *I. stilesi*.

SANITARY IMPORTANCE

B. chilensis, a genospecies belonging to the *Borrelia burgdorferi* sensu lato group, was isolated from *I. stilesi* ticks from Chile,[4] but the role of this tick as vector of human pathogenic *Borrelia* genospecies is unknown. In this sense, to date, there are no records of *I. stilesi* on humans.

DIAGNOSIS

Male (Fig. 2.111): total body length 2.7 mm, breadth 1.2 mm. Body outline oval, scapulae rounded; cervical grooves shallow, converging anteriorly then diverging as shallow depressions. Scutum inornate; punctuations few and small, larger on the antero-lateral field; setae short, scattered all over scutum but absent on centro-lateral field; presence of pseudoscutum barely perceptible; ventral surface covered by scattered setae, very few setae on anal plate; ventral plates as shown in Fig. 2.111. Basis capituli pentagonal dorsally; cornua small, triangular; presence of triangular and small auriculae; articles II and III of palps subequal in length, suture conspicuous. Hypostome short,

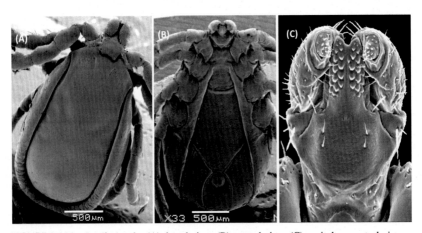

FIGURE 2.111   *I. stilesi* male: (A) dorsal view, (B) ventral view, (C) capitulum ventral view.

notched, dental formula 2/2 with an additional row of denticles in the anterior third. Genital aperture located between coxa II. Legs: coxa I with two triangular spurs, the internal longer than the external and reaching the coxa II; coxa II with two spurs, subequal in size; coxae III−IV with two spurs each, the external longer than the internal. Spiracular plates oval.

Female (Fig. 2.112): total body length from a slightly engorged specimen 5.3 mm, breadth 3.1 mm. Body outline oval, scapulae pointed; cervical grooves shallow, converging anteriorly then diverging as shallow depressions. Scutum length 1.4 mm, breadth 1.1 mm, inornate, with sinuous outline and presence of few and scattered setae; punctuations few, small, larger on the lateral field. Notum with numerous whitish setae. Ventral surface covered by scattered setae. Basis capituli subtriangular dorsally, cornua prominent and triangular, porose area rounded, auriculae rounded; article II of the palp longer than article III, suture conspicuous. Hypostome pointed, dental formula 3/3. Genital aperture located between coxae III and IV. Legs: coxa I with two subequal spurs, the internal slightly longer than the external spur; coxae II−IV each with a single small triangular external spur. Spiracular plates oval.

Nymph (Fig. 2.113): total body length from a slightly engorged specimen 2.3 mm, breadth 1.3 mm. Body outline oval, scapulae rounded; cervical grooves shallow, converging anteriorly then diverging as shallow depressions. Scutum inornate, outline oval with few setae and punctuations. Notum with scattered whitish setae. Ventral surface covered by scattered setae. Basis capituli subtriangular dorsally; presence of prominent triangular cornua and distinctive lateral protuberances, giving it a winged appearance; article I of palp with large ventral anterior and posterior processes which end bluntly

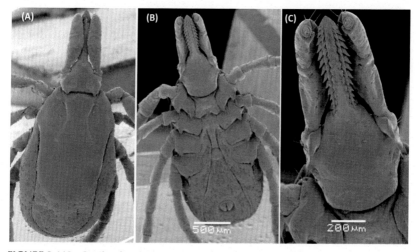

FIGURE 2.112   *I. stilesi* female: (A) dorsal view, (B) ventral view, (C) capitulum ventral view.

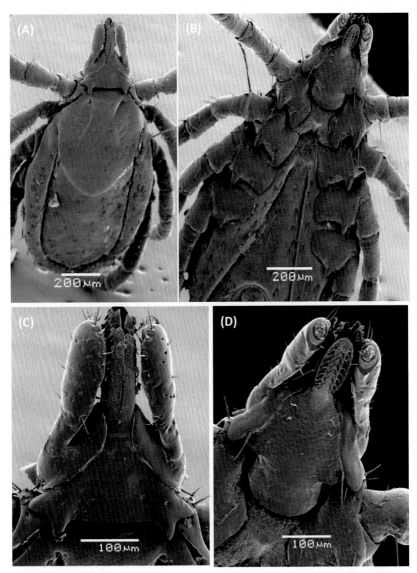

FIGURE 2.113 *I. stilesi* nymph: (A) dorsal view, (B) ventral view, (C) capitulum dorsal view, (D) capitulum ventral view.

at both extremities; auriculae absent. Hypostome blunt, dental formula 2/2 for most of the length. Legs: coxa I with two subequal triangular spurs; coxae II–IV each with single, small triangular external spur. Spiracular plates rounded.

Taxonomic notes: the female of *I. stilesi* was redescribed by Guglielmone et al.[1] using a partly engorged female. These authors stated that the female tick was characterized by a blunt hypostome, rectangular and blunt cornua, and genital aperture situated between coxae II and III. Later inspection of unfed females of *I. stilesi* showed that those previous morphological characters were wrong because of engorgement plus broken cornua and hypostome. Therefore Guglielmone et al.[2] amended the incorrect redescription of Guglielmone et al.,[1] where the female of *I. stilesi* is characterized with a cornua prominent and triangular, hypostome pointed, and genital aperture located between coxae III and IV. See also Taxonomic notes of *I. abrocomae* Lahille, 1916.

DNA sequences with relevance for tick identification and phylogeny: there are just three sequences of the mitochondrial 16S rRNA gene of *I. stilesi* in the Gen Bank (accession numbers DQ061292, DQ061293, EF362757).

REFERENCES

1. Guglielmone AA, Venzal JM, González-Acuña D, Nava S, Hinojosa A, Mangold AJ. The phylogenetic position of *Ixodes stilesi* Neumann, 1911 (Acari: Ixodidae): morphological and preliminary molecular evidences from 16S rDNA sequences. *Syst Parasitol* 2006;**65**:1−11.
2. Guglielmone AA, Nava S, González-Acuña D, Mangold AJ, Robbins RG. Additional observations on the morphology and host of *Ixodes stilesi* Neumann, 1911 (Acari: Ixodidae). *Syst Appl Acarol* 2007;**12**:135−9.
3. González-Acuña D, Guglielmone AA. Ticks (Acari: Ixodoidea: Argasidae, Ixodidae) of Chile. *Exp Appl Acarol* 2005;**35**:147−63.
4. Ivanova LB, Tomova A, González-Acuña D, Murúa R, Moreno CX, Hernández C, et al. *Borrelia chilensis*, a new member of the *Borrelia burgdorferi* sensu lato complex that extends the range of this genospecies in the Southern Hemisphere. *Environ Microbiol* 2014;**16**:1069−80.

Ixodes taglei
Kohls, 1969

Kohls, G.M. (1969) *Ixodes taglei* n. sp. (Acarina: Ixodidae) a parasite of the deer, *Pudu pudu* (Wol.), in Chile. *Journal of Medical Entomology*, 6, 280−283.

DISTRIBUTION

Chile.[1]

Biogeographic distribution in the Southern Cone of America: Maule and Valdivian Forest (Andean Region) (see Fig. 1 in the introduction).

HOSTS

The only host recorded for *I. taglei* adults is *Pudu puda* (Artiodactyla: Cervidae).[2]

ECOLOGY

There is no information on ecological aspects of *I. taglei*.

SANITARY IMPORTANCE

The capacity of *I. taglei* to transmit pathogens has not been investigated to date, and no records on humans and domestic animals have been found.

DIAGNOSIS

Male (Fig. 2.114): total body length 2.5 mm, breadth 1.5 mm. Body outline oval, scapulae rounded; cervical grooves shallow, converging anteriorly then diverging as shallow depressions. Scutum inornate; presence of pseudoscutum; pseudoscutal area with few punctuations, no setae; numerous, large, whitish setae elsewhere; ventral surface covered by scattered setae, very few setae on anal plate; ventral plates as showed in Fig. 2.114. Basis capituli pentagonal dorsally; cornua small, triangular; presence of triangular and small auriculae; articles II and III of palps subequal in length, suture conspicuous. Hypostome short, notched, dental formula 2/2 with an additional row of denticles in the anterior third. Genital aperture located between coxae II and III. Legs: coxa I with two triangular spurs, the internal longer than the external and reaching the coxa II; coxae II–IV with a short triangular

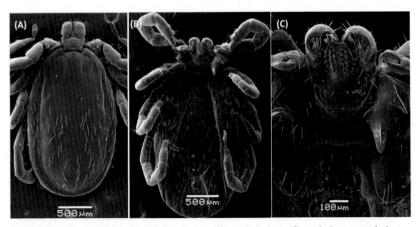

FIGURE 2.114   *I. taglei* male: (A) dorsal view, (B) ventral view, (C) capitulum ventral view.

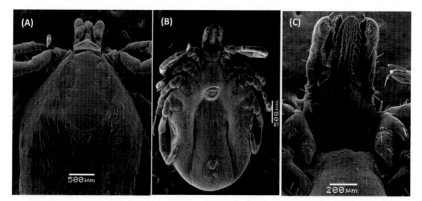

FIGURE 2.115   *I. taglei* female: (A) dorsal view, (B) ventral view, (C) capitulum ventral view.

external spur each; coxa IV with an internal spur as a small tubercle. Spiracular plates oval.

Female (Fig. 2.115): total body length from an slightly engorged specimen 2.8 mm, breadth 1.8 mm. Body outline oval, scapulae pointed; cervical grooves shallow, converging anteriorly then diverging as shallow depressions. Scutum inornate, with sinuous outline; punctuations few, small, larger on the lateral field. Notum with numerous, whitish setae. Ventral surface covered by scattered setae. Basis capituli subrectangular dorsally, cornua short and rounded, porose area oval, auriculae small; article II of the palp longer than article III, suture conspicuous. Hypostome blunt, dental formula 3/3 in the anterior third, then 2/2 to base. Genital aperture located between coxa III. Legs: coxa I with two subequal spurs, the internal slightly longer than the external spur; coxae II—IV each with a single, small triangular external spur. Spiracular plates rounded.

Taxonomic notes: in the phylogenetic tree constructed with mitochondrial 16S rDNA sequences (Fig. 2.90) it can be observed that *I. taglei* represents an independent lineage within Neotropical species of the genus *Ixodes*. The larva and nymph of *I. taglei* are unknown.

DNA sequences with relevance for tick identification and phylogeny: the only DNA sequence of *I. taglei* available in the Gen Bank corresponds to a fragment of the mitochondrial 16S rRNA gene (accession number KU729880).

REFERENCES

1. Guglielmone AA, Estrada-Peña A, Keirans JE, Robbins RG. Ticks (Acari: Ixodida) of the neotropical zoogeographic region. Special publication of the integrated consortium on ticks and tick-borne diseases-2. Houten (The Netherlands): Atalanta; 2003.

2. Kohls GM. *Ixodes taglei* n. sp. (Acarina: Ixodidae) a parasite of the deer, *Pudu pudu* (Wol.), in Chile. *J Med Entomol* 1969;**6**:280—3.

Ixodes uriae
White, 1852

White, A. (1852) Insects and Aptera. *In: Journal of the voyage in Baffin's Bay and Barrow Straite, in the years 1850–1851*, 2 (appendix), 208–211.

DISTRIBUTION

I. uriae has a circumpolar distribution in both southern and northern hemispheres, around the poles and adjacent areas of all zoogeographic regions of the world but the Oriental Region.[1,2] In the Southern Cone of America, *I. uriae* was found in Argentina, Chile, and their South Atlantic islands.

Biogeographic distribution in the Southern Cone of America: Central Patagonia, Magellanic Paramo, Magellanic Forest, Malvinas Islands (Andean Region), and Pampa (Neotropical Region) (see Fig. 1 in the introduction).

HOSTS

Seabird species of different orders are the principal hosts for all stages of *I. uriae*, while mammals are considered exceptional hosts for this tick species.[2,3] The hosts recorded for I. uriae in the Southern Hemisphere are listed in Table 2.36.

ECOLOGY

I. uriae is adapted to the extreme climatic conditions of areas close to both south and north poles. This tick species congregates in the substrate near the host nesting area.[4] *I. uriae* has a long life cycle span, which varies from 2 to 7 years depending on geographic location.[3,5–7]

TABLE 2.36 Hosts of Adults (A), Nymphs (N), and Larvae (L) of *I. uriae* in the Southern Cone of America

AVES		*P. antarctica*	ANL
CHARADRIIFORMES		*P. papua*	ANL
Haematopodidae		*Spheniscus magellanicus*	ANL
Haematopus ater	N	**SULIFORMES**	
SPHENISCFORMES		Phalacrocoracidae	
Spheniscidae		*Phalacrocorax* sp.	ANL
Pygoscelis adeliae	ANL		

SANITARY IMPORTANCE

Several viruses (families Reoviridae, Bunyaviridae, and Flaviviridae) and bacteria (families Spirochataceae, Rickettsiaceae, and Coxiellaceae) were detected in *I. uriae* (see the revision of Muñoz-Leal and González-Acuña[3]), and there is evidence suggesting that *I. uriae* is involved in the enzootic cycle of some viruses and bacteria from the families named above.[3] *I. uriae* was recorded biting humans.[8]

DIAGNOSIS

Male (Fig. 2.116): total length 3.4 mm, breadth 2.5 mm. Body outline oval elongated, with moderately convex lateral margins and posterior margin straight, scapulae pointed; cervical grooves conspicuous, diverging posteriorly. Scutum inornate, punctuations numerous and uneven in size. Ventral surface covered by short setae; ventral plates as shown in Fig. 2.116; posterior margin of ventral plates with numerous and long spines. Basis capituli rectangular dorsally, without cornua, auriculae absent, palps short and pointed, suture between articles II and III of palps inconspicuous. Hypostome short, notched, rudimentary, dental formula 1/1. Genital aperture located between coxae I and II. Legs: all coxae about equal in size and without spurs. Spiracular plates oval.

Female (Fig. 2.117): total length 3.0 mm, breadth 2.4 mm but see Taxonomic notes. Body outline oval elongated, with moderately convex lateral margins and posterior margin straight, scapulae rounded, cervical grooves straight and reaching the postero-lateral margins of the scutum. Scutum length 1.4 mm, breadth 1.1 mm, inornate; outline oval elongated; surface smooth; setae absent; punctuations numerous, fine, uniformly distributed. Notum with numerous, long, whitish setae. Ventral surface covered by numerous, long, whitish setae. Basis capituli subrectangular dorsally; cornua

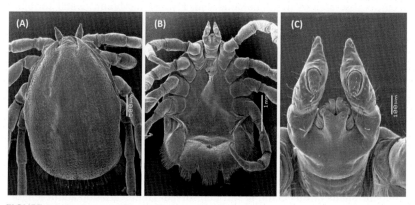

FIGURE 2.116   *I. uriae* male: (A) dorsal view, (B) ventral view, (C) capitulum ventral view.

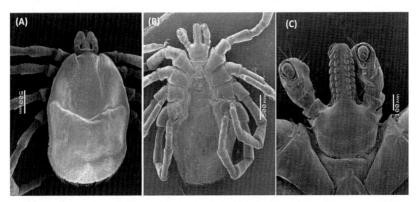

FIGURE 2.117   *I. uriae* female: (A) dorsal view, (B) ventral view, (C) capitulum ventral view.

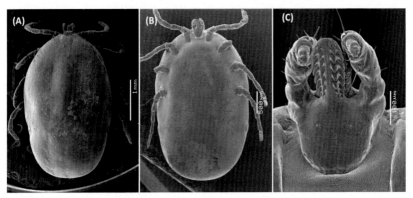

FIGURE 2.118   *I. uriae* nymph: (A) dorsal view, (B) ventral view, (C) capitulum ventral view. The specimens of the figure correspond to engorged nymphs.

absent, porose area oval; auriculae absent. Hypostome blunt, dental formula 2/2. Genital aperture located between coxae II and III. Legs: all coxae about equal in size and without spurs. Spiracular plates oval.

Nymph (Fig. 2.118): total body length 1.1 mm, breadth 1.0 mm. Body outline oval elongated, scapulae rounded, cervical grooves straight and reaching the postero-lateral margins of the scutum. Scutum inornate; outline oval elongated; surface smooth; setae absent; punctuations numerous, fine, uniformly distributed. Basis capituli subrectangular dorsally, cornua absent, auriculae absent. Hypostome blunt, dental formula 2/2. Legs: all coxae about equal in size and without spurs.

Taxonomic notes: evidence principally obtained from molecular data strongly suggests that *I. uriae* constitutes a species complex,[9,10−12] but the taxonomic status of this taxon is still unresolved. Huge variation about length and breadth of unfed females of *I. uriae* are presented by different authors being uncertain if this variation is related to the alleged different species contained under this name or typographical errors.

DNA sequences with relevance for tick identification and phylogeny: the following DNA sequences of *I. uriae* are available in the Gen Bank: mitochondrial 16S rDNA genes (accession numbers D88304, AB030016, AB030016, D88299, AF549859, U95908), mitochondrial 12S rDNA (accession numbers U95907, AM410581), mitochondrial cytochrome oxidase III (accession numbers KF586551−KF586646, EU849503−EU849550), and 18S rDNA (accession number AF115369), 28S rDNA (accession number AF120296), elongation factor 1-alpha gene (accession number GU392110), sequences of the fragment 5.8S ribosomal RNA gene-internal transcribed spacer 2-28S ribosomal RNA gene (accession numbers D88297−D88307, AB030284−AB030292), and microsatellites sequences (accession numbers AF293324−AF293331). The entire mitochondrial genome of *I. uriae* was sequenced.[13]

REFERENCES

1. Guglielmone AA, Estrada-Peña A, Keirans JE, Robbins RG. Ticks (Acari: Ixodida) of the neotropical zoogeographic region. Special publication of the integrated consortium on ticks and tick-borne diseases-2. Houten (The Netherlands): Atalanta; 2003.
2. Guglielmone AA, Robbins RG, Apanaskevich DA, Petney TN, Estrada-Peña A, Horak IG. *The hard ticks of the world (Acari: Ixodida: Ixodidae)*. London: Springer; 2014.
3. Muñoz-Leal S, González-Acuña D. The tick *Ixodes uriae* (Acari: Ixodidae): hosts, geographical distribution, and vector roles. *Ticks Tick-borne Dis* 2015;**6**:843−8.
4. Benoit JB, Yoder JA, Lopes-Martinez G, Elnitsky MA, Lee RE, Denlinger DL. Habitat requirements of the seabird tick, *Ixodes uriae* (Acari: Ixodidae), from the Antarctic Peninsula in relation to water balance characteristics of eggs, nonfed and engorged stages. *J Comp Physiol B* 2007;**177**:205−15.
5. Eveleigh ES, Threlfall W. The biology of *Ixodes* (*Ceratixodes*) *uriae* White, 1852 in Newfoundland. *Acarologia* 1974;**16**:621−35.
6. Barton TR, Harris MP, Wanless S, Elston DA. The activity periods and life-cycle of the tick *Ixodes uriae* (Acari: Ixodidae) in relation to host breeding strategies. *Parasitology* 1996;**112**:571−80.
7. Frenot Y, De Oliveira E, Gauthier-Clerc M, Deunff J, Bellido A, Vernon P. Life cycle of the tick *Ixodes uriae* in penguin colonies: reltionships with host breeding activity. *Int J Parasitol* 2001;**31**:1040−7.
8. Jaenson TGT, Jensen JK. Records of ticks (Acari, Ixodidae) from the Faroe Islands. *Norw J Entomol* 2007;**54**:11−15.
9. McCoy KD, Tirard C. Isolation and characterization of microsatellites in the seabird ectoparasite *Ixodes uriae*. *Mol Ecol* 2000;**9**:2213−14.
10. McCoy KD, Chapuis E, Tirard C, Boulinier T, Michalakis Y, Le Bohec C, et al. Recurrent evolution of host-specialized races in a globally distributed parasite. *Proc R Soc B* 2005;**272**:2389−95.
11. McCoy KD, Beis P, Barbosa A, Cuervo JJ, Fraser WR, González-Solís J, et al. Population genetic structure and colonisation of the western Antarctic Peninsula by the seabird tick *Ixodes uriae*. *Mar Ecol Prog Ser* 2012;**459**:109−20.

12. Dietrich M, Beati L, Elguero E, Boulinier T, McCoy K. Body size and shape evolution in host races of the tick *Ixodes uriae*. *Biol J Linn Soc* 2013;**108**:323−34.
13. Shao R, Barker SC, Mitani H, Aoki Y, Fukunaga M. Evolution of duplicate control regions in the mitochondrial genome of metazoa: a case study with Australasian *Ixodes* ticks. *Mol Biol Evol* 2005;**22**:620−9.

Genus *Rhipicephalus*

Rhipicephalus contains 85 species representing 12% of the total for Ixodidae. The great majority of the species are three-host ticks of Afrotropical origin (Gondwanian), but a few have a two-host or one-host cycle. The *Rhipicephalus* species with a one-host tick parasitic cycle belong to the subgenus *Boophilus* (formerly considered a genus independent of *Rhipicephalus*) which is of enormous importance for cattle industry in tropical and subtropical areas worldwide. One species, *R. microplus*, is firmly established in the Southern Cone of America. There are controversies about the correct taxonomic identification of the second species established in the region, *R. sanguineus* a name that surely represents a species complex with a worldwide distribution and of importance for animal and human health. *R. sanguineus* sensu lato is firmly established in Argentina, Chile, Paraguay, and Uruguay. These ticks have been extensively studied throughout the region.

Rhipicephalus microplus (Canestrini, 1888)

Haemophysalis [sic] *micropla*. Canestrini, G. (1888) Intorno da alcuni Acari ed Opilonidi dell'America. *Atti della Società Veneto-Trentina di Scienze Naturali Residente in Padova*, 11, 100−109.

DISTRIBUTION

R. microplus is chiefly distributed in tropical and subtropical areas worldwide belonging to different zoogeographic regions.[1] In the Southern Cone of America, *R. microplus* is present in Argentina, Paraguay, and Uruguay.

Biogeographic distribution in the Southern Cone of America: Parana Forest, Cerrado, Chaco, Pampa, and Yungas (Neotropical Region) (see Fig. 1 in the introduction).

HOSTS

There are records of *R. microplus* on mammals, birds, and amphibians but this tick species is strongly associated to cattle.[1] Table 2.37 contains records from Central America, South America, and southern Mexico.

TABLE 2.37 Hosts of *Rhipicephalus microplus* a One-Host Tick Species (Parasitic Stages not Shown)

AMBHIBIA	Primates
ANURA	Hominidae
Bufonidae	Human
Rhinella marina	RODENTIA
MAMMALIA	Caviidae
ARTIODACTYLA	*Hydrochoerus hydrochaeris*
Bovidae	CARNIVORA
Cattle	Canidae
Domestic buffalo	*Cerdocyon thous*
Goat	*Chrysocyon brachyurus*
Sheep	Domestic dog
Cervidae	*Lycalopex gymnocercus*
Blastoceros dichotomus	*L. vetulus*
Cervus elaphus	*Urocyon cinereoargenteus*
Mazama gouazoubira	Felidae
Odocoileus virginianus	*Leopardus colocolo*
Ozotoceros bezoarticus	*L. pardalis*
Suidae	*L. tigrinus*
Pig (domestic and feral)	*Panthera onca*
Tayassuidae	*Puma concolor*
Catagonus wagneri	LAGOMORPHA
Tayassu pecari	Leporidae
Pecari tajacu	*Lepus europaeus*
PERISSODACTYLA	AVES
Equidae	PELECANIFORMES
Donkey	Ardeidae
Horse	*Bubulcus ibis*

(*Continued*)

TABLE 2.37 (Continued)		
Tapiridae		
Tapirus terrestres		
T. bairdii		

Main host underlined (records from the Afrotropical, Australasian, and Nearctic Zoogeographic Regions excluded).

ECOLOGY

R. microplus has been introduced to many tropical and subtropical areas of the Americas with livestock of the early European colonialists.[2] This tick species has a one-host life cycle divided in parasitic and nonparasitic phases.[3] The parasitic phase occurs entirely on the same individual hosts and is relatively constant with a mode of approximately 23 days.[3] The nonparasitic phase includes preovipositional development and oviposition of engorged females, incubation of eggs, and host-seeking of larvae. The duration of the nonparasitic phase, which is strongly influenced by environmental factors as climate and vegetation, determines the number of generations per year of *R. microplus*. In this sense, *R. microplus* can complete from three to six generations yearly depending mainly on abiotic conditions for a given geographic area. Temperature and saturation deficit are deterministic factors for the presence and abundance of *R. microplus*, respectively.[4,5]

SANITARY IMPORTANCE

R. microplus is the major tick pest of cattle in tropical and subtropical areas of the world.[6,7] Parasitism by *R. microplus* ticks and the haemoparasites they transmit constitute a relevant constraint for cattle production due to depression on weight gain and milk production, hide damage, mortality, morbidity, and control costs.[8] *R. microplus* is the unique vector of the causative agents of bovine babesiosis, the protozoan *Babesia bovis* and *B. bigemina*, in Central and South America.[5] This tick is also involved in the transmission of *Anaplasma marginale*, the agent of another severe disease of cattle, but its relevance in the epidemiology of anaplasmosis is uncertain.[5] High levels of *R. microplus* infestation facilitate the occurrence of screwworm myiasis in cattle.[9] The records of *R. microplus* biting humans are occasional.

DIAGNOSIS

Male (Fig. 2.119): total body length 1.8 mm, breadth 1.1 mm. Body outline oval, scapulae rounded, cervical grooves shallow and barely perceptible,

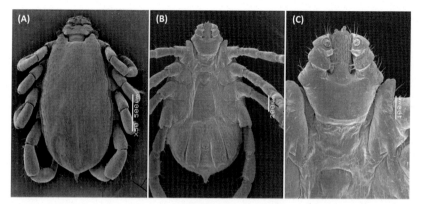

FIGURE 2.119   *R. microplus* male: (A) dorsal view, (B) ventral view, (C) capitulum ventral view.

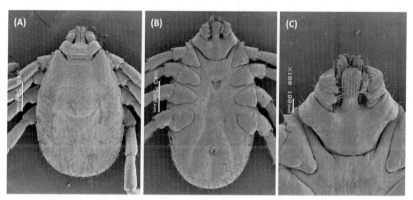

FIGURE 2.120   *R. microplus* female: (A) dorsal view, (B) ventral view, (C) capitulum ventral view.

presence of caudal appendage. Scutum inornate; marginal groove absent; dorsal surface covered by numerous, long, pale setae, arranged in rows. Ventral surface covered by numerous setae; presence of two pairs of adanal plates variable in shape, usually subrectangular, the posterior margin variable but usually with the postero-internal angle produced as a blunt spur. Eyes small and flat. Basis capituli hexagonal dorsally, cornua small, palp shorter than hypostome, both articles II and III ventrally with a short retrograde and internal process. Hypostome short, blunt, dental formula 4/4. Genital aperture located between coxa II. Legs: coxa I with two short, triangular spurs, subequal in size; coxa II−III each with a single, short external spur; coxa IV without spurs with a rounded salience. Spiracular plates rounded.

Female (Fig. 2.120): total body length 2.2 mm, breadth 1.1 mm. Body outline oval, scapulae rounded, cervical grooves shallow and barely perceptible. Scutum length 1.0 mm, breadth 0.9 mm, inornate, widest before the level

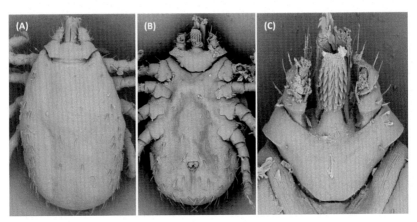

FIGURE 2.121 *R. microplus* nymph: (A) dorsal view, (B) ventral view, (C) capitulum ventral view.

of eyes, with setae concentrated in the anterior field. Notum and ventral surface covered by numerous setae. Eyes small and flat. Basis capituli hexagonal dorsally, porose area oval, palp shorter than hypostome, both articles II and III ventrally with a short retrograde and internal process. Hypostome short, blunt, dental formula 4/4. Genital aperture located between coxa II. Legs: coxa I with two short blunt spurs; coxae II−IV each with a single small blunt external spur. Spiracular plates rounded.

Nymph (Fig. 2.121): total body length 0.9 mm, breadth 0.7 mm. Body outline oval, scapulae rounded, cervical grooves shallow and barely perceptible. Scutum inornate, widest before the level of eyes, with few setae, more numerous in the anterior field. Notum and ventral surface covered by long whitish setae. Eyes small and flat. Basis capituli hexagonal dorsally, palp shorter than hypostome, both articles II and III ventrally with a short retrograde and internal process. Hypostome short, blunt, dental formula 3/3. Legs: coxa I with two short blunt spurs; coxae II−IV each with a single small blunt external spur. Spiracular plates rounded.

Taxonomic notes: *R. microplus* was traditionally described as a tick species distributed in tropical and subtropical areas of Asia and America, north-eastern Australia, New Caledonia, Madagascar, South Africa, and West Africa,[2,10] but after the reinstatement of *R. australis* Fuller, 1899, by Estrada-Peña et al.[11] it is recognized that the current distribution of *R. microplus* comprises America, Africa, and south-eastern Asia, while *R. australis* is distributed in Australia, New Caledonia, and also in south-eastern Asia.[11,12] The separation of *R. microplus* and *R. australis* as two different specific entities has been strongly supported by morphological, molecular, and biological evidence.[11−14]

DNA sequences with relevance for tick identification and phylogeny: there are many DNA sequences of *R. microplus* available in the Gen Bank. However, after the reinstatement of *R. australis* and the evidence of the existence of closely related or cryptic species in southern Asia,[11,12,15] it is difficult in many cases to assign a sequence named as belonging to *R. microplus* in Gen Bank to this taxon itself. The complete mitochondrion genome of bona fide *R. microplus* from different countries is available in the Gen Bank (accession numbers KP143546, KC503261, KC503260, KC503259). There are also DNA sequences of nuclear markers of bona fide *R. microplus* as follows: internal transcribed spacer 2 (accession numbers KC503272, KC503273, KC5032724), 18S rDNA gene (accession number KC769619), and 28S rDNA gene (accession number KC769640). The mitochondrial 16S and 12S rDNA sequences and the microsatellite loci sequences of *R. microplus* from South America and Africa presented in Labruna et al.[13] belong to bona fide *R. microplus*.

REFERENCES

1. Guglielmone AA, Robbins RG, Apanaskevich DA, Petney TN, Estrada-Peña A, Horak IG. *The hard ticks of the world (Acari: Ixodida: Ixodidae)*. London: Springer; 2014.
2. Estrada-Peña A, Bouattour A, Camicas JL, Guglielmone AA, Horak I, Jongejan F. The known distribution and ecological preferences of the tick subgenus *Boophilus* (Acari: Ixodidae) in Africa and Latin America. *Exp Appl Acarol* 2006;**38**:219−35.
3. Nuñez JL, Muñoz Cobeñas M, Moltedo H. *Boophilus microplus: la garrapata común del ganado vacuno*. Buenos Aires: Hemisferio Sur; 1982.
4. Guglielmone AA. The level of infestation with the vector of cattle babesiosis in Argentina. *Mem Inst Oswaldo Cruz* 1992;**87**(Suppl. III):133−7.
5. Guglielmone AA. Epidemiology of babesiosis and anaplasmosis in South and Central America. *Vet Parasitol* 1995;**57**:109−19.
6. Guglielmone AA, Estrada-Peña A, Keirans JE, Robbins RG. Ticks (Acari: Ixodida) of the neotropical zoogeographic region. Special publication of the integrated consortium on ticks and tick-borne diseases-2. Houten (The Netherlands): Atalanta; 2003.
7. Jongejan F, Uilenberg G. The global importance of ticks. *Parasitology* 2004;**129**:1−12.
8. Spath EJA, Guglielmone AA, Signorini AR, Mangold AJ. Estimación de las pérdidas económicas directas producidas por la garrapata *Boophilus microplus* y las enfermedades asociadas en la Argentina, 1a parte. *Therios* 1994;**23**:341−60.
9. Reck J, Marks FS, Rodrigues RO, Souza UA, Webster A, Leite RC, et al. Does *Rhipicephalus microplus* tick infestation increase the risk for myiasis caused by *Conchliomyia hominivorax* in cattle? *Prev Vet Med* 2014;**113**:59−62.
10. Madder M, Adehan S, de Deken R, Adehan R, Lokossou R. New foci of *Rhipicephalus microplus* in West Africa. *Exp Appl Acarol* 2012;**56**:385−90.
11. Estrada-Peña A, Venzal AJ, Nava S, Mangold A, Guglielmone AA, Labruna MB, et al. Reinstatement of *Rhipicephalus (Boophilus) australis* Fuller, the Australian cattle tick (Acari: Ixodidae) with redescription of the adult and larval stages. *J Med Entomol* 2012;**49**:794−802.
12. Burger TD, Shao R, Barker SC. Phylogenetic analysis of mitochondrial genome sequences indicates that the cattle tick, *Rhipicephalus (Boophilus) microplus*, contains a cryptic species. *Mol Phylogen Evol* 2014;**76**:241−53.
13. Labruna MB, Naranjo V, Mangold AJ, Thompson C, Estrada-Peña A, Guglielmone AA, et al. Allopatric speciation in ticks: genetic and reproductive divergence between geographic strains of *Rhipicephalus (Boophilus) microplus*. *BMC Evol Biol* 2009;**9**(article 46):12.

14. Mccooke JK, Guerrero FD, Barrero RA, Black M, Hunter A, Bell C, et al. The mitochondrial genome of a Texas outbreak strain of the cattle tick, *Rhipicephalus* (*Boophilus*) *microplus*, derived from whole genome sequencing Pacific Biosciences and Illumina reads. *Gene* 2015;**571**:135−41.
15. Low VL, Tay ST, Kho KL, Kho FX, Tan TK, Lim YAL, et al. Molecular characterization of the tick *Rhipicephalus microplus* in Malaysia: new insights into the cryptic diversity and distinct genetic assemblages throughout the world. *Parasit Vectors* 2015;**8**(article 341):10 p.

Rhipicephalus sanguineus sensu lato

DISTRIBUTION

R. sanguineus sensu lato has a cosmopolitan distribution, extending over different continents. In the Southern Cone of America, *R. sanguineus* sensu lato is present in Argentina, Chile, Paraguay, and Uruguay. See also Taxonomic notes.

Biogeographic distribution in the Southern Cone of America: *R. sanguineus* sensu lato is distributed in almost all biogeographic provinces of the Southern Cone of America.

HOSTS

R. sanguineus sensu lato was occasionally recorded on mammals of different orders in the Southern Cone of America (*Cerdocyon thous*, *Rattus norvegicus*, *Didelphis albiventris*, horse, goat, cattle, domestic cat, domestic rabbit), but all parasitic stages of this taxon are strongly associated to domestic dogs. See also Taxonomic notes.

ECOLOGY

It is assumed that *R. sanguineus* sensu lato ticks were introduced from the old world to America with dogs accompanying human migrations or displacements. The populations grouped under the name *R. sanguineus* sensu lato contain endophilic ticks, principally distributed in urban and periurban areas associated to dogs' resting places inside anthropic constructions. In the Southern Cone of America, all parasitic stages of *R. sanguineus* sensu lato are found active throughout the year, although they reach their peaks of abundance in spring and summer.[1−3]

SANITARY IMPORTANCE

R. sanguineus sensu lato ticks are among the most important ectoparasites of domestic dogs in the world, and they have also found parasitizing humans in several countries.[4,5] *R. sanguineus* sensu lato ticks are involved as potential vector of pathogenic agents to both humans and dogs, such as *Ehrlichia canis, Anaplasma platys, Babesia canis vogeli, B. gibsoni, Hepatozoon canis, Rickettsia rickettsii, R. conorii,* and *R. massiliae,* and also in cases of paralysis in dogs.[4-11] Moraes-Filho et al.[12] have found that *R. sanguineus* sensu lato ticks of the tropical lineage were highly competent vectors of *E. canis,* but not *R. sanguineus* sensu lato of the temperate lineage. It is unknown whether this difference in the vectorial competence occurs for other tick-borne pathogens.

DIAGNOSIS

Male (Fig. 2.122): body outline oval elongated, narrower in the anterior part; scapulae rounded; cervical grooves deep, short, comma-shaped; marginal groove incomplete, delimiting the first two festoons and extending anteriorly, ending before the eye level. Scutum inornate; punctuations unequal in size, larger and more densely distributed in the anterior field. Presence of one pair of adanal plates, subtriangular in shape, rounded posteriorly; presence of accessory adanal plates, smaller than adanal plates, narrow posteriorly. Eyes flat. Basis capituli hexagonal dorsally, cornua triangular, palps short and rounded apically. Hypostome short, blunt, dental formula 3/3. Genital aperture located between coxa II. Legs: coxa I with two long triangular spurs, the external narrower than the internal spur; coxae II−IV with a single short external spur each. Spiracular plates elongated, with a narrow dorsal prolongation.

   Female (Fig. 2.123): body outline oval; scapulae rounded; cervical grooves deep anteriorly, shallow posteriorly, sigmoid in shape. Scutum inornate; punctuations unequal in size, larger and more densely distributed in the

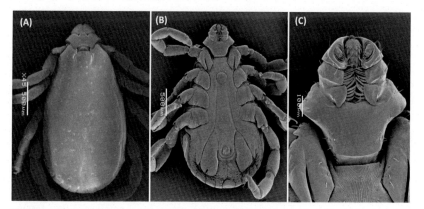

FIGURE 2.122   *R. sanguineus* sensu lato male: (A) dorsal view, (B) ventral view, (C) capitulum ventral view.

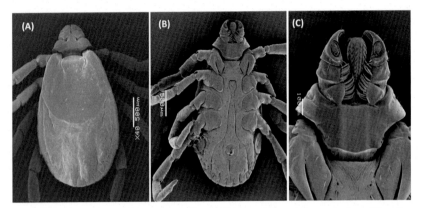

FIGURE 2.123 *R. sanguineus* sensu lato female: (A) dorsal view, (B) ventral view, (C) capitulum ventral view.

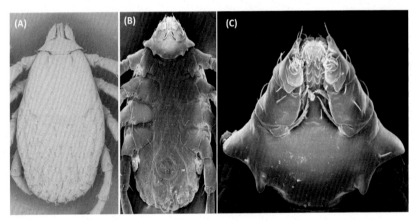

FIGURE 2.124 *R. sanguineus* sensu lato nymph: (A) dorsal view, (B) ventral view, (C) capitulum ventral view.

anterior field; posterior margin sinuous. Eyes flat. Basis capituli hexagonal dorsally, cornua small, porose area oval, palps short and rounded apically. Hypostome short, blunt, dental formula 3/3. Genital aperture located between coxa II, U-shaped. Legs: coxa I with two long triangular spurs, the external narrower than the internal spur; coxae II—IV with a single short external spur each. Spiracular plates elongated, with a narrow dorsal prolongation.

Nymph (Fig. 2.124): body outline oval; scapulae rounded; cervical grooves deep anteriorly, shallow posteriorly, sigmoid in shape. Scutum inornate, lateral margins nearly straight, posterior margin rounded. Notum and ventral surface covered by scattered whitish setae. Eyes small and flat. Basis capituli subtriangular dorsally, cornua absent, presence of ventral process, palps short and apically acute. Hypostome short, blunt, dental formula 2/2. Legs: coxa I with two triangular spurs, the external longer than the internal spur; coxae II–IV each with a single, short, external spur.

Taxonomic notes: the cosmopolitan taxon traditionally named as *R. sanguineus* (Latreille, 1806), ordinarily known *as the brown dog tick, constitute a species complex.*[13] *The name R. sanguineus* has been applied to a diversity of tick populations prone to feed on dogs, but, at the current state of the knowledge, the name R. sanguineus cannot be assigned to any tick population, and the term "*R. sanguineus* species group" or "*R. sanguineus* sensu lato" should be employed instead of "*R. sanguineus*" or "*R. sanguineus* sensu stricto."[13,14] The original description of Latreille[15] is not informative and the type is lost. In the specific case of the American continent, the tick populations previously determined as *R. sanguineus* belong to two different lineages, tropical and temperate, with reproductive incompatibility and significant genetic divergence.[16–20] The tropical lineage is distributed from northern Argentina to Mexico, while the temperate lineage is associated to temperate and cold localities from Argentina, Brazil, Chile, Uruguay, and United States.[11,13,18,19,21] In South America, the geographical boundary limiting the distributional range of these two lineages of *R. sanguineus* sensu lato appear to be an ecotonal zone situated between 24 and 25 degrees of south latitude.[19] The morphological diagnoses above presented are representative of ticks belonging to the temperate lineage. Differences for females of both lineages have been shown by Oliveira et al.[22] where the main features to separate them are body size and shape of genital aperture. Sanches et al.[23] measured unfed specimens of both lineages and found that body size of males (length 2.9 mm, breadth 1.9 mm) and females (length 3.1 mm, breadth 2.3 mm) of temperate lineage were larger than males (length 2.3 mm, breadth 1.3 mm) and females (length 2.3 mm, 1.5 mm breadth) from the tropical lineage. Further morphological comparison of all stages is needed to separate ticks of both lineages, a task currently in progress.

DNA sequences with relevance for tick identification and phylogeny: there are many sequences of *R. sanguineus* sensu lato in the Gen Bank. However, taking into account the current taxonomic status of *R. sanguineus* sensu lato (see Taxonomic notes and Nava et al.[13]), it is not possible to assign these sequences to any formally well-described taxon of a *Rhipicephalus* tick. Sequences of the mitochondrial genes 16S rRNA, 12S rRNA, and cytochrome oxidase I belonging to the tropical and temperate lineages of *R. sanguineus* sensu lato from Central and South America are available in the Gen Bank under the following accession numbers: (I) tropical lineage: 16S (JX206980, JX206981, GU553074–GU553076, KR909458, KR909459), 12S (JX206968–JX206971, JX206976, AY559842), and COI (KC243873, KC243876, KC243877, KC243879); (II) temperate lineage: 16S (JX195167–JX195172, GU553077, GU553084, GU553078, KR909454–KR909457) and 12S (AY559841, AY559843, JX206972–JX206975).

REFERENCES

1. González A, Castro D, González S. Ectoparasitic species from *Canis familiaris* (Linné) in Buenos Aires Province, Argentina. *Vet Parasitol* 2004;**120**:123−9.
2. Venzal JM, Estrada-Peña A, Castro O, De Souza CG, Portillo A, Oteo JA. Study on seasonal activity in dogs and ehrlichial infection in *Rhipicephalus sanguineus* (Latreille, 1806) (Acari: Ixodidae) from southern Uruguay. *Parasitol Latinoam* 2007;**62**:23−6.
3. Debárbora VN, Oscherov EB, Guglielmone AA, Nava S. Garrapatas (Acari: Ixodidae) asociadas a perros en diferentes ambientes de la provincia de Corrientes, Argentina. *InVet* 2011;**13**:45−51.
4. Walker JB, Keirans JE, Horak IG. *The genus Rhipicephalus (Acari: Ixodidae): a guide to the brown ticks of the world.* Cambridge: Cambridge University Press; 2000.
5. Guglielmone AA, Estrada-Peña A, Keirans JE, Robbins RG. Ticks (Acari: Ixodida) of the neotropical zoogeographic region. Special publication of the integrated consortium on ticks and tick-borne diseases-2. Houten (The Netherlands): Atalanta; 2003.
6. Parola P, Paddock CD, Raoult D. Tick-borne rickettsioses around the world: emerging diseases challenging old concepts. *Clin Microbiol Rev* 2005;**18**:719−56.
7. Otranto D, Dantas-Torres F, Breitschwerdt EB. Managing canine vector-borne diseases of zoonotic concern: part one. *Trends Parasitol* 2009;**25**:157−63.
8. Bowman DD. Introduction to the Alpha-proteobacteria: Wolbachia and Bartonella, Rickettsia, Brucella, Ehrlichia, and Anaplasma. *Top Companion Anim Med* 2011;**26**:173−7.
9. Labruna MB, Mattar S, Nava S, Bermúdez S, Venzal JM, Dolz G, et al. Rickettsioses in Latin America, Caribbean, Spain and Portugal. *Rev MVZ Córdoba* 2011;**16**:2435−57.
10. Cicuttin GL, Brambati DF, Rodríguez Eugui JI, Lebrero CG, De Salvo MN, Beltrán FJ, et al. Molecular characterization of *Rickettsia massiliae* and *Anaplasma platys* infecting *Rhipicephalus sanguineus* ticks and domestic dogs, Buenos Aires (Argentina). *Ticks Tick-borne Dis* 2014;**5**:484−8.
11. Cicuttin GL, Tarragona EL, De Salvo MN, Mangold AJ, Nava S. Infection with *Ehrlichia canis* and *Anaplasma platys* (Rickettsiales: Anaplasmataceae) in two lineages of *Rhipicephalus sanguineus* sensu lato (Acari: Ixodidae) from Argentina. *Ticks Tick-borne Dis* 2015;**6**:724−9.
12. Moraes-Filho J, Krawczak FS, Costa FB, Soares JF, Labruna MB. Comparative evaluation of the vector competence of four South American populations of the *Rhipicephalus sanguineus* group for the bacterium *Ehrlichia canis*, the agent of canine monocytic ehrlichiosis. *Plos One* 2015;**10**(article e0139386):16 p.
13. Nava S, Estrada-Peña A, Petney T, Beati L, Labruna MB, Szabó MPJ, et al. The taxonomic status of *Rhipicephalus sanguineus* (Latreille, 1806). *Vet Parasitol* 2015;**208**:2−8.
14. Guglielmone AA, Nava S. Names for Ixodidae (Acari: Ixodoidea): valid, synonyms, *incertae sedis, nomina dubia, nomina nuda, lapsus,* incorrect and suppressed names − with notes on confusions and misidentifications. *Zootaxa* 2014;**3767**:1−256.
15. Latreille PA. *Genera crustaceorum et insectorum secundum ordinem naturalem in familia disposita, iconibus exemplisque plurimis explicata.* Parisiis: Koenig; 1806.
16. Szabó MPJ, Mangold AJ, Joao CF, Bechara GH, Guglielmone AA. Biological and DNA evidence of two dissimilar populations of the *Rhipicephalus sanguineus* tick group (Acari: Ixodidae) in South America. *Vet Parasitol* 2005;**130**:131−40.
17. Burlini L, Texeira KRS, Szabó MPJ, Famadas KM. Molecular dissimilarities of *Rhipicephalus sanguineus* (Acari: Ixodidae) in Brazil and its relation with samples throughout the world: is there a geographical pattern? *Exp Appl Acarol* 2010;**50**:361−74.

18. Moraes-Filho J, Marcili A, Nieri-Bastos F, Richtzenhain LJ, Labruna MB. Genetic analysis of ticks belonging to the *Rhipicephalus sanguineus* group in Latin America. *Acta Tropica* 2011;**117**:51−5.

19. Nava S, Mastropaolo M, Venzal JM, Mangold AJ, Guglielmone AA. Mitochondrial DNA analysis of *Rhipicephalus sanguineus* sensu lato (Acari: Ixodidae) in the Southern Cone of South America. *Vet Parasitol* 2012;**190**:547−55.

20. Dantas-Torres F, Latrofa MS, Annoscia G, Gianelli A, Parisi A, Otranto D. Morphological and genetic diversity of *Rhipicephalus sanguineus* sensu lato from the New and Old Worlds. *Parasit Vectors* 2013;**6**(article 213):7 p.

21. Levin ML, Studer E, Killmaster L, Zemtsova G, Mumcuoglu KY. Crossbreeding between different geographical populations of the brown dog tick, *Rhipicephalus sanguineus* (Acari: Ixodidae). *Exp Appl Acarol* 2012;**58**:51−68.

22. Oliveira PR, Bechara GH, Denardi SE, Saito CS, Nunes ET, Szabó MPJ, et al. Comparison of the external morphology of *Rhipicephalus sanguineus* (Latreille, 1806) (Acari: Ixodidae) ticks from Brazil and Argentina. *Vet Parasitol* 2005;**129**:139−47.

23. Sanches GS, Evora PM, Mangold AJ, Jittapalapong S, Rodríguez-Mallon A, Guzmán PEE, et al. Molecular, biological, and morphometric comparisons between different geographical populations of Rhipicephalus sanguineus sensu lato (Acari: Ixodidae). *Vet Parasitol* 2016;**215**:78−87.

Chapter 3

Genera and Species of Argasidae

There are five extant genera of Argasidae around the world; all of them are found in the Neotropical Zoogeographic Region, but only three (*Argas*, *Ornithodoros*, and *Otobius*) are found in the Southern Cone of America. A total of 18 species of Argasidae are determined for Argentina, Chile, Paraguay, and Uruguay, which represents 8% of the species worldwide, and information about all of them are presented in the paragraphs below.

Genus *Argas*

There are 61 species of *Argas* worldwide which represents 29% of Argasidae. Five taxa are established in the Southern Cone of America and all of them feed on Aves using a multihost feeding cycle characteristic of most Argasidae. Main information for the five species for the region is presented in this chapter.

Argas keiransi Estrada-Peña, Venzal, and González-Acuña, 2003

Estrada Peña, A., Venzal, J.M., González Acuña, D. & Guglielmone, A.A. (2003). *Argas* (*Persicargas*) *keiransi* n. sp. (Acari: Argasidae), a parasite of the chimango, *Milvago c. chimango* (Aves: Falconiformes) in Chile. *Journal of Medical Entomology*, 40, 766–769.

DISTRIBUTION

Argentina and Chile.[1,2]

    Biogeographic distribution in the Southern Cone of America: Pampa (Neotropical Region) and Santiago (Andean Region) (see Fig. I1.1).

Ticks of the Southern Cone of America. DOI: http://dx.doi.org/10.1016/B978-0-12-811075-1.00003-0
269

HOSTS

To date, the only host—tick association recorded for *A. keiransi* corresponds to larvae collected on the bird *Milvago chimango* (Falconiformes: Falconidae).[1–3]

ECOLOGY

Besides the little data on host association and geographical distribution, there is no information on the ecology of *A. keiransi*.

SANITARY IMPORTANCE

The capacity of *A. keiransi* to transmit pathogens has not been investigated, and no records on humans and domestic animals have been found.

DIAGNOSIS

Female (Fig. 3.1): body outline oval, total length 7.2 mm, breadth 4.7 mm (measures are from the only unfed tick collected). Lateral margins subparallel, posterior margin broadly rounded, anterior margin narrower than posterior. Dorsum with discs medium to large in size, arranged in rows as showed in Fig. 3.1. Most dorsal peripheral cells with a single small setiferous pit not occupying most of the total surface area, a few cells with two setiferous pits. Capitulum: camerostome between coxa I; basis capituli ventrally subquadrangular; postpalpal setae present plus seven to eight pairs of ventral setae; posthypostomal setae arising in a straight line immediately posterior to the level of palpal insertion. Hypostome blunt, dental formula 2/2. Legs: coxae I and II separated; coxae II—IV contiguous; tarsus slightly humped subapically.

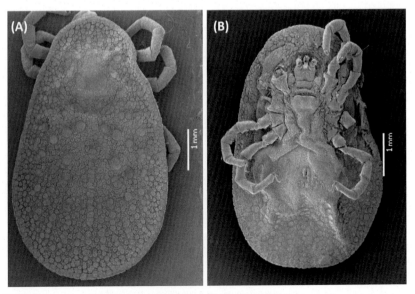

FIGURE 3.1   *A. keiransi* female: (A) dorsal view, (B) ventral view.

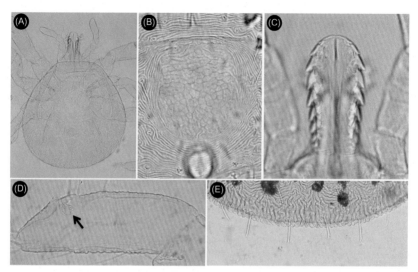

FIGURE 3.2 *A. keiransi* larva: (A) dorsal view, (B) dorsal shield, (C) hypostome, (D) tarsus I with capsule of Haller's organ indicated by the *arrow*, (E) setae in posterior margin of the body.

Larva (Fig. 3.2): body outline circular, body length (including capitulum) 0.90 mm (0.75−1.00), breadth 0.69 mm (0.62−0.75) (unengorged specimens). Dorsal shield oval in shape, length 0.20 mm (0.18−0.24), breadth 0.17 mm (0.15−0.19). Dorsum with 24 pairs of setae—6 anterolateral pairs, 8 posterolateral pairs, and 10 central pairs; venter with seven setal pairs— three sternal, three circumanal, and one pair on anal valves. Hypostome blunt, dental formula 2/2 or 3/3 in corona, then 2/2 at base; palpal segment IV as long as or longer than the other palpal segments. Legs: three pairs of ventral setae in tarsus I, absence of a trumpet-shaped sensillum extending from the capsule of Haller's organ into the lumen of the tarsus.

Taxonomic notes: Venzal et al.[2] found a genetic divergence from 1.9% to 2.2% between the 16S rDNA sequences of *A. keiransi* from Argentina and Chile. The nymphs and male of this tick species are unknown.

DNA sequences with relevance for tick identification and phylogeny: there are sequences of the mitochondrial 16S rDNA gene of *A. keiransi* in the Gen Bank as follows: accession numbers DQ295778, KJ465098−KJ465101.

REFERENCES

1. Estrada-Peña A, Venzal JM, González-Acuña D, Mangold AJ, Guglielmone AA. Notes on new world *Persicargas* ticks (Acari: Argasidae) with description of female *Argas* (*P.*) *keiransi*. *J Med Entomol* 2006;**43**:801−9.
2. Venzal JM, Flores FS, Solaro C, Santillán MA, Mangold AJ, Nava S. The presence of *Argas keiransi* Estrada-Peña, Venzal & González-Acuña, 2003 (Acari: Argasidae) in Argentina. *Syst Appl Acarol* 2014;**19**:399−403.
3. Estrada Peña A, Venzal JM, González Acuña D, Guglielmone AA. *Argas* (*Persicargas*) *keiransi* n. sp. (Acari: Argasidae), a parasite of the chimango, *Milvago c. chimango* (Aves: Falconiformes) in Chile. *J Med Entomol* 2003;**40**:766−9.

Argas miniatus Koch, 1844

Koch, C.L. (1844) Systematische Übersicht über die Ordnung der Zecken. *Archive fur Naturgeschichte*, 10, 217–239.

DISTRIBUTION

Argentina, Brazil, Colombia, Cuba, Guyana, Jamaica, Panama, Paraguay, Puerto Rico, Trinidad and Tobago, and Venezuela[1–4] (see also Taxonomic notes). *A. miniatus* is also found in the Nearctic region.

Biogeographic distribution in the Southern Cone of America: Chaco (see Fig. 1 in the introduction). The presence and distribution of *A. miniatus* in Argentina is subject to confirmation; Guglielmone & Nava[4] believe that several records of argasids from chicken and chicken houses may correspond to *A. miniatus*. Therefore its exact biogeographic distribution in the Southern Cone of America cannot be accurately determined (see also Taxonomic notes).

HOSTS

All findings of *A. miniatus* in the Southern Cone of America were made on chickens or chicken houses.

ECOLOGY

Besides the little data on host association and geographical distribution, there is no information on the ecology of *A. miniatus*.

SANITARY IMPORTANCE

A. miniatus is a vector of *Borrelia anserina*, the etiologic agent of the avian spirochaetosis, a grave diseases of poultry,[5,6] and it has been also involved as causing paralysis of Aves.[7,8]

DIAGNOSIS

Female (Fig. 3.3): body outline oval, total length 6.0 mm (4.3–7.7), breadth 3.8 mm (2.8–4.9), anterior margin narrow. Most of the dorsal peripheral cells have a single small setiferous pit but some peripheral cells have two setiferous pits. Discs conspicuous, arranged in radiating rows; posteromedian discs irregular in size and spacing, discs on the anterior margin smaller than those of the posterior margin. Capitulum: basis capituli rectangular ventrally, postpalpal setae present, 9–12 pairs of ventral setae. Hypostome blunt, dental formula 2/2. Legs: coxae II–IV contiguous, coxae I and II separated;

tarsus moderately stout, dorsal surfaces flat, very slightly elevated but not humped subapically.

Male (Fig. 3.3): all morphological characteristics are similar to those of the female, with the exception of the shape of genital aperture.

Larva (Fig. 3.4): body outline oval rounded, length (including capitulum) 0.81 mm (0.76−0.87), breadth 0.63 mm (0.53−0.73) (unengorged

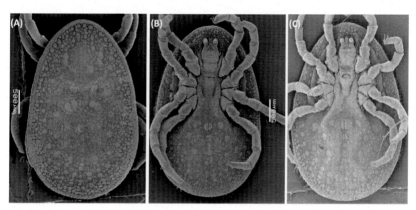

FIGURE 3.3   *A. miniatus*: (A) female dorsal view, (B) female ventral view, (C) male ventral view.

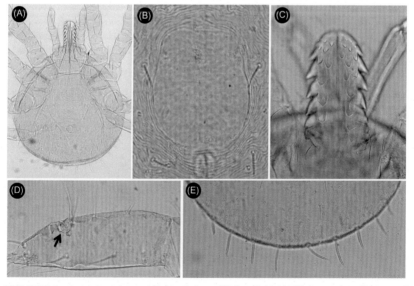

FIGURE 3.4   *A. miniatus* larva: (A) dorsal view, (B) dorsal shield, (C) hypostome, (D) tarsus I with capsule of Haller's organ indicated by the *arrow*, (E) setae on the posterior margin of the body.

specimens). Dorsal shield oval elongated, length 0.20 mm (0.19−0.22), breadth 0.16 mm (0.15−0.18). Dorsum with 25−31 pairs of setae—14 dorsolateral pairs of setae (posterolateral setae length 0.065−0.075 mm) and 11−17 central pairs of setae; venter with seven pairs of setae—three sternal, three circumanal, and one pair on anal valves. Hypostome blunt, dental formula 3/3 to 4/4 in corona, then 3/3 in the anterior third and 2/2 at base. Legs: three pairs of ventral setae on tarsus I; absence of a trumpet-shaped sensillum extending from the capsule of Haller's organ into the lumen of the tarsus.

Taxonomic notes: the geographic distribution of *A. miniatus* is probably greater than indicated above because many reports of *A. persicus* on poultry in South America may correspond to *A. miniatus*.[2−4] The authors consider this species as established in the Southern Cone of America but additional studies are needed to further confirm its presence in the region.

DNA sequences with relevance for tick identification and phylogeny: there are DNA sequences of *A. miniatus* in Gen Bank as follows: nuclear 18 S rDNA (KC769610) and 28 S rDNA genes (KC769631). The complete mitochondrion genome of *A. miniatus* is available in Gen Bank (KC769590).

REFERENCES

1. Aragão HB. Ixodidas brasileiros e de algunos paízes limitrophes. *Mem Inst Oswaldo Cruz* 1936;**31**:759−843.

2. Kohls GM, Hoogstraal H, Clifford CM, Kaiser MN. The subgenus *Persicargas* (Ixodoidea, Argasidae, *Argas*). 9. Redescription and New World records of *Argas* (*P.*) *persicus* (Oken), and resurrection, redescription, and records of A. (*P.*) *radiatus* Railliet, A. (*P.*) *sanchezi* Dugès, and A. (*P.*) *miniatus* Koch, New World ticks misidentified as A. (*P.*) *persicus. Ann Entomol Soc Am* 1970;**63**:590−606.

3. Guglielmone AA, Estrada-Peña A, Keirans JE, Robbins RG. Ticks (Acari: Ixodida) of the neotropical zoogeographic region. Special publication of the integrated consortium on ticks and tick-borne diseases. Houten (The Netherlands): Atalanta, 2003.

4. Guglielmone AA, Nava S. Las garrapatas de la familia Argasidae y de los géneros *Dermacentor, Haemaphysalis, Ixodes* y *Rhipicephalus* (Ixodidae) de la Argentina: distribución y hospedadores. *Rev Invest Agropec* 2005;**34**(2):123−41.

5. Hoogstraal H. Argasid and Nuttalliellid ticks as parasites and vectors. *Adv Parasitol* 1985;**24**:135−238.

6. Lisboa RS, Teixeira RC, Rangel CP, Santos HA, Massard CL, Fonseca AH. Avian spirochaetosis in chickens following experimental transmission of *Borrelia anserina* by *Argas* (*Persicargas*) *miniatus. Avian Dis* 2009;**53**:166−8.

7. Magalhães FEP, Massard CL, Serra Freire NM. Paralysis in *Gallus gallus* and *Cairina moschata* induced by larvae of *Argas* (*Persicargas*) *miniatus. Pesq Vet Bras* 1987;**7**:47−9.

8. Capriles JM, Gaud SM. The ticks of Puerto Rico (Arachnida: Acarina). *J Agric Univ Puerto Rico* 1977;**61**:402−4.

Argas monachus Keirans, Radovsky, and Clifford, 1973

Keirans, J.E., Radovsky, F.J. & Clifford, C.M. (1973). *Argas (Argas) monachus*, new species (Ixodoidea: Argasidae), from nests of the monk parakeet, *Myiopsitta monachus*, in Argentina. *Journal of Medical Entomology*, 5, 511−516.

DISTRIBUTION

Argentina and Paraguay[1] but Keirans et al.[2] alerted that the distribution of *A. monachus* may encompass the range of its host that also includes Bolivia, Brazil, and Uruguay.

Biogeographic distribution in the Southern Cone of America: Chaco (Neotropical Region) and Monte (South American Transition Zone) (see Fig. I1.1).

HOSTS

The monk parakeet *M. monachus* (Psittaciformes: Psittacidae) is the only host recorded for *A. monachus*.[1,2]

ECOLOGY

Besides the little data on host association and geographical distribution, there is not information on the ecology of *A. monachus*.

SANITARY IMPORTANCE

The capacity of *A. monachus* to transmit pathogens has not been investigated to date, and no records on humans and domestic animals have been found.

DIAGNOSIS

Female (Fig. 3.5): body outline oval, total length 6.0 mm (5.2−6.3), breadth 3.8 mm (3.4−4.2). Lateral margins gradually converging anteriorly, posterior margin more broadly rounded than anterior margin, widest at level of anus. Discs distinct, somewhat radially arranged, increasing in size toward middle of dorsum. Integumental striations of the same size on dorsal and ventral surface. Capitulum: camerostome between coxa I; basis capituli broadly rectangular ventrally; posterior border straight; presence of about eight pairs of lateral and sublateral setae, and one pair of posthypostomal setae. Hypostome: apex broadly rounded, dental formula 2/2. Legs: coxae I and II

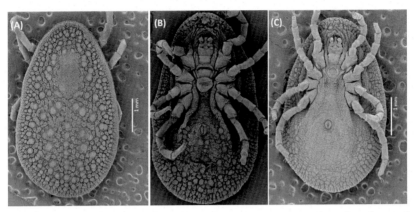

FIGURE 3.5 *A. monachus*: (A) female dorsal view, (B) female ventral view, (C) male ventral view.

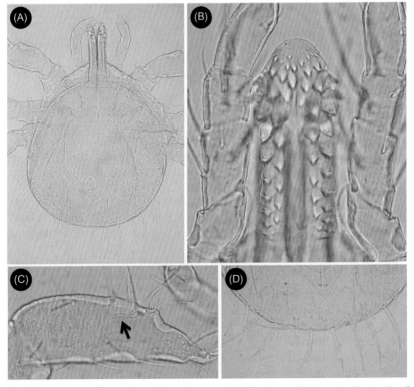

FIGURE 3.6 *A. monachus* larva: (A) dorsal view, (B) hypostome, (C) tarsus I with capsule of Haller's organ and trumpet-shaped sensillum indicated by the *arrow*, (D) setae in posterior margin of the body.

separated, coxae II–IV contiguous; presence of a prominent subapical dorsal protuberance on tarsus I only.

Male (Fig. 3.5): all morphological characteristics are similar to those of the female, with the exception of the shape of genital aperture.

Larva (Fig. 3.6): body outline circular, body length (excluding capitulum) 0.66 mm, breadth 0.66 mm (measures are from the only unengorged specimen collected). Dorsal shield oval, length 0.28 mm, breadth 0.18 mm. Dorsum with 24 pairs of setae—seven anterolateral pairs, eight posterolateral pairs, and nine central pairs; venter with seven pairs of setae—three sternal, three circumanal, and one pair on anal valves. Hypostome blunt, dental formula 4/4 or 5/5 in the first row of the corona, followed by 3/3 dentition in the two following rows, then 2/2 from mid-line to base; palpal segments I and II of equal size, segment III slightly shorter than IV. Legs: three or four pairs of ventral setae on tarsus I, presence of a trumpet-shaped sensillum extending from the capsule of Haller's organ into the lumen of the tarsus.

DNA sequences with relevance for tick identification and phylogeny: there are DNA sequences of *A. monachus* in the Gen Bank as follows: mitochondrial 16S rDNA gene (accession numbers EU283344, JF443858–JF443860) and nuclear 18S rDNA gene (KC769609).

REFERENCES

1. Mastropaolo M, Turienzo P, Di Iorio O, Nava S, Venzal JM, Guglielmone AA, et al. Distribution and 16s rDNA sequences of *Argas monachus* (Acari: Argasidae), a soft tick parasite of *Myiopsitta monachus* (Aves: Psittacidae). *Exp Appl Acarol* 2011;**55**:283–91.
2. Keirans JE, Radovsky FJ, Clifford CM. *Argas (Argas) monachus*, new species (Ixodoidea: Argasidae), from nests of the monk parakeet, *Myiopsitta monachus*, in Argentina. *J Med Entomol* 1973;**5**:511–16.

Argas neghmei Kohls and Hoogstraal, 1961

Kohls, G.M. & Hoogstraal, H. 1961. Observations on the subgenus *Argas* (Ixodoidea, Argasidae, *Argas*). 4. *A. neghmei*, new species, from poultry houses and human habitations in northern Chile. *Annals of the Entomological Society of America*, 54, 844–851.

DISTRIBUTION

Argentina and Chile but its presence in Peru is quite probable.[1]

Biogeographic distribution in the Southern Cone of America: Atacama, Monte, and Puna (South American Transition Zone) (see Fig. I1.1).

HOSTS

The hosts recorded for *A. neghmei* are the chicken, *Asthenes dorbignyi* (Passeriformes: Furnariidae), and humans.[2–4]

ECOLOGY

Besides the little data on host association and geographical distribution, there is no information on the ecology of *A. neghmei*.

SANITARY IMPORTANCE

A. neghmei was recorded on humans, producing erythematous nodular lesions with a central hemorrhagic point and intense pruritus.[3] This tick can be found in chicken coops in rural areas. The capacity of *A. neghmei* to transmit pathogens has not been investigated to date but up to 24% of houses were found infested with this tick in northern Chile.[5]

DIAGNOSIS

Female (Fig. 3.7): body outline oval elongated, narrower in the anterior part, total length 8.2 mm (5.5–10.8), breadth 5.1 mm (3.3–6.4). Discs large and distinct, arranged in radiating rows as showed in Fig. 3.7. Capitulum: camerostome between coxa I; basis capituli broadly rectangular ventrally; posterior border straight; approximately 10 pairs of lateral and sublateral setae on

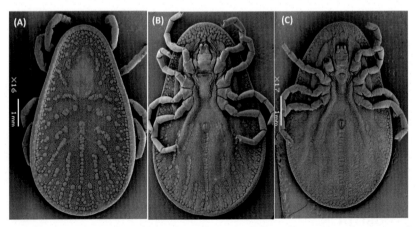

FIGURE 3.7 *A. neghmei*: (A) female dorsal view, (B) female ventral view, (C) male ventral view.

ventral surface; one pair of posthypostomal setae. Hypostome blunt, dental formula 2/2. Legs: coxae I and II separated, coxae II−IV contiguous; tarsus with an elevated subapical dorsal protuberance, with apical margin distinctly retrograde forming a triangular process dorsally.

Male (Fig. 3.7): all morphological characteristics similar to those of the female, with the exception of the shape of genital aperture.

Larva (Fig. 3.8): body outline subcircular, length (excluding capitulum) 0.75 mm (0.63−0.80), breadth 0.72 mm (0.62−0.78) (unengorged specimens). Dorsal shield oval elongated in shape, with a squamous surface, length 0.36 mm (0.34−0.39), breadth 0.28 mm (0.25−0.30). Setae of body fringed on apical half. Dorsum with 28 pairs of setae—20 dorsolateral pairs and 8 central pairs (2 pairs lateral to dorsal plate and 6 pairs posterior to dorsal plate); venter with seven pairs of setae—three sternal, three circumanal, and one pair on anal valves. Hypostome blunt, dental formula 2/2; palpal segments I−III subequal in length, segment IV approximately 1.5 times longer than segment III. Legs: four pairs of ventral setae on tarsus I; presence of a trumpet-shaped sensillum extending from the capsule of Haller's organ into the lumen of the tarsus.

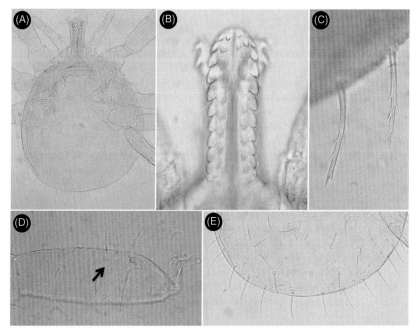

FIGURE 3.8   *A. neghmei* larva: (A) dorsal view, (B) hypostome, (C) detail of the body setae (fringed), (D) tarsus I with capsule of Haller's organ and trumpet-shaped sensillum indicated by the *arrow*, (E) setae in posterior margin of the body.

Taxonomic notes: as other species of *Argas* from the Neotropical Region *A. neghmei* has been confused with *A. persicus* (Oken, 1818).[1] Kohls and Hoogstraal[2] stated that the females of *A. neghmei* and *A. reflexus* (Fabricius, 1794) are morphologically similar but larvae of both species differ greatly. The diagnosis of adults of *A. neghmei* and related species is sometimes difficult; Clifford et al.[6] provided morphological details to separate the females of *A. cucumerinus* Neumann, 1901, *A. dalei* Clifford, Keirans, Hoogstraal, and Corwin, 1976, *A. dulus* Keirans, Clifford and Capriles, 1971, *A. magnus* Neumann, 1896, *A. monachus* Keirans, Radovsky, and Clifford, 1973, *A. neghmei*, and *Argas* species No. 1, which was later described as *A. moreli* Keirans, Hoogstraal, and Clifford, 1979, by Keirans et al.,[7] but difficulties in morphologically separation of these species are not resolved properly with the information provided by these authors.[6] This situation was somewhat improved by Keirans et al.[7] who presented dichotomous keys to separate adults and larvae of the species named above using external morphological characters.

DNA sequences with relevance for tick identification and phylogeny: there are mitochondrial 16S rDNA gene sequences of *A. neghmei* in the Gen Bank as follows: accession numbers DQ295781, FJ853598.

REFERENCES

1. Guglielmone AA, Estrada-Peña A, Keirans JE, Robbins RG. Ticks (Acari: Ixodida) of the neotropical zoogeographic region. Special publication of the integrated consortium on ticks and tick-borne diseases. Houten (The Netherlands): Atalanta, 2003.

2. Kohls GM, Hoogstraal H. Observations on the subgenus *Argas* (Ixodoidea, Argasidae, *Argas*). 4. *A. neghmei*, new species, from poultry houses and human habitations in northern Chile. *Ann Entomol Soc Am* 1961;**54**:844−51.

3. Aguirre DH, Gaido AB, Viñabal AE, Guglielmone AA, Estrada-Peña A. First detection of Argas (Argas) neghmei (Acari: Argasidae) in Argentina. *Medicina (Buenos Aires)* 1997;**57**:445−6.

4. Di Iorio O, Turienzo P, Nava S, Mastropaolo M, Mangold AJ, González Acuña D, et al. *Asthenes dorbignyi* (Reinchenbach) (Passeriformes: Furnariidae) host of *Argas neghmei* Kohls & Hoogstraal, 1961 (Acari: Argasidae). *Exp Appl Acarol* 2010;**51**:419−22.

5. Burchard L. Infestación de viviendas por garrapatas de la especie *Argas neghmei* en Calama, Chile. *Bol Chil Parasitol* 1985;**40**:45−6.

6. Clifford CM, Hoogstraal H, Keirans JE, Rice RCA, Dale WE. Observations on the subgenus *Argas* (Ixodoidea: Argasidae: *Argas*). 14. Identity and biological observations of *Argas* (A.) *cucumerinus* from Peruvian seaside cliffs and a summary of the status of the subgenus in the Neotropical faunal region. *J Med Entomol* 1978;**15**:57−73.

7. Keirans JE, Hoogstraal H, Clifford CM. Observations on the subgenus *Argas* (Ixodoidea: Argasidae: *Argas*). 16. *Argas* (A.) *moreli*, new species, and keys to Neotropical species of the subgenus. *J Med Entomol* 1979;**15**:246−52.

Argas persicus (Oken, 1818)

Rhynchoprion persicum. Oken, L. (1818) Sogenannte giftige Wanze in Persie. *Isis*, 1567−1570.

DISTRIBUTION

A. persicus is considered to be originated in Central Asia, but its presence in America, Africa, Asia, Australia, and Europe is well documented.[1] In the Southern Cone of America, *A. persicus* was reported in Argentina and Paraguay,[2] and its presence in Chile is currently under investigation. See also Taxonomic notes below.

Biogeographic distribution in the Southern Cone of America: Chaco (Neotropical Region) (see Fig. I1.1).

HOSTS

A. persicus is a common parasite of chicken, but it was also found on some species of wild birds[1,3] but in the Southern Cone of America this tick species was only found on chicken or chicken houses.[4,5]

ECOLOGY

A. persicus is a xerophilic species which inhabits desert, steppe, and forested steppe zones where its usual habitats are cracks and crevices in trees and wooden structures, and in periurban and rural areas this tick is associated to premises for domestic fowls.[1,3]

SANITARY IMPORTANCE

A. persicus is thought to be vector of different bacterial, protozoan, and viral agents infecting domestic fowl. Nevertheless, this tick species is mostly recognized as vector of two important avian diseases such as aegyptianellosis caused by *Aegyptianella pullorum* and avian spirochaetosis by *Borrelia anserina*. In addition, tick paralysis has been reported in chickens heavily infested with larvae of *A. persicus*.[3,6] Authentic records of *A. persicus* parasitizing man are rare, especially in view of the wide distribution of this tick and its close association with human activities.[6]

DIAGNOSIS

Female (see Taxonomic notes): body outline oval elongated, narrower in the anterior part, total length 8.4 mm (5.1−9.7), breadth 4.8 mm (3.4−6.1).

Peripheral dorsal cells with a single large setiferous pit occupying most of the surface cell area. Discs conspicuous and arranged in radiating rows; posteromedian discs irregular in size, discs on the anterior part of the body smaller than those present posteriorly. Capitulum: camerostome between coxa I; basis capituli rectangular ventrally, eight pairs of posterolateral setae, posthypostomal and postpalpal setae almost equidistantly spaced. Hypostome blunt, dental formula 2/2. Legs: coxae I and II separated, coxae II−IV contiguous. Tarsus slightly humped subapically.

Male (see Taxonomic notes): all morphological characteristics are similar to those of the female, with the exception of the shape of genital aperture.

Larva (Fig. 3.9): body length including capitulum 0.96 mm (0.95−0.98), breadth 0.71 mm (0.68−0.74) (unengorged specimens). Dorsal shield large, length 0.22 mm (0.20−0.24), breadth 0.19 mm (0.18−0.20). Dorsum with 26−29 pairs of setae—14−16 dorsolateral pairs and 12−13 central pairs; venter with seven pairs of setae—three sternal, three circumanal, and one pair on anal valves. Hypostome blunt dental formula 4/4 in apex, 3/3 in anterior third, then 2/2 posteriorly at base. Legs: three pairs of ventral setae on tarsus I; absence of a trumpet-shaped sensillum extending from the capsule of Haller's organ into the lumen of the tarsus.

Taxonomic notes: *A. persicus* from the Palearctic and Neotropical Regions appear to be very similar morphologically and we consider that this tick is established in the Southern Cone of America. However, integrative studies including cross-mating experiments and comparison of morphology and DNA sequences are needed to accurately determine if all *A. persicus*

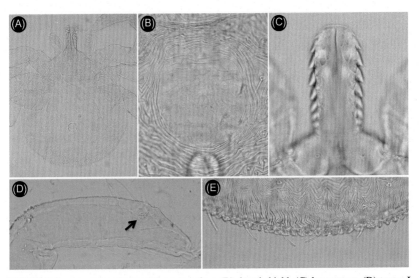

FIGURE 3.9 *A. persicus* larva: (A) ventral view, (B) dorsal shield, (C) hypostome, (D) tarsus I with capsule of Haller's organ indicated by the *arrow*, (E) setae on posterior margin of the body.

populations distributed around the world belong to the same specific entity. The authors were unable to obtain specimens of *A. persicus* adults to properly figure this stage; therefore, the diagnosis presented here is based on the descriptions and drawings of Kohls et al.,[4] and readers should consult this article to get access to figures of all stages of *A. persicus*.

DNA sequences with relevance for tick identification and phylogeny: there are DNA sequences of several genes of *A. persicus* in the Gen Bank as follows: mitochondrial 16S rDNA (accession numbers AF001402, AY436769, AY436770, AY436772, GU355920, GU451248, KR297209, KR297208, L34321), mitochondrial 12S rDNA (accession number APU95864), mitochondrial cytochrome oxidase subunit I (accession number FN394341), and nuclear 18S rDNA (accession number L76353).

REFERENCES

1. Pantaleoni RA, Baratti M, Barraco L, Contini C, Cossu CS, Filippelli MT, et al. *Argas (Persicargas) persicus* (Oken, 1818) (Ixodida: Argasidae) in Sicily with considerations about its Italian and West-Mediterranean distribution. *Parasite* 2010;**17**:349−55.
2. Guglielmone AA, Estrada-Peña A, Keirans JE, Robbins RG. Ticks (Acari: Ixodida) of the neotropical zoogeographic region. Special publication of the integrated consortium on ticks and tick-borne diseases. Houten (The Netherlands): Atalanta, 2003.
3. Nosek J, Hoogstraal H, Labuda M, Cyprich D. Bionomics and health importance of fowl tick *Argas (Persicargas) persicus* (Oken, 1818) (Ixodoidea: Argasidae). *Z Parasitenkd* 1980;**63**:209−12.
4. Kohls GM, Hoogstraal H, Clifford CM, Kaiser MN. The subgenus *Persicargas* (Ixodoidea, Argasidae, *Argas*). 9. Redescription and New World records of *Argas* (*P.*) *persicus* (Oken), and resurrection, redescription, and records of *A.* (*P.*) *radiatus* Railliet, *A.* (*P.*) *sanchezi* Dugès, and *A.* (*P.*) *miniatus* Koch, New World ticks misidentified as *A.* (*P.*) *persicus*. *Ann Entomol Soc Am* 1970;**63**:590−606.
5. Ivancovich JC, Luciani CA. *Las garrapatas de Argentina*. Buenos Aires: Asociación Argentina de Parasitología Veterinaria; 1992.
6. Hoogstraal H. Argasid and Nuttalliellid ticks as parasites and vectors. *Adv Parasitol* 1985;**24**:135−238.

Genus *Ornithodoros*

There are 126 species of *Ornithodoros* worldwide which represents 61% of Argasidae being also the most numerous genus in the Southern Cone of America with 12 taxa. The life cycle of many species is characterized by a multihost parasitic cycle but this should be considered cautiously because the cycle of several species remains unknown. More than 10 species of *Ornithodoros* were described for the region lately, and several new taxa will be described in the near future. Nevertheless, we were unable to obtain specimens to figure adults and larva of *Or. rudis* and adults of *Or. hasei*.

ABOUT THE PRESENCE OF *ORNITHODOROS CAPENSIS* IN CHILE

Recently Duron et al.[1] stated that *Or. capensis* has been found in the island of Pan de Azúcar in Chile. However, the ticks mentioned as *Or. capensis* were obtained by one of the author of this book (Daniel González-Acuña) and they are in fact *Or. spheniscus*. The tick *Or. capensis* was originally described by Neumann[2] in 1901 under the name *Or. talaje capensis*, from specimens collected in nests of penguins and guano in the southern coast of South Africa, not from specimens collected in Saint Paul Rocks (Atlantic Ocean, Brazil) as stated in Hoogstraal[3] and other workers. Hoogstraal[3] postulated *Or. capensis* as the progenitor of a group of several species with a huge range mostly in islands and coastal lines south of the Tropic of Cancer and included *Or. spheniscus* described by Hoogstraal et al.,[4] into this group. Here, we affirm that *Or. capensis* sensu stricto has not been found yet in Chile or any other country of the Southern Cone of America.

REFERENCES

1. Duron O, Noël V, McCoy KD, Bonazzi M, Sidi-Boumedine K, et al. The recent evolution of a maternally-inherited endosymbiont of ticks led to the emergence of the Q fever pathogen, *Coxiella burnetii*. *Plos Pathog* 2015;**11**(5) (article e1004892) 23 p.
2. Neumann LG. Révision de la famille des ixodidés (4a mémoire). *Mém Soc Zool Fr* 1901;**14**:249−372.
3. Hoogstraal H. Argasid and nuttalliellid ticks as parasites and vectors. *Adv Parasitol* 1985;**24**:135−238.
4. Hoogstraal H, Wassef HY, Hays A, Keirans JE. *Ornithodoros (Alectorobius) spheniscus* n. sp. [Acarina: Ixodoidea: Argasidae: *Ornithodoros (Alectorobius) capensis* group], a tick parasite of the Humboldt penguin in Peru. *J Parasitol* 1985;**71**:635−44.

Ornithodoros amblus Chamberlin, 1920

Chamberlin, R.V. (1920) South American Arachnida, chiefly from the guano islands of Peru. *Science Bulletin of the Museum of the Brooklyn Institute of Arts and Sciences*, 3, 35−44.

DISTRIBUTION

Chile and Peru.[1]

Biogeographic distribution in the Southern Cone of America: Atacama (South American Transition Zone) (see Fig. I1.1).

TABLE 3.1 Hosts of *Or. amblus*

MAMMALIA	PELECANIFORMES
Primates	Pelecanidae
Human	*Pelecanus thagus*
AVES	SULIFORMES
CHARADRIIFORMES	Phalacrocoracidae
Laridae	*Phalacrocorax bougainvilliorum*
Larus sp.	*P. gaimardi*
SPHENISCIFORMES	Sulidae
Spheniscidae	*Sula neuboxi*
Spheniscus humboldti	*S. variegata*

The species of hosts were mainly determined by tick records in breeding sites being probable that all parasitic stages (omitted) feed on these hosts.

HOSTS

Seabirds are the principal hosts for *Or. amblus*.[2] The complete list of hosts of *Or. amblus* is given in Table 3.1.

ECOLOGY

Besides the data on host association and geographical distribution of *Or. amblus*, Ianaccone and Ayala[3] found a direct relationship between density of Aves and tick numbers, and also a similar relation between wind and tick abundance, but this last association may be due in fact to avian host preference for breeding sites rather than tick choice.

SANITARY IMPORTANCE

Or. amblus is very aggressive to humans. The bites of this tick cause severe inflammation and itching.[2] *Orbivirus* (Kemerovo serogroup) and *Nairovirus* (Hughes serogroup) were isolated from this tick species in Peru.[4] Duffy[5] determined that severe discomfort from bites of *Or. amblus* caused birds to abandon their nests. Presence of *Coxiella* endosymbionts has been reported in this tick species.[6]

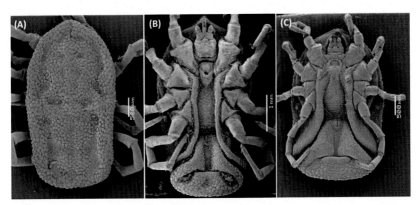

FIGURE 3.10   *Or. amblus*: (A) female dorsal view, (B) female ventral view, (C) male ventral view.

DIAGNOSIS

Female (Fig. 3.10): body outline oval, total length 7.5 mm (5.2−8.7), breadth 4.5 mm (3.3−5.7). Mammillae numerous, regularly spaced, subequal in size and shape. Discs conspicuous, irregular in size and shape. Mammillae present on ventral surface. Capitulum: basis capituli broader than long; cheeks small and narrow. Hypostome indented apically, dental formula 2/2. Legs long, micromammillated; coxae I and II separated, coxae II−IV contiguous; tarsus lacking humps and protuberances.

Male (Fig. 3.10): all morphological characteristics are similar to those of the female, with the exception of the shape of genital aperture.

Larva (Fig. 3.11): body outline subcircular, length (excluding capitulum) 0.658 mm (0.610−0.695), breadth 0.534 mm (0.476−0.610) (unengorged specimens). Dorsal shield pyriform, posterior margin rounded or concave, length 0.259 mm (0.280−0.312), breadth 0.200 mm (0.170−0.240). Dorsum with 15 pairs of setae—seven anterolateral pairs, four posterolateral pairs, and four central pairs. Venter surface with eight pairs of setae—one pair on anal valves, three pairs of sternal setae, three pairs of circumanal setae, and one pair of postcoxal setae; posteromedian seta absent. Hypostome pointed, dental formula 5/5 or 4/4 in the first quarter, then 3/3 in midportion, then 2/2 at base. Capsule of Haller's organ subcircular in shape from dorsal view.

Taxonomic notes: Hoogstraal et al.[7] stated that *Or. amblus* is morphologically related to *Or. spheniscus* Hoogstraal, Wassef, Hays, and Keirans, 1985, and *Or. yunkeri* Keirans, Clifford, and Hoogstraal, 1984. Guglielmone et al.[1] alerted that laboratory studies of *Or. amblus* from different localities yielded dissimilar results,[2,8,9] being uncertain whether these differences were due to intraspecific variation. Perhaps, wrong tick identification was the cause of the dissimilarities found in those studies.

DNA sequences with relevance for tick identification and phylogeny: there are no sequences of *Or. amblus* in the Gen Bank.

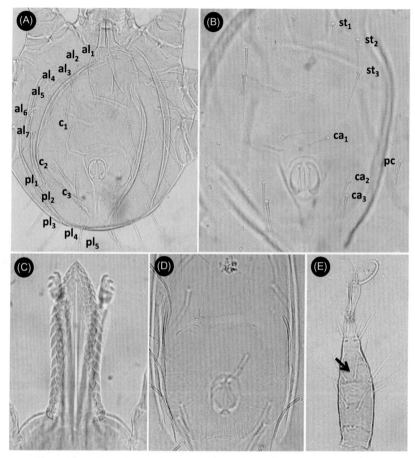

FIGURE 3.11   *Or. amblus* larva: (A) dorsal view, (B) ventral view, (C) hypostome, (D) dorsal shield, (E) tarsus I with capsule of Haller's organ indicated by the *arrow*. We acknowledge the contribution of Sebastián Muñoz-Leal for generously providing photos of *Or. amblus*.

REFERENCES

1. Guglielmone AA, Estrada-Peña A, Keirans JE, Robbins RG. Ticks (Acari: Ixodida) of the neotropical zoogeographic region. Special publication of the integrated consortium on ticks and tick-borne diseases. Houten (The Netherlands): Atalanta, 2003.
2. Clifford CM, Hoogstraal H, Radovsky FJ, Stiller D, Keirans JE. *Ornithodoros* (*Alectorobius*) *amblus* (Acarina: Ixodoidea: Argasidae): identity, marine bird and human hosts, virus infections, and distribution in Peru. *J Parasitol* 1980;**66**:312–13.
3. Ianaccone J, Ayala L. Censo de *Ornithodoros amblus* (Acarina: Argasidae) en la Isla Mazorca, Lima, Perú. *Parasitol Latinoam* 2004;**59**:56–60.
4. Hoogstraal H. Argasid and Nuttalliellid ticks as parasites and vectors. *Adv Parasitol* 1985;**24**:135–238.
5. Duffy DC. The ecology of tick parasitism on densely nesting Peruvian seabirds. *Ecology* 1983;**64**:110–19.

6. Duron O, Noël V, Mccoy KD, Bonazzi M, Sidi-Boumedine K, Morel O, et al. The recent evolution of a maternally-inherited endosymbiont of ticks led to the emergence of the Q Fever pathogen, *Coxiella burnetii*. *PLoS Pathog* 2015;**11**(5) (article e1004892) 23 p.
7. Hoogstraal H, Wassef HY, Hays A, Keirans JE. *Ornithodoros* (*Alectorobius*) *spheniscus* n. sp. [Acarina: Ixodoidea: Argasidae: *Ornithodoros* (*Alectorobius*) *capensis* group], a tick parasite of the Humboldt penguin in Peru. *J Parasitol* 1985;**71**:635−44.
8. Khalil GM, Hoogstraal H. The life cycle of *Ornithodoros* (*Alectorobius*) *amblus* (Acari: Ixodoidea: Argasidae) in the laboratory. *J Med Entomol* 1981;**18**:134−9.
9. Schumaker TTS, Mori CM, Ferreira CS. Experimental infestation of *Gallus gallus* with *Ornithodoros* (*Alectorobius*) *amblus* (Ixodoidea: Argasidae). *J Med Entomol* 1997;**34**:521−6.

Ornithodoros hasei (Schulze, 1935)

Argas hasei. Schulze, P. (1935) Zur Vergleichenden Anatomie der Zecken. *Zeitschrift für Morphologie und Ökologie der Tiere*, 30, 1−40.

DISTRIBUTION

Antigua and Barbuda, Argentina, Bolivia, Brazil, Colombia, Costa Rica, Dominican Republic, Guadeloupe, Guatemala, Guyana, Jamaica, Martinique, southern Mexico, Nicaragua, Panama, Paraguay, Peru, Trinidad and Tobago, Uruguay, Virgin Island, and Venezuela.[1−3]

Biogeographic distribution in the Southern Cone of America: Pampa and Yungas (Neotropical Region) (see Fig. I1.1).

HOSTS

Bats (Chiroptera) are the principal hosts for *Or. hasei* (Table 3.2).

ECOLOGY

Besides the little data on host association and geographical distribution, there is no information on the ecology of *Or. hasei*.

SANITARY IMPORTANCE

The capacity of *Or. hasei* to transmit pathogens has not been investigated to date.

DIAGNOSIS

Female (see Taxonomic notes): body outline oval, total length 3.4 mm long, breadth 2.0 mm. Mammillae relatively few in number, smaller on the margins. Discs large and conspicuous, occupying much of the dorsal surface.

TABLE 3.2 Hosts of *Or. hasei*

MAMMALIA	*Chiroderma salvini*
CHIROPTERA	*Desmodus rotundus*
Emballonuridae	*Glossophaga longirostris*
Peropteryx sp.	*Lonchorinha orinocensis*
Molossidae	*Mimon crenulatum*
Molossops mattogrossensis	*Phyllostomus hastatus*
M. temmincki	*Rhinophylla pumilio*
Molossus molossus	*Sturnira lilium*
M. rufus	*Uroderma magnirostrum*
Mormoopidae	**Vespertilionidae**
Mormoops megalophylla	*Histiotus laephotis*
Pteronotus daveyi	*Myotis albescens*
Noctilionidae	*M. nigricans*
Noctilio albiventris	*M. velifer*
N. leporinus	*Rhogeessa minutilla*
Phyllostomidae	**RODENTIA**
Artibeus jamaicensis	**Cricetidae**
A. lituratus	*Necromys urichi*
Brachyphylla cavernarum	

With few exceptions, records of this species are from larvae found on hosts, and parasitic stages are omitted.

Capitulum: basis capituli wider than long, with the surface granulated; cheeks small and oval. Hypostome notched, dental formula 2/2. Legs: all coxae contiguous; tarsus I with a subapical dorsal protuberance, dorsal humps absent.

Male (see Taxonomic notes): all morphological characteristics are similar to those of the female, with the exception of the shape of genital aperture.

Larva (Fig. 3.12): body outline oval, length (including capitulum) 0.684 mm (0.615–0.753), breadth 0.345 mm (0.300–0.390) (unengorged specimens). Dorsal shield pyriform, length 0.230 mm (0.220–0.240), breadth 0.138 mm (0.127–0.150). Dorsum with 17–20 pairs of setae (19 pairs in specimens from Argentina and Uruguay)—seven anterolateral pairs, seven to nine posterolateral pairs (typically eight), and three to five central pairs (typically four). Venter surface with eight pairs of setae—one pair on anal valves, three pairs of sternal setae, three pairs of circumanal setae, and one pair of

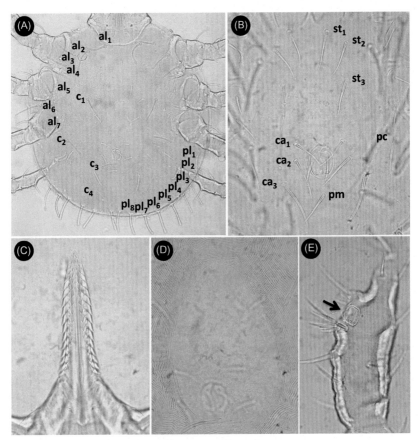

FIGURE 3.12 *Or. hasei* larva: (A) dorsal view, (B) ventral view, (C) hypostome, (D) dorsal shield, (E) tarsus I with capsule of Haller's organ indicated by the *arrow*.

postcoxal setae; posteromedian seta present. Hypostome pointed, dental formula 3/3 in the anterior two-thirds, 2/2 in the posterior third. Capsule of Haller's organ slightly oval.

Taxonomic notes: *Or. hasei* morphologically belongs to the *Or. talaje* species group. The evidence obtained with the analysis of the mitochondrial 16S rDNA sequences of the Neotropical species of *Ornithodoros* showed that *Or. hasei*, *Or. rioplatensis* Venzal, Estrada-Peña, and Mangold, 2008, *Or. puertoricensis* Fox, 1947, and *Or. guaporensis* Nava, Venzal, and Labruna, 2013, are phylogenetically closely related (Fig. 3.13). The great majority of records of this species are from larva parasitizing bats and the authors were unable to obtain adult ticks; therefore, the diagnosis presented here is based on the descriptions and drawings of Matheson[4] and Cooley and Kohls[5] (under the name *Or. dunni*). Readers should consult the articles of these authors to get access to figures of adults of *Or. hasei*. Nevertheless, it

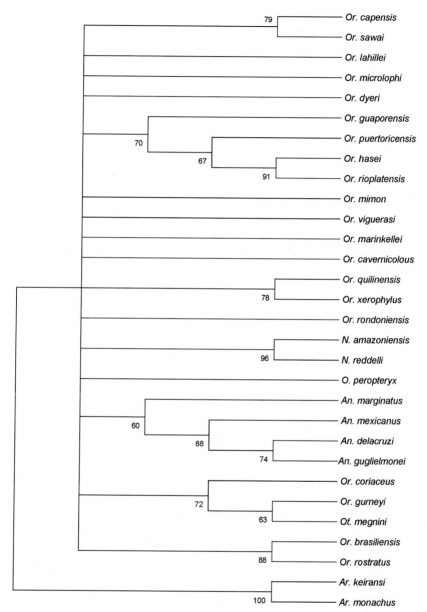

FIGURE 3.13 Maximum-likelihood analysis of the mitochondrial 16S rDNA sequences of species of Argasidae. The phylogenetic condensed tree was generated with the GTR model and a discrete Gamma-distribution.

should be considered that Jones et al.[6] found morphological variation among ticks from different localities and concluded that more than one species may be represented in material currently under this name, and Nava et al.[3] treated *Or. hasei* as a name for a species complex.

DNA sequences with relevance for tick identification and phylogeny: there are sequences of the mitochondrial 16S rDNA gene of *Or. hasei* in the Gen Bank as follows: accession numbers KT894588, DQ295779.

REFERENCES

1. Guglielmone AA, Estrada-Peña A, Keirans JE, Robbins RG. Ticks (Acari: Ixodida) of the neotropical zoogeographic region. Special publication of the integrated consortium on ticks and tick-borne diseases. Houten (The Netherlands): Atalanta, 2003.
2. Nava S, Lareschi M, Rebollo C, Benítez Usher C, Beati L, Robbins RG, et al. The ticks (Acari: Ixodida: Argasidae, Ixodidae) of Paraguay. *Ann Trop Med Parasitol* 2007;**101**:255−70.
3. Nava S, Venzal JM, Díaz MM, Mangold AJ, Guglielmone AA. The *Ornithodoros hasei* (Schulze, 1935) species group in Argentina. *Syst Appl Acarol* 2007;**12**:27−30.
4. Matheson R. Three new species of ticks, *Ornithodoros* (Acarina, Ixodoidea). *J Parasitol* 1935;**21**:347−53.
5. Cooley RA, Kohls GM. The Argasidae of North America, Central America and Cuba. *Am Midl Natur Monogr* 1944;**1**:152.
6. Jones EK, Clifford CM, Keirans JE, Kohls GM. The ticks of Venezuela (Acarina: Ixodoidea) with a key to the species of *Amblyomma* in the Western Hemisphere. *Brigham Young Univ Sci Bull Biol Ser* 1972;**17**(4):40.

Ornithodoros lahillei Venzal, González-Acuña, and Nava, 2015

Venzal J.M., González-Acuña D., Muñoz-Leal S., Mangold A.J. & Nava S. (2015) Two new species of *Ornithodoros* (Ixodida: Argasidae) from the Southern Cone of South America. *Experimental and Applied Acarology*, 66, 127−139.

DISTRIBUTION

Chile.[1]

Biogeographic distribution in the Southern Cone of America: Atacama (Andean Region) (see Fig. I1.1).

HOSTS

Larvae of *Or. lahillei* were collected on *Philodryas chamissonis* (Serpentes: Colubridae) and *Callopistes maculatus* (Sauria: Teiidae).[1]

ECOLOGY

Besides the little data on host association and geographical distribution, there is no information on the ecology of *Or. lahillei*.

SANITARY IMPORTANCE

The capacity of *Or. lahillei* to transmit pathogens has not been investigated to date.

DIAGNOSIS

Larva (Fig. 3.14): body outline oval, length (including capitulum) 1.023 mm (0.931−1.166), breadth 0.653 mm (0.568−0.745). Dorsal shield subtriangular,

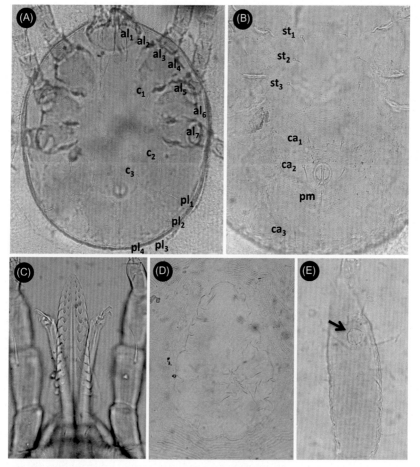

FIGURE 3.14   *Or. lahillei* larva: (A) dorsal view, (B) ventral view, (C) hypostome, (D) dorsal shield, (E) tarsus I with capsule of Haller's organ indicated by the *arrow*.

surface smooth in some portions with margins corrugated, length 0.207 mm (0.195–0.219), breadth 0.146 mm (0.134–0.166). Dorsum with 14 pairs of setae—seven anterolateral pairs, four posterolateral pairs, and three central pairs. Venter surface with seven pairs of setae—one pair on anal valves, three pairs of sternal setae, three pairs of circumanal setae; posteromedian seta present, postcoxal setae absent. Hypostome pointed, dental formula 3/3 in anterior third, then 2/2 at base. Capsule of Haller's organ oval in shape from dorsal view.

Taxonomic notes: adults and nymphs of *Or. lahillei* are unknown. At least with the phylogenetic analysis performed with mitochondrial 16S rDNA sequences, the phylogenetic position of *Or. lahillei* remains unresolved (Fig. 3.13).

DNA sequences with relevance for tick identification and phylogeny: there is just one sequence of the mitochondrial 16S rDNA gene of *Or. lahillei* in the Gen Bank (accession number KP403288).

REFERENCES

1. Venzal JM, González-Acuña D, Muñoz-Leal S, Mangold AJ, Nava S. Two new species of *Ornithodoros* (Ixodida; Argasidae) from the Southern Cone of South America. *Exp Appl Acarol* 2015;**66**:127–39.

Ornithodoros microlophi Venzal, Nava, and González-Acuña, 2013

Venzal J.M., Nava S., González-Acuña D., Mangold A.J., Muñoz-Leal S., Lado P. & Guglielmone A.A. (2013) A new species of *Ornithodoros* (Acari: Argasidae), parasite of *Microlophus* spp. (Reptilia: Tropiduridae) from northern Chile. *Ticks and Tick-borne Diseases*, 4, 128–132.

DISTRIBUTION

Chile.[1]

Biogeographic distribution in the Southern Cone of America: Atacama (South American Transition Zone) (see Fig. I1.1).

HOSTS

Or. microlophi was collected on *Microlophus atacamensis* and *M. quadrivittatus* (Squamata: Tropiduridae).[1]

ECOLOGY

Besides the little data on host association and geographical distribution, there is no information on the ecology of *Or. microlophi*.

SANITARY IMPORTANCE

The capacity of *Or. microlophi* to transmit pathogens has not been investigated to date.

DIAGNOSIS

Larva (Fig. 3.15): body outline oval, length (including capitulum) 0.670 mm, breadth 0.549 mm (based on slightly engorged specimens). Dorsal shield pyriform, length 0.295 mm (0.284−0.313), breadth 0.217 mm (0.196−0.245). Dorsum with 19−21 pairs of setae (typically

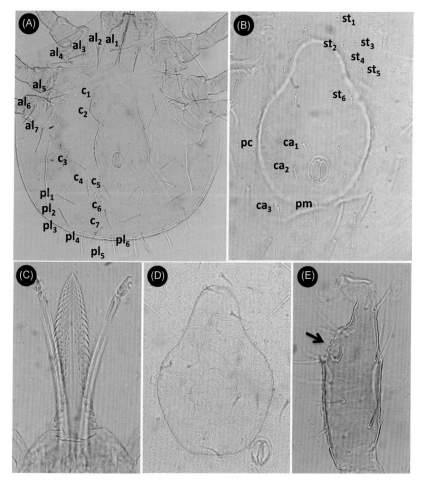

FIGURE 3.15 *Or. microlophi* larva: (A) dorsal view, (B) ventral view, (C) hypostome, (D) dorsal shield, (E) tarsus I with capsule of Haller's organ indicated by the *arrow*.

20)—seven anterolateral pairs of setae, six to seven posterolateral pairs of setae, and six to eight central pairs of setae (typically seven). Venter surface provided with 11 pairs of setae—one pair on anal valves, six pairs of sternal setae, three pairs of circumanal setae, and one pair of postcoxal setae; posteromedian seta present. Hypostome pointed, dental formula 2/2 or 3/3 apically, 4/4 in the midportion, then 2/2 at base. Capsule of Haller's organ irregular in shape.

Taxonomic notes: Adults and nymphs of *Or. microlophi* are unknown.

Phylogenetic analysis of 16S rDNA sequences suggests that *Or. microlophi* represents an independent lineage within Neotropical species of the Argasidae (Fig. 3.13).

DNA sequences with relevance for tick identification and phylogeny: there are sequences of the mitochondrial 16S rDNA gene of *Or. microlophi* in the Gen Bank (accession numbers JX455897−JX455899).

REFERENCES

1. Venzal JM, Nava S, González-Acuña D, Mangold AJ, Muñoz-Leal S, Lado P, et al. A new species of *Ornithodoros* (Acari: Argasidae), parasite of *Microlophus* spp. (Reptilia: Tropiduridae) from northern Chile. *Ticks Tick-borne Dis* 2013;4:128−32.

Ornithodoros mimon Kohls, Clifford, and Jones, 1969

Kohls G.M., Clifford C.M. & Jones E.K. (1969) The systematics of the subfamily Ornithodorinae (Acarina: Argasidae). IV. Eight new species of *Ornithodoros* from the Western Hemisphere. *Annals of the Entomological Society of America*, 62, 1035−1043.

DISTRIBUTION

Argentina, Bolivia, Brazil, and Uruguay.[1−3]

Biogeographic distribution in the Southern Cone of America: Pampa and Yungas (Neotropical Region) (see Fig. I1.1).

HOSTS

Most of the records of larvae of *Or. mimon* were made on bats, but there also are findings on opossums (Didelphidae), rodents (Cricetidae), and birds[2,4,5] (Table 3.3).

TABLE 3.3 Hosts of *Or. mimon*

MAMMALIA	PRIMATES
CHIROPTERA	Hominidae
Phyllostomidae	Human
Mimon crenulatum	RODENTIA
Vespertilionidae	Cricetidae
Eptesicus brasiliensis	*Nectomys rattus*
E. diminutus	AVES
E. furinalis	PASSERIFORMES
Histiotus macrotus	Dendrocolaptidae
DIDELPHIMORPHIA	*Dendroplex picus*
Didelphidae	*Lepidocolaptes angustirostris*
Didelphis albiventris	*Xiphorhynchus guttatus*
Gracilinanus agilis	Thraupidae
Thylamis macrurus	*Eucometis penicillata*

With few exceptions, most of the records of this species correspond to larvae.

ECOLOGY

Besides the little data on host association and geographical distribution, there is no information on the ecology of *Or. mimon*.

SANITARY IMPORTANCE

The capacity of *Or. mimon* to transmit pathogens has not been investigated to date, but bites of this species on humans cause severe inflammatory reactions.[4]

DIAGNOSIS

Female (Fig. 3.16): body outline oval, total length 4.5 mm (3.9−5.0), breadth 2.5 mm (2.3−2.9). Mammillae large and with similar size through the dorsum. Discs with different sizes. Mammillae present on ventral surface. Capitulum: basis capituli slightly wider than long, cheeks large. Hypostome blunt, dental formula 2/2. Legs: coxae I and II separated, coxae II−IV contiguous; tarsi with dorsal humps absent.

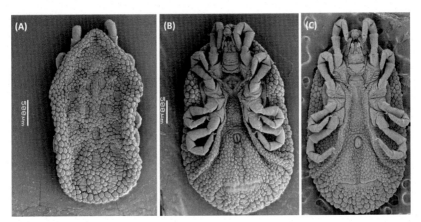

FIGURE 3.16   *Or. mimon*: (A) female dorsal view, (B) female ventral view, (C) male ventral view.

Male (Fig. 3.16): all morphological characteristics are similar to those of the female, with the exception of the shape of genital aperture.

Larva (Fig. 3.17): body outline subcircular, length (including capitulum) 0.740 mm (0.710−0.780), breadth 0.450 mm (0.430−0.460) (unengorged specimens). Dorsal shield pyriform, length 0.170 mm (0.150−0.226), breadth 0.110 mm (0.109−0.188). Dorsum with 14 pairs of setae—seven anterolateral pairs of setae, four posterolateral pairs of setae, and three central pairs of setae. Venter surface with eight pairs of setae—one pair on anal valves, three pairs of sternal setae, three pairs of circumanal setae, and one pair of postcoxal setae; posteromedian seta present. Hypostome blunt, dental formula 4/4 or 3/3 in the anterior third, then 2/2 at base. Capsule of Haller's organ slightly oval and placed laterally.

Taxonomic notes: at least with the phylogenetic analysis performed with mitochondrial 16S rDNA sequences, the phylogenetic position of *Or. mimon* remains unresolved (Fig. 3.13). Barros-Battesti et al.[3] described the adults and redescribed the larva of this species under the name *Carios mimon* but we consider that this species belongs to the genus *Ornithodoros*.

DNA sequences with relevance for tick identification and phylogeny: there are sequences of genes of *Or. mimon* in the Gen Bank as follows: mitochondrial 16S rDNA (accession numbers GU198362, KC677675−KC677680, KR869153, KP739868, KJ739864) and nuclear 18S rDNA (KC769599) and 28S rDNA (KC769627).

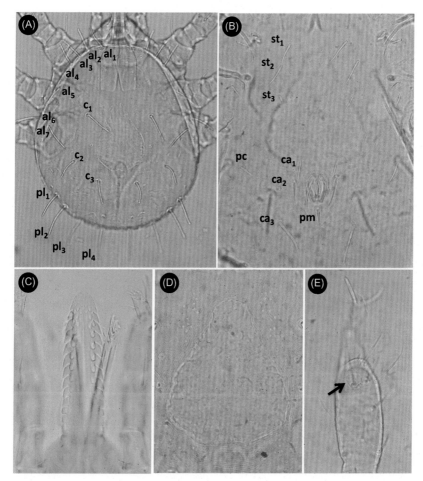

FIGURE 3.17   *Or. mimon* larva: (A) dorsal view, (B) ventral view, (C) hypostome, (D) dorsal shield, (E) tarsus I with capsule of Haller's organ indicated by the *arrow*.

REFERENCES

1. Kohls GM, Clifford CM, Jones EK. The systematics of the subfamily Ornithodorinae (Acarina: Argasidae). IV. Eight new species of *Ornithodoros* from the Western Hemisphere. *Ann Entomol Soc Am* 1969;**62**:1035−43.
2. Venzal JM, Autino AG, Nava S, Guglielmone AA. *Ornithodoros mimon* Kohls, Clifford & Jones, 1969 (Acari: Argasidae) on Argentinean bats, and new records from Uruguay. *Syst Appl Acarol* 2004;**9**:37−9.
3. Barros-Battesti DM, Landulfo GA, Onofrio VC, Faccini JLH, Marcili A, Nieri-Bastos FA, et al. *Carios mimon* (Acari: Argasidae): description of adults and redescription of larva. *Exp Appl Acarol* 2011;**54**:93−104.

4. Labruna MB, Marcili A, Ogrzewalska M, Barros-Battesti DM, Dantas-Torres F, Fernandes AA, et al. New records and human parasitism by *Ornithodoros mimon* (Acari: Argasidae) in Brazil. *J Med Entomol* 2014;**51**:283−7.
5. Ramos DG, Melo AL, Martins TF, Alves Ada S, Pacheco Tdos A, Pinto LB, et al. Rickettsial infection in ticks from wild birds from Cerrado and the Pantanal region of Mato Grosso, midwestern Brazil. *Ticks Tick-borne Dis* 2015;**6**:836−42.

Ornithodoros peruvianus Kohls, Clifford, and Jones, 1969

Kohls G.M., Clifford C.M. & Jones E.K. (1969) The systematics of the subfamily Ornithodorinae (Acarina: Argasidae). IV. Eight new species of *Ornithodoros* from the Western Hemisphere. *Annals of the Entomological Society of America*, 62, 1035−1043.

DISTRIBUTION

Chile and Peru.[1,2]

    Biogeographic distribution in the Southern Cone of America: Atacama (South American Transition Zone) (see Fig. I1.1).

HOSTS

Bats (Chiroptera) are the hosts of *Or. peruvianus*[1,2] (Table 3.4).

ECOLOGY

Besides the little data on host association and geographical distribution, there is no information on the ecology of *Or. peruvianus*.

SANITARY IMPORTANCE

There are no data involving *Or. peruvianus* as a parasite with medical and veterinary importance. Presence of *Coxiella* endosymbionts has been reported in this tick species.[3]

TABLE 3.4 Hosts of *Or. peruvianus* that is Known Only by Its Larval Stage

MAMMALIA	Phyllostomidae
CHIROPTERA	*Desmodus rotundus*
Molossidae	*Glossophaga* sp.
Molossus molossus	

DIAGNOSIS

Larva (Fig. 3.18): body outline oval, length (including capitulum) 1.52−2.20 mm, breadth 1.30−1.68 mm (measures are from only two specimens, one engorged and the other slightly engorged). Dorsal shield pyriform, length 0.330 mm (0.320−0.340), breadth 0.210 mm (0.170−0.250). Dorsum with 14 pairs of setae—seven anterolateral pairs, four posterolateral pairs, and three central pairs. Venter surface with eight pairs of setae—one pair on anal valves, three pairs of sternal setae, three pairs of circumanal setae, and one pair of postcoxal setae; posteromedian seta present. Hypostome pointed, dental formula 3/3

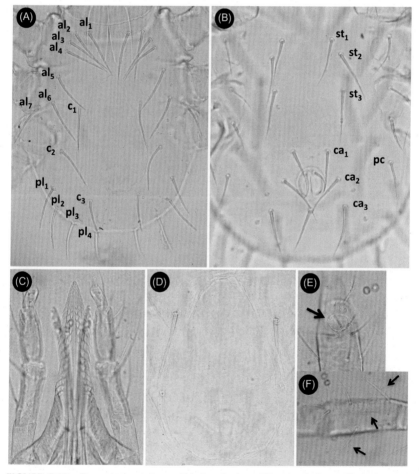

FIGURE 3.18 *Or. peruvianus* larva: (A) dorsal view, (B) ventral view, (C) hypostome, (D) dorsal shield, (E) tarsus I with capsule of Haller's organ indicated by the *arrow*, (F) tarsus I with fringed setae indicated by *arrows*. We acknowledge the contribution of Sebastián Muñoz-Leal for generously providing photos of *Or. peruvianus*.

in anterior part and 2/2 posteriorly. Capsule of Haller's organ subcircular in shape from dorsal view. Some setae of tarsus I fringed.

Taxonomic notes: adults and nymphs of *Or. peruvianus* are unknown.

DNA sequences with relevance for tick identification and phylogeny: there is a sequence of the mitochondrial 16S rDNA gene of *Or. peruvianus* in the Gen Bank (accession number HQ111351).

REFERENCES

1. Kohls GM, Clifford CM, Jones EK. The systematics of the subfamily Ornithodorinae (Acarina: Argasidae). IV. Eight new species of *Ornithodoros* from the Western Hemisphere. *Ann Entomol Soc Am* 1969;**62**:1035—43.
2. Venzal JM, González-Acuña D, Mangold AJ, Guglielmone AA. *Ornithodoros peruvianus* Kohls, Clifford & Jones, 1969 in Chile, a tentative diagnosis. *Neotr Entomol* 2012;**41**:74—8.
3. Duron O, Noël V, McCoy KD, Bonazzi M, Sidi-Boumedine K, Morel O, et al. The recent evolution of a maternally-inherited endosymbiont of ticks led to the emergence of the Q Fever pathogen, *Coxiella burnetii*. *PLoS Pathog* 2015;**11**(5) (article e1004892) 23 p.

Ornithodoros quilinensis Venzal, Nava, and Mangold, 2012

Venzal J.M., Nava S., Mangold A.J., Mastropaolo M., Casás G., Guglielmone A.A. (2012) *Ornithodoros quilinensis* sp. nov. (Acari, Argasidae), a new tick species from the Chacoan region in Argentina. *Acta Parasitologica*, 57, 329—336.

DISTRIBUTION

Argentina.[1]

Biogeographic distribution in the Southern Cone of America: Chaco (Neotropical Region) (see Fig. I1.1).

HOSTS

Or. quilinensis is known only from the rodent *Graomys centralis* (type host) from which the larvae to describe the species were collected.[1]

ECOLOGY

Besides the little data on host association and geographical distribution, there is no information on the ecology of *Or. quilinensis*.

SANITARY IMPORTANCE

The capacity of *Or. quilinensis* to transmit pathogens has not been investigated to date. However it is important to remark that DNA of a novel tick-borne relapsing fever *Borrelia* was detected in *Ornithodoros* sp. from Bolivia,[2] a taxon closely related to *Or. quilinensis*.

DIAGNOSIS

Larva (Fig. 3.19): body outline oval, length (including capitulum) 0.525 mm (0.503−0.579), breadth 0.358 mm (0.247−0.380). Dorsal shield oval, length 0.200 mm (0.172−0.221), breadth 0.152 mm (0.141−0.168). Dorsum with

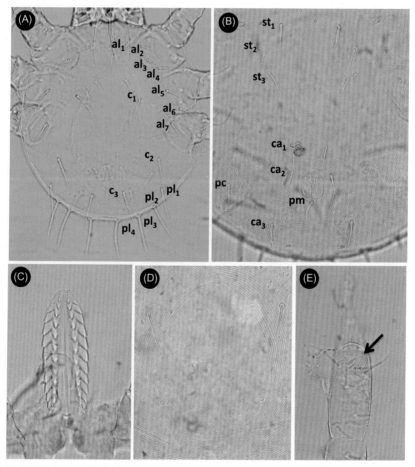

FIGURE 3.19   *Or. quilinensis* larva: (A) dorsal view, (B) ventral view, (C) hypostome, (D) dorsal shield, (E) tarsus I with capsule of Haller's organ indicated by the *arrow*.

14 pairs of setae—seven anterolateral pairs of setae, four posterolateral pairs of setae, and three central pairs of setae. Venter surface with eight pairs of setae—one pair of setae on anal valves, three pairs of sternal setae, three pairs of circumanal setae, and one pair of postcoxal setae; posteromedian seta present. Hypostome blunt, dental formula 2/2 throughout entire length. Capsule of Haller's organ irregular in shape.

Taxonomic notes: adults and nymphs of *Or. quilinensis* are unknown. *Or. quilinensis* and *Or. xerophylus* Venzal, Nava, and Mangold, 2015, are two sympatric species morphologically and phylogenetically closely related as stated by Venzal et al.[3] and shown in Fig. 3.13.

DNA sequences with relevance for tick identification and phylogeny: there are sequences of the mitochondrial 16S rDNA gene of *Or. quilinensis* in the Gen Bank (accession numbers JN255574 and JN255575).

REFERENCES

1. Venzal JM, Nava S, Mangold AJ, Mastropaolo M, Casás G, Guglielmone AA. *Ornithodoros quilinensis* sp. nov. (Acari, Argasidae), a new tick species from the Chacoan region in Argentina. *Acta Parasitol* 2012;**57**:329−36.
2. Parola P, Ryelandt J, Mangold AJ, Mediannikov O, Guglielmone AA, Raoult D. Relapsing fever *Borrelia* in *Ornithodoros* ticks from Bolivia. *Ann Trop Med Parasitol* 2011;**105**:407−11.
3. Venzal JM, González-Acuña D, Muñoz-Leal S, Mangold AJ, Nava S. Two new species of *Ornithodoros* (Ixodida; Argasidae) from the Southern Cone of South America. *Exp Appl Acarol* 2015;**66**:127−39.

Ornithodoros rioplatensis Venzal, Estrada-Peña, and Mangold, 2008

Venzal J.M., Estrada-Peña A., Mangold A.J., González-Acuña D. & Guglielmone A.A. (2008) The *Ornithodoros* (*Alectorobius*) *talaje* species group (Acari: Ixodida: Argasidae): description of *Ornithodoros* (*Alectorobius*) *rioplatensis* n. sp. from southern South America. *Journal of Medical Entomology*, 45, 832−840.

DISTRIBUTION

Argentina, Chile, Paraguay, and Uruguay.[1]

Biogeographic distribution in the Southern Cone of America: Chaco, Pampa (Neotropical Region), and Santiago (Andean Region) (see Fig. I1.1).

HOSTS

Mammals and reptiles are the hosts for *Or. rioplatensis*.[1] The complete list of hosts of *Or. rioplatensis* is given in Table 3.5.

TABLE 3.5 Hosts of *Or. rioplatensis*

MAMMALIA	REPTILIA
CARNIVORA	**SQUAMATA**
Mephitidae	**Liolaemidae**
Conepatus chinga	*Liolaemus jamesi*
RODENTIA	*Phymaturus palluma*
Caviidae	**Phyllodactylidae**
Cavia sp.	*Homonota uruguayensis*
Cricetidae	
Oryzomys sp.	
Reithrodon typicus	

The larva is the only parasitic stage found on hosts.

ECOLOGY

Besides the little data on host association and geographical distribution, there is no information on the ecology of *Or. rioplatensis*.

SANITARY IMPORTANCE

The capacity of *Or. rioplatensis* to transmit pathogens has not been investigated to date, but it is known that larvae of *Or. rioplatensis* (mentioned as *Or. aff. puertoricensis*) have the capacity to produce clinical symptoms compatible with a toxicosis in laboratory mice, including some deaths.[2]

DIAGNOSIS

Female (Fig. 3.20): body outline subrectangular, total length 5.2 mm (3.0−7.5), breadth 4.0 mm (3.4−4.7). Lateral margins subparallel, posterior margin broadly rounded, pointed anteriorly. Mammillae with similar size through the dorsum, abundant in the margins of the body, few mammillae in the center. Discs numerous. Capitulum: basis capituli slightly wider than long, cheeks large. Hypostome blunt, dental formula 2/2. Legs: coxae I and II separated, coxae II−IV contiguous; dorsal humps on tarsus absent.

Male (Fig. 3.20): all morphological characteristics are similar to those of the female, with the exception of the shape of genital aperture.

Larva (Fig. 3.21): body outline subcircular, length (including capitulum) 0.760 mm (0.740−0.780), breadth 0.408 mm (0.380−0.450) (unengorged specimens). Dorsal shield pyriform, length 0.262 mm (0.242−0.275), breadth

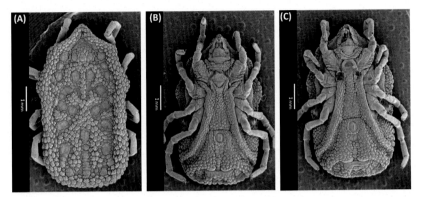

FIGURE 3.20 *Or. rioplatensis*: (A) female dorsal view, (B) female ventral view, (C) male ventral view.

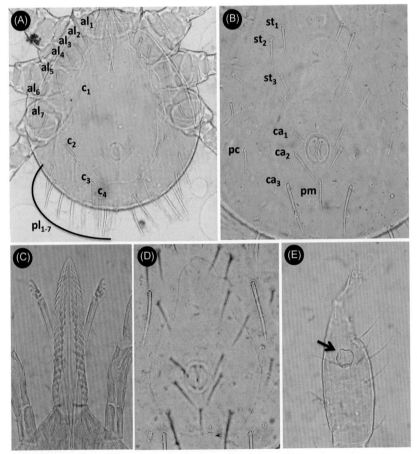

FIGURE 3.21 *Or. rioplatensis* larva: (A) dorsal view, (B) ventral view, (C) hypostome, (D) dorsal shield, (E) tarsus I with capsule of Haller's organ indicated by the *arrow*.

0.179 mm (0.167−0.187). Dorsum with 20 pairs of setae—seven anterolateral pairs, nine posterolateral pairs, and four central pairs. Venter surface with eight pairs of setae—one pair on anal valves, three pairs of sternal setae, three pairs of circumanal setae, and one pair of postcoxal setae; posteromedian seta present. Hypostome pointed, dental formula 3/3 in the anterior half, then 2/2 at base. Capsule of Haller's organ irregular in shape.

Taxonomic notes: *Or. talaje* (Guérin-Méneville, 1849) and/or *Or. puertoricensis* Fox, 1947, have been considered established in the Southern Cone of America[3−6] but never confirmed. Most of those records might be in fact records of *Or. rioplatensis*[1] (that originally was called *Or. aff. puertoricensis*[2]) or related species. See also Taxonomic notes of *Or. hasei* (Schulze, 1935).

DNA sequences with relevance for tick identification and phylogeny: there is just one sequence of 16S rRNA gene of *Or. rioplatensis* in the Gen Bank (accession number EU283343).

REFERENCES

1. Venzal JM, Estrada-Peña A, Mangold AJ, González-Acuña D, Guglielmone AA. The *Ornithodoros* (*Alectorobius*) *talaje* species group (Acari: Ixodida: Argasidae): description of *Ornithodoros* (*Alectorobius*) *rioplatensis* n. sp. from southern South America. *J Med Entomol* 2008;**45**:832−40.

2. Venzal JM, Estrada-Peña A, Fernández de Luco D. Effects produced by the feeding of larvae of *Ornithodoros* aff. *puertoricensis* Fox, 1947 (Acari: Argasidae) on laboratory mice. *Exp Appl Acarol* 2007;**42**:217−23.

3. Porter CE. Un caso de otocariasis en el norte de Chile. *An Zool Apl* 1917;**4**:30.

4. Cordero EH, Vogelsang EG, Cossio V. *Ornithodoros talaje* (Guérin-Méneville) y su presencia en el Paraguay y en el Uruguay. *Physis* 1928;**9**:125−7.

5. Capriles JM, Gaud SM. The ticks of Puerto Rico (Arachnida: Acarina). *J. Agric. Univ. Puerto Rico* 1977;**61**:402−4.

6. Nava S, Lareschi M, Rebollo C, Benítez Usher C, Beati L, Robbins RG, et al. The ticks (Acari: Ixodida: Argasidae, Ixodidae) of Paraguay. *Ann Trop Med Parasitol* 2007;**101**:255−70.

Ornithodoros rostratus Aragão, 1911

Aragão H. de B. (1911) Notas sobre ixódidas brazileiros. *Memórias do Instituto Oswaldo Cruz*, 3, 145−195.

DISTRIBUTION

Argentina, Bolivia, Brazil, and Paraguay.[1]

Biogeographic distribution in the Southern Cone of America: Chaco (Neotropical Region) (see Fig. I1.1).

HOSTS

Domestic animals (pigs, goats, and dogs), humans, and eventually wild animals are the target of this species; nevertheless most of these hosts are inferred because this tick is usually found as adults in the environment (see below).

ECOLOGY

Or. rostratus it is usually found under the soil or sand in holes, burrows, and roosting on wild and domestic animals, and also in human dwellings of rural and periurban areas. This tick species is associated to arid and semiarid environments.

SANITARY IMPORTANCE

Or. rostratus is very aggressive to humans. Their bites are very painful and itchy and scratching can lead to secondary infections,[2] and Hoogstraal[3] speculated that *Or. rostratus* may play a role in the maintenance of spotted fever rickettsiae in nature but this fact remains to be confirmed. It has been found that *Or. rostratus* has a *Coxiella* symbiont present in 100% of tick analyzed.[4]

DIAGNOSIS

Female (Fig. 3.22): body outline oval, total length 8.4 mm (4.9−9.3), breadth 3.4 mm (3.0−4.2). Mammillae numerous and regular in size; long, fine, and numerous hairs implanted on mammillae. Circular and prominent peritremes located laterally between the coxae III−IV. Cheeks absent. Hypostome blunt, dental formula 2/2. Legs: coxae contiguous, progressively decreasing in size from coxa I to coxa IV, tarsus I−III with two dorsal humps (proximal and distal), tarsus IV with one dorsal hump (proximal hump absent).

Male (Fig. 3.22): all morphological characteristics are similar to those of the female, with the exception of the shape of genital aperture.

Larva (Fig. 3.23): body outline oval, length (including capitulum) 0.865 mm (0.810−0.920), breadth 0.557 mm (0.510−0.605) (unengorged specimens). Dorsal shield broadly quadrangular, slightly longer than wide, concave anteriorly, length 0.168 mm (0.151−0.185), breadth 0.136 mm (0.127−0.146). Dorsum with 13 pairs of setae—seven anterolateral pairs, three posterolateral pairs, and three central pairs. Venter surface with eight pairs of setae—one pair on anal valves, three pairs of sternal setae, three pairs of circumanal setae, and one pair of postcoxal setae; posteromedian seta present. Ventral grooves presents. Hypostome notched, dental formula 2/2, with crenulations. Area adjacent to capsule of Haller's organ spinose.

Taxonomic notes: *Or. rostratus* is morphologically and phylogenetically related to *Or. brasiliensis* Aragão, 1923, as stated by Barros-Battesti et al.[5]

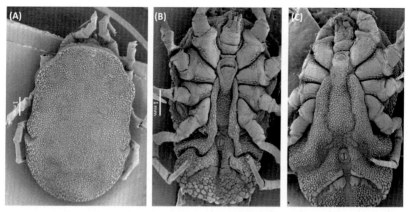

FIGURE 3.22   *Or. rostratus*: (A) female dorsal view, (B) female ventral view, (C) male ventral view.

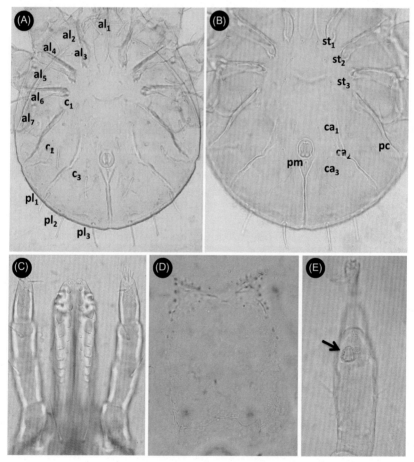

FIGURE 3.23   *Or. rostratus* larva: (A) dorsal view, (B) ventral view, (C) hypostome, (D) dorsal shield, (E) tarsus I with spinose area adjacent to capsule of Haller's organ indicated by the *arrow*.

and this phylogenetic relationship is also shown in Fig. 3.13. According to Aragão[6] *Or. rostratus* has been confused in Argentina with *Or. turicata* (Dugès, 1876) by Barbará and Dios[7] and Dios and Knopoff.[8]

DNA sequences with relevance for tick identification and phylogeny: there are sequences of *Or. rostratus* genes in the Gen Bank as follows: mitochondrial 16S rDNA (accession numbers JN887882, DQ295780) and nuclear 18S rDNA (KC769605) and 28S rDNA (KC769628).

REFERENCES

1. Guglielmone AA, Estrada-Peña A, Keirans JE, Robbins RG. Ticks (Acari: Ixodida) of the neotropical zoogeographic region. Special publication of the integrated consortium on ticks and tick-borne diseases. Houten (The Netherlands): Atalanta, 2003.

2. Boero JJ. *Las garrapatas de la República Argentina (Acarina: Ixodoidea).* Buenos Aires: Universidad de Buenos Aires; 1957.

3. Hoogstraal H. Argasid and Nuttalliellid ticks as parasites and vectors. *Adv Parasitol* 1985;**24**:135−238.

4. Almeida AP, Marcili A, Leite RC, Nieri-Bastos FA, Domingues LN, Martins JR, et al. *Coxiella* symbiont in the tick *Ornithodoros rostratus* (Acari: Argasidae). *Ticks Tick-borne Dis* 2012;**3**:203−6.

5. Barros-Battesti DM, Onofrio VC, Nieri-Bastos FA, Soares JF, Marcili A, Famadas KM, et al. *Ornithodoros brasiliensis* Aragão (Acari: Argasidae): description of the larva, redescription of male and female, and neotype designation. *Zootaxa* 2012;**3178**:22−32.

6. Aragão HB. Observações sobre os ixodideos da República Argentina. *Mem Inst Oswaldo Cruz* 1935;**30**:519−33.

7. Barbará B, Dios RL. Contribución al estudio de la sistemática y la biología de los Ixodidae de la República Argentina y de algunos países vecinos. *Rev Inst Bacteriol. Dep Nac Hig* 1918;**1**:285−322.

8. Dios RL, Knopoff R. Sobre Ixodoidea de la República Argentina. *Rev Inst Bacteriol Dep Nac Hyge* 1934;**6**:359−412.

Ornithodoros rudis Karsch, 1880

Karsch F. (1880) Vier neue Ixodidae des Berliner Museums. *Mittheilungen des Münchener Entomologischen Vereins*, 4, 141−142.

DISTRIBUTION

Brazil, Colombia, Ecuador, Panama, Paraguay, Peru, and Venezuela.[1]

Biogeographic distribution in the Southern Cone of America: Chaco (Neotropical Region) (see Fig. I1.1).

HOSTS

Or. rudis is usually found in the environments associated to humans and wild and domestic animals; nevertheless most of these hosts are inferred because this tick is usually found as nonparasitic adults.

ECOLOGY

Besides the little data on host association and geographical distribution, there is no information on the ecology of *Or. rudis*.

SANITARY IMPORTANCE

Or. rudis is aggressive to humans and domestic animals. This species has been taken from beds in houses of Rocky Mountain spotted fever patients in Colombia, and Hoogstraal[2] speculated that *Or. rudis* may play a role in the maintenance of spotted fever rickettsiae in nature but this fact remains to be confirmed. *Borrelia venezuelensis* infects *Or. rudis* and causes mild to severe human relapsing fever in Central America and northern South America.[2]

DIAGNOSIS

Female (see Taxonomic notes): body outline oval, total length 4.7 mm (3.9−5.5), breadth 2.5 mm (2.0−3.0), pointed anteriorly. Mammillae moderate in number. Discs superficial and faintly differentiated from the mammillae, arranged in asymmetrical pattern on the dorsum. Capitulum: basis capitulum slightly wider than long, cheeks reniform with the free edges irregular. Hypostome notched, dental formula 2/2. Legs: coxae I and II separated, coxae II−IV contiguous; tarsus I with subapical dorsal protuberance, small on tarsus I, very small or absent on tarsus IV; dorsal humps absent.

Male (see Taxonomic notes): all morphological characteristics are similar to those of the female, with the exception of the shape of genital aperture.

Larva (see Taxonomic notes): body outline oval, length (including capitulum) 0.710 mm (0.680−0.740), breadth 0.490 mm (0.455−0.525) (unengorged specimens). Dorsal shield small, wider than long, length 0.073 mm (0.068−0.078), breadth 0.131 mm (0.125−0.138). Dorsum with 16−21 pairs of setae—seven anterolateral pairs, five to seven posterolateral pairs, and four to seven central pairs (typically five). Venter surface with seven pairs of setae—one pair on anal valves, three pairs of sternal setae, three pairs of circumanal setae; posteromedian setae present, postcoxal setae absent. Hypostome blunt, dental formula 3/3 in the anterior portion, then 2/2 at base.

Taxonomic notes: *Or. rudis* was cited frequently in tick literature from about 1900 to 1975 to decline abruptly thereafter. The identification of this species has been controversial, and Osorno Mesa[3] in 1940 stated that it was confused in Colombia with *O. turicata* (Dugès, 1876) and *Or. talaje* (Guérin-Méneville, 1849) and this situation is probably not exclusive of Colombia. It was not possible to obtain specimens of bona fide *Or. rudis* for this study. The diagnosis of this species presented here is based on the descriptions and drawings of Cooley and Kohls[4] and Kohls et al.,[5] and readers should consult these articles to get access to figures of adults and larva of *Or. rudis*.

DNA sequences with relevance for tick identification and phylogeny: there are no DNA sequences of *Or. rudis* in the Gen Bank.

REFERENCES

1. Guglielmone AA, Estrada-Peña A, Keirans JE, Robbins RG. Ticks (Acari: Ixodida) of the neotropical zoogeographic region. Special publication of the integrated consortium on ticks and tick-borne diseases. Houten (The Netherlands): Atalanta, 2003.
2. Hoogstraal H. Argasid and Nuttalliellid ticks as parasites and vectors. *Adv Parasitol* 1985;24:135−238.
3. Osorno Mesa E. Las garrapatas de la República de Colombia. *Rev Acad Colomb Cienc Exact Físic Nat* 1940;4:6−24.
4. Cooley RA, Kohls GM. The Argasidae of North America, Central America and Cuba. *Am Midl Natur Monogr* 1944;1:152.
5. Kohls GM, Sonenshine DE, Clifford CM. The systematics of the subfamily Ornithodorinae (Acarina: Argasidae). II. Identification of the larvae of the Western Hemisphere and description of three new species. *Ann Entomol Soc Am* 1965;58:331−64.

Ornithodoros spheniscus Hoogstraal, Wassef, Hays, and Keirans, 1985

Hoogstraal H., Wassef H.Y., Hays A. & Keirans J.E. (1985) *Ornithodoros* (*Alectorobius*) *spheniscus* n. sp. [Acarina: Ixodoidea: Argasidae: *Ornithodoros* (*Alectorobius*) *capensis* group], a tick parasite of the Humboldt penguin in Peru. *Journal of Parasitology*, 71, 635−644.

DISTRIBUTION

Chile and Peru.[1,2]

Biogeographic distribution in the Southern Cone of America: Atacama (South American transition zone) (see Fig. I1.1).

HOSTS

The Humboldt Penguin, *Spheniscus humboldti* (Aves: Spheniscidae) is the only recorded host for *Or. spheniscus*.[2,3]

ECOLOGY

Besides the little data on host association and geographical distribution, there is no information on the ecology of *Or. spheniscus*.

SANITARY IMPORTANCE

Or. spheniscus is aggressive to humans. Bites of these ticks on humans have been often accompanied by a piercing and burning sensation.[3] Hoogstraal et al.[3] stated that infestation of *Or. spheniscus* may be a cause to abandon their nests by the Humboldt Penguin. Presence of *Coxiella* endosymbionts has been reported in this tick species.[4]

DIAGNOSIS

Female (Fig. 3.24): body outline oval, total length 5.8 mm (5.3−6.8), breadth 3.1 mm (2.6−4.0). Mammillae numerous, closely spaced, outlines irregular. Discs conspicuous, size and shape variable; Mammillae present on ventral surface. Capitulum: basis capituli about 1.6 times as broad as long, cheeks thin. Hypostome indented apically, dental formula 2/2. Legs: coxae I and II separated, coxae II−IV contiguous; tarsus I mildly elevated dorsally, tarsus II−IV without humps and protuberances.

Male (Fig. 3.24): all morphological characteristics are similar to those of the female, with the exception of the shape of genital aperture.

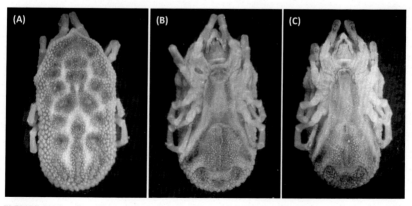

FIGURE 3.24   *Or. spheniscus*: (A) female dorsal view, (B) female ventral view, (C) male ventral view.

Larva (Fig. 3.25): body outline subcircular, length (excluding capitulum) 0.453 mm (0.420−0.490), breadth 0.410 mm (0.374−0.439) (unengorged specimens). Dorsal shield pyriform, length 0.215 mm (0.196−0.234), breadth 0.188 mm (0.168−0.196). Dorsum with 15−16 pairs of setae (typically 15)— seven anterolateral pairs, four to five posterolateral pairs (typically four), and four central pairs. Venter surface with eight pairs of setae—one pair on anal valves, three pairs of sternal setae, three pairs of circumanal setae, and one pair of postcoxal setae; posteromedian setae absent. Hypostome pointed, dental formula 4/4 in the first quarter, 3/3 in midportion, then 2/2 at base. Capsule of Haller's organ subcircular in shape from dorsal view.

Taxonomic notes: *Or. spheniscus* is morphologically close to *Or. amblus* Chamberlin, 1920, and *Or. yunkeri* Keirans, Clifford, and Hoogstraal, 1984, and

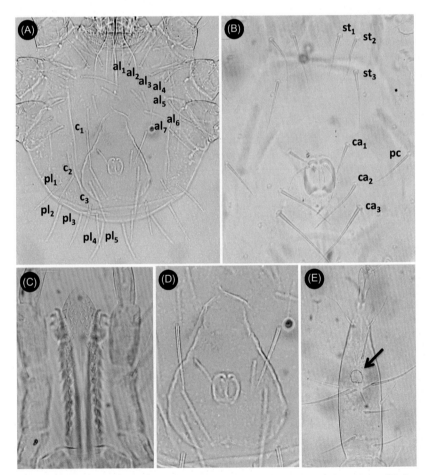

FIGURE 3.25 *Or. spheniscus* larva: (A) dorsal view, (B) ventral view, (C) hypostome, (D) dorsal shield, (E) tarsus I with capsule of Haller's organ indicated by the *arrow*. We acknowledge the contribution of Sebastián Muñoz-Leal for generously providing photos of *Or. spheniscus*.

Hoogstraal et al.[3] provided characters to differentiate these species. Duron et al.[4] stated that *Or. capensis* has been found in the island of Pan de Azúcar in Chile. However, the ticks mentioned as *Or. capensis* were obtained by one of the author of this book (Daniel González-Acuña) and they are in fact *Or. spheniscus*.

DNA sequences with relevance for tick identification and phylogeny: there are no DNA sequences of *Or. spheniscus* in the Gen Bank.

REFERENCES

1. Guglielmone AA, Estrada-Peña A, Keirans JE, Robbins RG. Ticks (Acari: Ixodida) of the neotropical zoogeographic region. Special publication of the integrated consortium on ticks and tick-borne diseases. Houten (The Netherlands): Atalanta, 2003.
2. González-Acuña D, Moreno L, Guglielmone AA. First report of *Ornithodoros spheniscus* (Acari: Ixodoidea: Argasidae) from the Humboldt Penguin in Chile. *Syst Appl Acarol* 2008;**13**:120−2.
3. Hoogstraal H, Wassef HY, Hays A, Keirans JE. *Ornithodoros* (*Alectorobius*) *spheniscus* n. sp. [Acarina: Ixodoidea: Argasidae: *Ornithodoros* (*Alectorobius*) *capensis* group], a tick parasite of the Humboldt penguin in Peru. *J Parasitol* 1985;**71**:635−44.
4. Duron O, Noël V, McCoy KD, Bonazzi M, Sidi-Boumedine K, et al. The recent evolution of a maternally-inherited endosymbiont of ticks led to the emergence of the Q fever pathogen, *Coxiella burnetii. Plos Pathog* 2015;**11**(5) (article e1004892) 23 p.

Ornithodoros xerophylus Venzal, Mangold, and Nava, 2015

Venzal J.M., González-Acuña D., Muñoz-Leal S., Mangold A.J. & Nava S. 2015. Two new species of *Ornithodoros* (Ixodida; Argasidae) from the Southern Cone of South America. *Experimental and Applied Acarology*, 66, 127−139.

DISTRIBUTION

Argentina.[1]

Biogeographic distribution in the Southern Cone of America: Chaco (Neotropical Region) (see Fig. I1.1).

HOSTS

Or. xerophylus is known only from the rodent *Graomys centralis* (type host) from which the larvae used to describe this species were collected.[1]

ECOLOGY

Besides the little data on host association and geographical distribution, there is no information on the ecology of *Or. xerophylus*.

SANITARY IMPORTANCE

The capacity of *Or. xerophylus* to transmit pathogens has not been investigated to date. However, it is important to remark that DNA of a novel tick-borne relapsing fever *Borrelia* was detected in *Ornithodoros* sp. from Bolivia,[2] a taxon closely related to *Or. xerophylus*.

DIAGNOSIS

Larva (Fig. 3.26): body outline subcircular, length (including capitulum) 0.906 mm (0.813−0.970), breadth 0.637 mm (0.689−0.720). Dorsal shield oval, length 0.243 mm (0.232−0.260), breadth 0.207 mm (0.185−0.225).

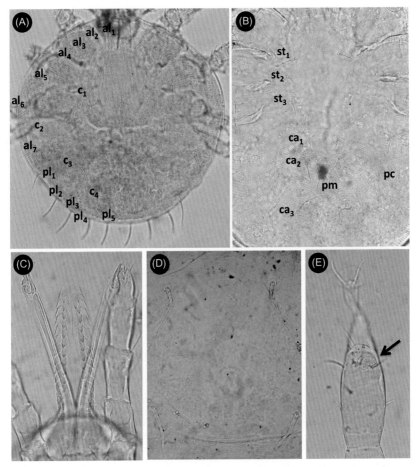

FIGURE 3.26   *Or. xerophylus* larva: (A) dorsal view, (B) ventral view, (C) hypostome, (D) dorsal shield, (E) tarsus I with capsule of Haller's organ indicated by the *arrow*.

Dorsum with 16 pairs of setae—seven anterolateral pairs, five posterolateral pairs, and four central pairs. Venter surface with eight pairs of setae—one pair on anal valves, three pairs of sternal setae, three pairs of circumanal setae, and one pair of postcoxal setae; posteromedian setae present. Hypostome blunt, dental formula 3/3 in apex, then 2/2 at base. Capsule of Haller's organ oval in shape.

Taxonomic notes: see Taxonomic notes of *Or. quilinensis* Venzal, Nava, and Mangold, 2012.

DNA sequences with relevance for tick identification and phylogeny: there is just one sequence of the 16S rDNA mitochondrial gene of *Or. xerophylus* in the Gen Bank (accession number KP403287).

REFERENCES

1. Venzal JM, González-Acuña D, Muñoz-Leal S, Mangold AJ, Nava S. Two new species of *Ornithodoros* (Ixodida; Argasidae) from the Southern Cone of South America. *Experimental and Applied Acarology* 2015;**66**:127–39.
2. Parola P, Ryelandt J, Mangold AJ, Mediannikov O, Guglielmone AA, Raoult D. Relapsing fever *Borrelia* in *Ornithodoros* ticks from Bolivia. *Ann Trop Med Parasitol* 2011;**105**:407–11.

Genus *Otobius*

There are only two species of *Otobius* worldwide which represents less than 1% of the species of Argasidae and one of them is established in the Southern Cone of America as well as in many countries throughout the world and is of considerable medical and veterinary importance. The life cycle is characterized by nonparasitic adults that lay eggs autogenically.

Otobius megnini (Dugès, 1883)

Argas megnini. Dugès, A. (1883). Turicata y garrapata de Guanajuato. *Naturaleza*, 6, 195–198.

DISTRIBUTION

Argentina, Bolivia, Chile, Guatemala, Mexico, Peru, and Venezuela.[1,2] It is uncertain whether *Ot. megnini* is established in Brazil.[1] *Ot. megnini* is also present in Europe, North America, Africa, Asia, and Australia.[1–4]

Biogeographic distribution in the Southern Cone of America: Chaco (Neotropical Region), Atacama, Monte, Prepuna, Puna (South American transition zone), and Santiago (Andean Region) (see Fig. I1.1).

HOSTS

The complete list of hosts of *Ot. megnini* is given in Table 3.6.

ECOLOGY

Ot. megnini, the spinose ear tick, is an argasid thought to be of Nearctic origin that has been able to colonize other parts of the world.[5] *Ot. megnini* ticks are usually found deep in the external ear canal of their hosts. This tick species has a single-host life cycle where the parasitic stages are larva and nymph (there is no agreement among authors about the number of nymphal stages), while male and female are nonparasitic stages (females are autogenic).[6,7] *Ot. megnini* is generally treated as a tick species adapted to arid and semiarid environments; nevertheless it has also been reported frequently in temperate and humid regions,[2,7,8] indicating that additional studies are needed to properly identify the abiotic factors governing its cycle.

SANITARY IMPORTANCE

Ot. megnini is a species with sanitary importance because it is a parasite of cattle, sheep, goats, and horses, and reports of human infestation are also

TABLE 3.6 Hosts of *Ot. megnini*, a One-Host Tick and Parasitic Stages Omitted

MAMMALIA	PERISSODACTYLA
ARTIODACTYLA	Equidae
Bovidae	Donkey
Cattle	Horse
Goat	Mule
Sheep	PRIMATES
Camelidae	Hominidae
Lama glama	Human
CARNIVORA	
Canidae	
Domestic dog	

frequent.[3] *Ot. megnini* may play a role in the maintenance of the agent of Q fever in nature.[9] Besides deleterious effect on domestic animals (otitis, decrease of milk production in cows, and neurological problems in horses), *Ot. megnini* can cause otitis in humans and paralysis in a child that may have been caused by this tick species has also been reported.[9–15]

DIAGNOSIS

Female (Fig. 3.27): body with a constriction behind level of coxa IV (panduriform), total length 6.2 mm (4.0−8.5), breadth 4.5 mm (3.0−6.0). Mammillae absent, integument of dorsal and ventral surface granulated with numerous circular depressions, each depression with a central tubercle; discs small. Capitulum: basis capituli broad and short; cheeks absent. Hypostome vestigial without denticles and with the apical margin curved and thin. Legs: coxae I and II separated, coxae II−IV contiguous; tarsus II−IV with a subapical dorsal protuberance, negligible on tarsus I.

Male: all morphological characteristics are similar to those of the female, with the exception of the shape of genital aperture.

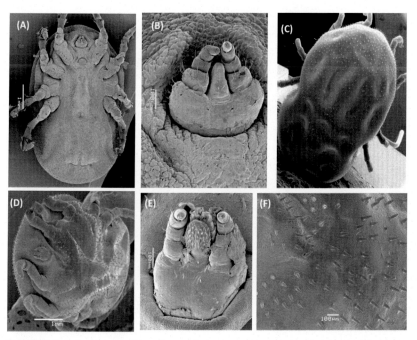

FIGURE 3.27   *Ot. megnini*: (A) male ventral view, (B) male capitulum and detail of the integument, (C) nymph dorsal view, (D) nymph ventral view, (E) nymph hypostome, (F) nymph detail of the integument.

Nymph (Fig. 3.27): general characteristic similar to those of adult ticks, but the integument presents numerous spines and the hypostome is functional with large and numerous denticles.

Larva (Fig. 3.28): body outline oval, length (including capitulum) 0.696 mm (0.675−0.717) mm, breadth 0.365 mm (0.345−0.385). Two pairs of eyes present, the larger anterior pair located dorsally to coxa I, the smaller posterior pair located dorsally to coxa III. Dorsal shield elongated, widest anteriorly, length 0.250 mm (0.239−0.260), breadth 0.150 mm (0.140−0.161). Dorsum with 9 or 10 pairs of setae—four anterolateral pairs of setae, three or four posterolateral pairs of setae, and three central pairs of setae. Venter surface with five or six pairs of setae—one pair on anal valves, three pairs of sternal setae, one pair of circumanal setae; postcoxal pair of

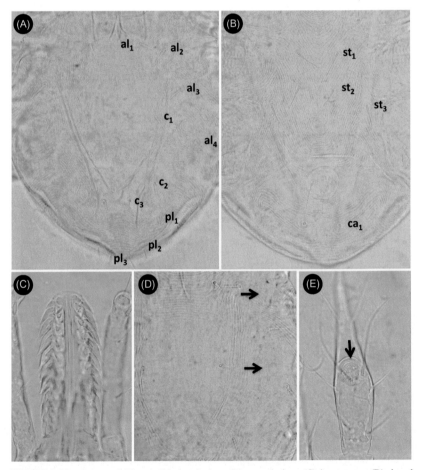

FIGURE 3.28 *Ot. megnini* larva: (A) dorsal view, (B) ventral view, (C) hypostome, (D) dorsal shield and eyes indicated by *arrows*, (E) tarsus I with capsule of Haller's organ indicated by the *arrow*.

setae and posteromedian seta absent. Hypostome blunt, dental formula 2/2. Capsule of Haller's organ irregular in shape.

DNA sequences with relevance for tick identification and phylogeny: there are DNA sequences of genes of *Ot. megnini* in the Gen Bank as follows: mitochondrial 16S rDNA (accession numbers L34325, EF120989, DQ159447) and 12S rDNA (U95913), nuclear 18S rDNA (KC769607, L76356) and 28S rDNA (KC769630, AF120297).

REFERENCES

1. Guglielmone AA, Estrada-Peña A, Keirans JE, Robbins RG. Ticks (Acari: Ixodida) of the neotropical zoogeographic region. Special publication of the integrated consortium on ticks and tick-borne diseases. Houten (The Netherlands): Atalanta, 2003.
2. Estrada-Peña A, Nava S, Horak IG, Guglielmone AA. Using ground-derived data to assess the environmental niche of the spinose ear tick. *Otobius megnini*. *Entomol Exp Appl* 2010;**137**:132−42.
3. Keirans JE, Pound JM. An annotated bibliography of the spinose ear tick, *Otobius megnini* (Dugès, 1883) (Acari: Ixodida: Argasidae) 1883−2000. *Syst Appl Acarol Spec Publ* 2003;**13**:1−68.
4. Bursali A, Keskin A, Tekin S. A Review of the ticks (Acari: Ixodida) of Turkey: species diversity, hosts and geographical distribution. *Exp Appl Acarol* 2012;**57**:91−104.
5. Keirans JE. Systematics of the Ixodida (Argasidae, Ixodidae, Nuttalliellidae): an overview and some problems. In: Fivaz B, Petney T, Horak I, editors. *Tick vector biology. Medical and veterinary aspects*. Berlin: Springer-Verlag; 1992. p. 1−21.
6. Jagannath MS, Lokesh YV. Lifecycle of *Otobius megnini* (Acari: Argasidae). In: Channabasavanna GP, Viraktamath CA, editors. *Progress in Acarology*. New Delhi: IBH Publishing Co; 1989. p. 91−4.
7. Nava S, Mangold AJ, Guglielmone AA. Field and laboratory studies in a Neotropical population of the spinose ear tick, *Otobius megnini*. *Med Vet Entomol* 2009;**23**:1−5.
8. Boero JJ. *Las garrapatas de la República Argentina (Acarina: Ixodoidea)*. Buenos Aires: Universidad de Buenos Aires; 1957.
9. Barbará B, Dios RL. Contribución al estudio de la sistemática y la biología de los Ixodidae de la República Argentina y de algunos países vecinos. *Rev Inst Bacteriol Dep Nac Hig* 1918;**1**:285−322.
10. Descazeaux MJ. Sur la présence au Chili de l'*Ornithodoros megnini*. *Bull Soc Path Exot* 1925;**18**:408−9.
11. Jellison WL, Bell EJ, Huebner RJ, Parker RR, Welsh HH. Q fever studies in southern California. IV. Occurrence of *Coxiella burnetti* in the spinose ear tick, *Otobius megnini*. *Publ Health Rep* 1948;**63**:1483−9.
12. Mucherl LM. Investigación de la enfermedad llamada meningo-encefalitis de la zona de Calama e interior fronterizo, identificada como otoacariasis a *Ornithodoros megnini* (Dugès, 1883). *Bol Inf Parasit Chil* 1952;**7**:8−9.
13. Peacock PB. Tick paralysis or poliomyelitis. *S Afr Med J* 1958;**32**:201−2.
14. Bulman GM, Walker JB. A previously unrecorded feeding site on cattle for the immature stages of the spinose ear tick, *Otobius megnini* (Dugès, 1844). *J S Afr Vet Assoc* 1979;**50**:107−8.
15. Madigan JE, Valberg SJ, Ragle C, Moody JL. Muscle spasms associated with ear tick (*Otobius megnini*) infestations in five horses. *J Am Vet Med Assoc* 1995;**207**:74−6.

Chapter 4

Morphological Keys for Genera and Species of Ixodidae and Argasidae

In the following pages several morphological keys for Ixodidae and Argasidae are presented. As already stated in the introduction, keys for genera of Ixodidae are for males, females, and nymphs but not for larvae because there is lack of information for this parasitic stage of several species, while larvae are of great diagnostic importance for Argasidae of the Southern Cone of America, especially for taxa of the genus *Ornithodoros*. Therefore, keys for Argasidae include keys for adult ticks (morphological differences of males and females are irrelevant) and larvae but not for nymphs because of their relative taxonomic value.

Morphological keys may aid to species identification but commonly other tools are needed for confirmation of closely related species. Nevertheless, this type of key should be useful to identify some species, but probably its main usefulness is grouping species of ticks sharing morphological characters as a first step before using more precise morphological diagnostic characters sometimes in combination with ecological factors and molecular identification.

MORPHOLOGICAL KEY FOR GENERA OF IXODIDAE

Numbers of figures for Ixodidae refer to those numbers used in Chapter 2, "Genera and Species of Ixodidae."

1.	Anal groove curving anteriorly to the anus; eyes and festoons absent	*Ixodes*
	Anal groove curving posteriorly to the anus or indistinct; eyes and festoons present or absent	**2**
2.	Eyes absent, scutum inornate, article II of palps extending laterally	*Haemaphysalis*[a]
	Eyes present, scutum ornate or inornate	**3**
3.	Scutum inornate, palps short and rounded apically, basis capituli hexagonal dorsally, ventral plates in males	*Rhipicephalus*[b]
	Scutum ornate or inornate, palps elongate and subcylindrical	**4**

Ticks of the Southern Cone of America. DOI: http://dx.doi.org/10.1016/B978-0-12-811075-1.00004-2

4.	Seven festoons, palps about as long as basis capituli, spiracular plates subcircular with few and very large globet cells, scutum inornate	***Dermacentor*[c]**
	Eleven festoons, palps longer than basis capituli, spiracular plates oval or comma-shaped with numerous and very small globet cells, scutum ornate or inornate	***Amblyomma***

[a]*A few species of the genus* Haemaphysalis *absent from the Neotropical Region lack the lateral extension of the article II of palps.*
[b]*These characters are representative of the* Rhipicephalus *species present in the Southern Cone of America.*
[c]*These characters are representative of* D. nitens, *the only species of the genus* Dermacentor *present in the Southern Cone of America.*

MORPHOLOGICAL KEY FOR SPECIES OF *AMBLYOMMA*

Morphological Key for Males

The male of *A. rotundatum* is included in the key below but this is a parthenogenetic species and the male is rarely found in nature.

1.	Marginal groove absent	**2**
	Marginal groove present	**9**
2.	Coxae II–IV with two distinct spurs	**3**
	Coxae II–IV with one spur	**5**
3.	External spur on coxa I longer than internal spur; scutum with punctuations numerous, areolate, interspersed with fine punctuations, absent in the anteromedian field	***A. dissimile*** (Fig. 2.25)
	Spurs on coxa I equal or subequal in size	**4**
4.	Scutum with small brownish spots; scutum with punctuations numerous, areolate, absent in the central and anteromedian fields	***A. argentinae*** (Fig. 2.1)
	Scutum with pale iridescent patches of orange ornamentation on lateral fields; scutum with punctuations numerous, more densely distributed on anterolateral fields	***A. rotundatum*** (Fig. 2.64)
5.	Carena present	**6**
	Carena absent	**7**
6.	Basis capituli dorsally quadrangular; all ventral plates incised, except the plate corresponding to the central festoon	***A. incisum*** (Fig. 2.34)
	Basis capituli dorsally subtriangular; ventral plates not incised	***A. aureolatum*** (Fig. 2.4)
7.	Coxa I with two distinct, short, subequal triangular spurs; limiting spots converging posteriorly toward the median line forming the outline of a pseudoscutum; lateral spots fused but distinct	***A. pacae*** (Fig. 2.49)
	Coxa I with two distinct, long, triangular, blunt spurs, subequal in size; scutum with irregular pale spots in the anterolateral fields; pseudoscutum absent	**8**
8.	Scutum with irregular pale spots Y-shaped in the anterolateral fields; coxa IV with one long, triangular sharp spur	***A. calcaratum*** (Fig. 2.19)
	Scutum with irregular pale spots J-shaped in the anterolateral fields; coxa IV with one short, triangular spur	***A. nodosum*** (Fig. 2.43)

9.	Marginal groove incomplete	**10**
	Marginal groove complete	**13**
10.	Coxae II–III with two triangular, short spurs; coxa IV with two spurs, internal spur long and triangular and the external spur very short; carena present	**A. brasiliense** (Fig. 2.16)
	Coxae II–IV with one spur	**11**
11.	Spines on the tibia of legs II–IV absent; presence of five ventral plates in the posterior field of the ventral surface	**A. longirostre** (Fig. 2.37)
	Presence of one or two spines on the tibia of legs II–IV; ventral plates absent	**12**
12.	Presence of one spine on the tibia of legs II–IV; eyes flat	**A. neumanni** (Fig. 2.40)
	Presence of two spines on the tibia of legs II–IV; eyes orbited	**A. parvitarsum** (Fig. 2.52)
13.	Trochanters with spurs; article I of palps ventrally with a large blunt spur directed posteriorly	**14**
	Trochanters without spurs	**17**
14.	Scutum ornate with small pale spots adjacent to the marginal groove; scutum with few punctuations; carena absent; basis capituli subquadrangular dorsally; cornua short	**A. pseudoconcolor** (Fig. 2.58)
	Scutum inornate	**15**
15.	Body outline oval rounded; scutum inornate with punctuations small and shallow; carena absent; basis capituli subquadrangular dorsally; cornua short	**A. auricularium** (Fig. 2.8)
	Body outline oval; scutum inornate with punctuations moderately deep; basis capituli subquadrangular dorsally; cornua long	**16**
16.	Scutum inornate with punctuations numerous, moderately deep, uniformly distributed; carena present	**A. parvum** (Fig. 2.55)
	Scutum inornate with punctuations sparse, moderately deep, uniformly distributed; carena absent	**A. pseudoparvum** (Fig. 2.61)
17.	Presence of one spine on the tibia of legs II–IV, body outline oval elongated, narrower in the anterior part	**18**
	Absence of one spine on the tibia of legs II–IV	**19**
18.	Scutum with posteromedian spot narrower than the area between posteromedian and posterolateral spots; tubercles absent; spiracular plates comma-shaped	**A. tigrinum** (Fig. 2.70)
	Scutum with posteromedian spot wider than the area between posteromedian and posterolateral spots; tubercles present; spiracular plates oval	**A. triste** (Fig. 2.76)
19.	Dental formula 2/2; eyes orbited; coxa IV with long, sickle-shaped, medially directed spur arising from its internal margin; scutum with coloration from light gray to very pale ivory, with single bilateral white stripe converging on the middle level of scutum and then diverging posteriorly	**A. boeroi** (Fig. 2.12)
	Dental formula 3/3; eyes not orbited	**20**
20.	Coxa I with two distinct, long, sharp spurs, equal in size, tip of the external spur with a slight curve outward; basis capituli dorsally subtriangular; body outline oval elongated; scutum with two narrow yellow stripes in the lateral fields	**A. ovale** (Fig. 2.46)
	Coxa I with two short, triangular spurs, unequal in size; basis capituli dorsally quadrangular	**21**

21.	Carena present	**22**
	Carena absent; body outline oval rounded; anterior extremity of limiting spots sometimes merging with anteroaccessory and ocular spot	**24**
22.	Spiracular plate oval; scutum with pale strips extending from the level of eyes to the level of the first festoons	**A. dubitatum** (Fig. 2.28)
	Spiracular plate comma-shaped	**23**
23.	Scutum with two bright red-orange patches in the scapular area, patches converge posteriorly forming the outline of a pseudoscutum; carena present, regular in shape and not incised	**A. coelebs** (Fig. 2.22)
	Scutum with limiting spots converging posteriorly toward the median line forming the outline of a pseudoscutum, lateral spots, posteroaccessory spots, and posteromedian spot small; carena present, irregular in shape, sometimes with a small incision on festoons 4, 5, and 6	**A. hadanii** (Fig. 2.31)
24.	Scutum with posteromedian spot narrower than adjacent enameled yellowish stripe	**A. sculptum** (Fig. 2.67)
	Scutum with posteromedian spot wider than adjacent enameled yellowish stripe	**A. tonelliae** (Fig. 2.73)

Morphological Key for Females

1.	Coxae II–III with two distinct spurs	**2**
	Coxae II–III with one spur	**5**
2.	Chitinous tubercles at the posterobody margin present; dental formula 4/4; scutum ornate with punctuations numerous, shallow, larger in the anterolateral fields	**A. brasiliense** (Fig. 2.17)
	Chitinous tubercles at the posterobody margin absent; dental formula 3/3	**3**
3.	Scutum ornate, extensively pale yellowish, cervical spots elongated touching posteriorly limiting spots; punctuations large, areolate, concentrated in anterolateral fields	**A. argentinae** (Fig. 2.2)
	Scutum ornate, with an irregular pale spot in each anterolateral field, and a pale spot in the posterior field	**4**
4.	Pale spot in the posterior field of the scutum large, extending from the posterior margin to the midlevel of the scutum	**A. dissimile** (Fig. 2.26)
	Pale spot in the posterior field of the scutum small, restricted to the posterior margin of the scutum	**A. rotundatum** (Fig. 2.65)
5.	Trochanters with spurs	**6**
	Trochanters without spurs	**9**
6.	Scutum ornate with small pale spots on a yellowish-brown ground; punctuations of the scutum small, moderately deep; notum glabrous	**A. pseudoconcolor** (Fig. 2.59)
	Scutum inornate	**7**
7.	Scutum inornate with punctuations small and shallow; notum glabrous	**A. auricularium** (Fig. 2.9)
	Scutum inornate with punctuations deep or moderately deep; notum with setae	**8**

8. Scutum with punctuations numerous, moderately deep,   ***A. parvum***
    uniformly distributed; notum with short setae, evenly distributed   (Fig. 2.56)
    Scutum with punctuations numerous and deep, forming an   ***A. pseudoparvum***
    evident depression on each lateral field; punctuations scarce   (Fig. 2.62)
    around eyes; notum with numerous, long, white setae,
    uniformly distributed
9. Presence of one or two spines on the tibia of legs II–IV   **10**
    Spines on the tibia of legs II–IV absent   **13**
10. Presence of two spines on the tibia of legs II–IV; eyes   ***A. parvitarsum***
    orbited; scutum ornate, extensively pale yellowish, central   (Fig. 2.53)
    area short and narrow; notum glabrous
    Presence of one spine on the tibia of legs II–IV; eyes not orbited   **11**
11. Notum with short coarse setae; coxa I with two distinct,   ***A. neumanni***
    triangular, short spurs, the external longer than the internal;   (Fig. 2.41)
    scutum ornate, extensively pale yellowish, cervical spots
    narrow and divergent posteriorly, central area short and narrow
    Notum glabrous; coxa I with two distinct spurs, the external spur   **12**
    long, narrow and sharp, the internal spur as a small tubercle
12. Chitinous tubercles at posterior body margin absent; scutum   ***A. tigrinum***
    with central area long and narrow, not reaching the posterior   (Fig. 2.71)
    margin of the scutum; spiracular plates comma-shaped
    Chitinous tubercles at posterior body margin present; scutum   ***A. triste*** (Fig. 2.77)
    with central area long and narrow, reaching the posterior
    margin of the scutum; spiracular plates oval
13. Chitinous tubercles at posterior body margin present   **14**
    Chitinous tubercles at posterior body margin absent   **16**
14. Dental formula 4/4; notum glabrous; scutum ornate,   ***A. incisum***
    extensively pale yellowish, cervical spots narrow, and   (Fig. 2.35)
    divergent posteriorly
    Dental formula 3/3; notum with setae; scutum ornate, with   **15**
    an irregular and diffuse central area not reaching the
    posterior margin of scutum
15. Notum with short setae, more densely distributed on central   ***A. sculptum***
    and posterior fields; genital aperture U-shaped   (Fig. 2.68)
    Notum with long and stout setae, and with three deep   ***A. tonelliae***
    grooves; genital aperture V-shaped   (Fig. 2.74)
16. Eyes orbited; dental formula 2/2; notum with setae; scutum   ***A. boeroi***
    ornate, central area reaching the posterior scutal margin, and   (Fig. 2.13)
    cervical spots externally concave
    Eyes not orbited; dental formula 3/3   **17**
17. Hypostome pointed, lanceolate; scutum ornate, with an   ***A. longirostre***
    irregular longitudinal pale patch in the median field   (Fig. 2.38)
    Hypostome spatulate   **18**
18. Spurs of coxa I longer than width of the article   **19**
    Spurs of coxa I as long as or shorter than width of the article   **20**
19. Scutum ornate, extensively pale yellowish, cervical spots   ***A. aureolatum***
    narrow and elongated; the external spur slightly longer than   (Fig. 2.5)
    the internal spur

	Scutum ornate, with an yellowish posterior spot without central stripe; the external spur slightly longer than the internal spur, tip of the external spur with a slight curve outward	*A. ovale* (Fig. 2.47)
20.	Basis capituli dorsally subtriangular	**21**
	Basis capituli dorsally subrectangular	**23**
21.	Coxa I with two distinct, short, triangular spurs, unequal in size, external spur longer than internal spur; scutum ornate, with pale marking in the posterior angle	*A. pacae* (Fig. 2.50)
	Coxa I with two distinct, triangular, blunt spurs, equal in size; punctuations numerous, uniformly distributed	**22**
22.	Scutum ornate, with an irregular pale spot in the posterior field	*A. calcaratum* (Fig. 2.20)
	Scutum ornate, with a Y-shaped pale spot in the anterolateral field, and a small and irregular pale spot in the posterior field	*A. nodosum* (Fig. 2.44)
23.	Scutum ornate, with a broad and longitudinal central area, extending to the posterior angle of the scutum; spiracular plate oval; posterolateral margins of the scutum sinuous	*A. dubitatum* (Fig. 2.29)
	Lacking this combination of characters	**24**
24.	Scutum ornate with an enameled yellowish posterior spot without central stripe	*A. coelebs* (Fig. 2.23)
	Scutum ornate with a patchy enameled yellowish posterior spot, with a central stripe reaching posterior scutal margin, central stripe become narrower toward posterior scutal margin	*A. hadanii* (Fig. 2.32)

Morphological Key for Nymphs

1.	Pseudoauricula present	**2**
	Pseudoauricula absent	**6**
2.	Coxa I with one spur; ventral process present	**3**
	Coxa I with two spurs; ventral process absent	**4**
3.	Coxa I with a long, narrow sharp spur; scutum lacking a sinuous posterolateral margin	*A. tigrinum* (Fig. 2.72)
	Coxa I with a triangular, robust spur; scutum with a sinuous posterolateral margin	*A. triste* (Fig. 2.78)
4.	Ventral cornua present	**5**
	Ventral cornua absent; coxa I with stout spurs; eyes located at the level of the scutal midlength; idiosoma longilinear; hypostome rounded apically	*A. aureolatum* (Fig. 2.6)
5.	Coxa I with two short spurs; eyes located at the level of the scutal midlength; scutal surface extensively shagreened (rugose) (this character is visualized under optical microscopy better than under scanning electron microscopy); hypostome sharply pointed	*A. longirostre* (Fig. 2.39)
	Coxa I with two medium spurs; eyes located at the level of the posterior third of the scutum; idiosoma longilinear; hypostome rounded apically	*A. ovale* (Fig. 2.48)
6.	Cornua present	**7**
	Cornua absent	**10**

7.	Tubercles present at posterointernal angles of festoons, pronounced triangular cornua	**8**
	Tubercles absent at posterointernal angles of festoons, minute triangular cornua	**9**
8.	Scutum length <0.75 mm	***A. brasiliense*** (Fig. 2.18)
	Scutum length >0.75 mm	***A. incisum*** (Fig. 2.36)
9.	Dorsum of basis capituli subrectangular; eyes elongate	***A. parvum*** (Fig. 2.57)
	Dorsum of basis capituli rectangular; eyes not elongate	***A. pseudoparvum*** (Fig. 2.63)
10.	Eyes not orbited	**11**
	Eyes orbited; coxa I with triangular, large, external spur and minute, rounded, internal spur; palpi short and wide (robust); idiosoma longilinear	***A. boeroi*** (Fig. 2.14)
11.	Hypostomal dentition 3/3 for most of the length, 2/2 at the base	**12**
	Hypostomal dentition 2/2	**14**
12.	Coxa II with only one spur	**13**
	Coxa II with two spurs, the internal very small	***A. rotundatum*** (Fig. 2.66)
13.	Scutum with deep punctuations evenly distributed, larger and deeper laterally	***A. argentinae*** (Fig. 2.3)
	Scutum with deep punctuations concentrated in the lateral fields	***A. dissimile*** (Fig. 2.27)
14.	Coxa I with two unequal spurs, the external longer than the internal	**15**
	Coxa I with two subequal rounded short spurs, the external slightly stouter; scutum with few punctuations; dorsal of basis capituli rectangular; palpal article IV projecting apically	***A. parvitarsum*** (Fig. 2.54)
15.	Deep punctuations rarely present on scutum or when present they are concentrated in the lateral fields	**16**
	Scutum moderately punctate, deep punctuations evenly distributed	**20**
16.	Dorsum of basis capituli subtriangular or trapezoidal; scutal surface extensively shagreened (rugose) (this character is visualized under optical microscopy better than under scanning electron microscopy)	**17**
	Dorsum of basis capituli broadly rectangular; scutal surface slightly shagreened	**19**
17.	Dorsum of basis capituli subtriangular	**18**
	Dorsum of basis capituli trapezoidal; palpi short and wide (robust); cervical groove long and deep	***A. neumanni*** (Fig. 2.42)
18.	Scutum breadth/length ratio <1.3, deep punctuations concentrated in the lateral fields	***A. calcaratum*** (Fig. 2.21)
	Scutum breadth/length ratio >1.3, few deep punctuations	***A. nodosum*** (Fig. 2.45)
19.	Scutum breadth/length ratio <1.3, cervical groove extending to the level of scutal midlength	***A. auricularium*** (Fig. 2.10)
	Scutum breadth/length ratio >1.3, cervical groove extending to the scutal posterior border	***A. pseudoconcolor*** (Fig. 2.60)

20.	Coxa I with external spur twice as long as internal spur; dorsum of basis capituli rectangular	**21**
	Coxa I with external spur longer than internal, however less than twice as long	**22**
21.	Scutal surface smooth; coxa I with anterolateral seta shorter than the length of coxa I external spur	*A. sculptum* (Fig. 2.69)
	Scutal surface extensively (rugose) (this character is visualized under optical microscopy better than under scanning electron microscopy); coxa I anterolateral seta longer than the length of coxa I external spur	*A. tonelliae* (Fig. 2.75)
22.	Cervical grooves short, ending as a small shallow depression at the level of the posterior margin of the eyes	**23**
	Cervical grooves long, well surpassing the level of the posterior margin of the eyes	**24**
23.	Dorsum of basis capituli broadly hexagonal; scutum with large and deep punctuations in both lateral and central fields	*A. coelebs* (Fig. 2.24)
	Dorsum of basis capituli rectangular; scutum with large punctuations in the lateral fields, smaller punctuations in the central field	*A. hadanii* (Fig. 2.33)
24.	Cervical grooves deep throughout; lacking a shallow and large depression in its posterior divergent half; scutum breadth/length ratio <1.3	*A. dubitatum* (Fig. 2.58)
	Cervical grooves deep in its anterior convergent half, and as a shallow large depression on its posterior divergent half; scutum breadth/length ratio >1.3	*A. pacae* (Fig. 2.50)

MORPHOLOGICAL KEY FOR SPECIES OF *HAEMAPHYSALIS*

Morphological Key for Males

1.	Hypostome spatulate, dental formula 4/4; article III of palps with a long, retrograde, ventral spur; ventral process absent	*H. juxtakochi* (Fig. 2.82)
	Hypostome spatulate, dental formula 3/3; article III of palps with a short, retrograde, ventral spur; ventral process present	*H. leporispalustris* (Fig. 2.85)

Morphological Key for Females

1.	Hypostome spatulate, dental formula 4/4; article III of palps with a long, retrograde, ventral spur; ventral process absent	*H. juxtakochi* (Fig. 2.83)
	Hypostome spatulate, dental formula 3/3; article III of palps with a short, retrograde, ventral spur; ventral process present	*H. leporispalustris* (Fig. 2.86)

Morphological Key for Nymphs

1.	Article III of palps with a long, retrograde, ventral spur; internal spur on coxa I long, reaching the coxa II; ventral process absent	*H. juxtakochi* (Fig. 2.84)
	Article III of palps with a short, retrograde, ventral spur; internal spur on coxa I short, not reaching the coxa II; ventral process present	*H. leporispalustris* (Fig. 2.87)

MORPHOLOGICAL KEY FOR SPECIES OF *IXODES*

Morphological Key for Males

The males of *I. chilensis, I. cornuae, I. longiscutatus, I. neuquenensis, I. schulzei,* and *I. sigelos* are unknown.

1.	Suture between articles II and III of the palps inconspicuous	2
	Suture between articles II and III of the palps conspicuous	3
2.	Hypostome short, notched, dental formula 3/3 in the anterior third, 6/6 in the median portion; coxae I–IV with a short, external spur, increasing in size from coxa I to coxa IV; scutum with surface smooth	*I. auritulus* (no figure available)
	Hypostome short, notched, dental formula 1/1; all coxae without spurs; scutum with punctuations numerous, uneven in size	*I. uriae* (Fig. 2.116)
3.	Ventral outline of basis capituli with trilobed transverse ridge, two laterals lobes and one median lobe triangular and displaced posteriorly; hypostome short with lateral teeth pointed and longer than median teeth; coxa I with two spurs, internal spur long, pointed, reaching the coxa II, external spur short and rounded	*I. aragaoi/I. pararicinus* (Figs. 2.91 and 2.105)
	Ventral outline of basis capituli without trilobed transverse ridge	4
4.	Coxae II–IV with one spur; basis capituli subtriangular dorsally; hypostome blunt; auriculae rounded	5
	Coxae II–IV with two spurs; basis capituli pentagonal dorsally; hypostome notched	6
5.	Coxa I with two short, blunt spurs, subequal in size, the internal spur slightly narrower than external spur	*I. loricatus* (Fig. 2.97)
	Coxa I with two spurs, the external spur long, narrow, and sharp, reaching the coxa II, and the internal spur triangular and short	*I. luciae* (Fig. 2.100)
6.	No indication of pseudoscutum; scutum with punctuations numerous, fine, uniformly distributed; setae scattered all over the scutum but absent on posterocentral field	*I. abrocomae* (Fig. 2.88)
	Presence of pseudoscutum	7
7.	Scutum with numerous, large, whitish setae, except in the anterior part corresponding to the pseudoscutum	*I. taglei* (Fig. 2.114)
	Scutum with few and short setae	8
8.	Pseudoscutum faint	*I. stilesi* (Fig. 2.111)
	Pseudoscutum notorious	*I. nuttalli* (no figure available)

Morphological Key for Females

1.	Legs with spurs on trochanters I–IV; auriculae large, horn-like	2
	Legs without spurs on trochanters I–IV	3
2.	Article I of the palp with a long internal projection	*I. auritulus* (Fig. 2.93)
	Article I of the palp with a long internal projection, and presence of a strong meso-dorsal spur	*I. cornuae* (no figure available)

3.	Legs with all coxae without spurs; cornua and auriculae absent; dental formula 2/2	*I. uriae* (Fig. 2.117)
	Legs with coxae with spurs	4
4.	Coxa I with one spur	5
	Coxa I with two spurs	6
5.	Coxae I–III with a single, short, external spur; coxa IV with a single, short, central spur	*I. abrocomae* (Fig. 2.89)
	Coxae I–IV with an external spur as a small tubercle	*I. chilensis* (no figure available)
6.	Coxae II–IV each with two triangular spurs	*I. neuquenensis* (Fig. 2.103)
	Coxae II–IV without two spurs	7
7.	Coxa I with a short internal spur, and two posteroexternal spurs, one dorsal and one ventral; coxae II–III with two posteroexternal spurs, one dorsal one ventral	*I. nuttalli* (no figure available)
	Coxae I–III each with one-posteroexternal spur	8
8.	Coxa I with one spur pointed and much longer than the other spur	9
	Coxa I with two spurs equal or subequal in size	10
9.	Coxa I with two spurs, the external spur long, narrow, and sharp, reaching the coxa II, and the internal spur triangular and short	*I. luciae* (Fig. 2.101)
	Coxa I with two spurs, the internal spur long, narrow, and sharp, reaching the coxa II, and the external spur short and rounded	*I. aragaoi/I. pararicinus* (Figs. 2.92 and 2.106)
10.	Hypostome blunt	11
	Hypostome pointed	12
11.	Scutum with numerous punctuations, larger on the median-posterior field	*I. loricatus* (Fig. 2.98)
	Scutum with few punctuations, larger on the lateral field	*I. taglei* (Fig. 2.115)
12.	Auriculae pointed and extended laterally; hypostome pointed, dental formula 2/2	*I. schulzei* (Fig. 2.107)
	Auriculae short and small, not extended laterally	13
13.	Scutum with few, short, scattered setae; few punctuations on the scutum, larger on the lateral field; dental formula 3/3	*I. stilesi* (Fig. 2.112)
	Scutum with long and numerous setae	14
14.	Lateral margins in the midportion of the scutum straight, almost parallel; cervical grooves short	*I. sigelos* (Fig. 2.109)
	Lateral margins of the scutum convex; cervical grooves long	*I. longiscutatus* (Fig. 2.95)

Morphological Key for Nymphs

The nymphs of *I. abrocomae*, *I. chilensis*, and *I. taglei* are unknown.

1.	Legs with spurs on trochanters I–IV; auriculae large, horn-like	2
	Legs without spurs on trochanters I–IV	3
2.	Article I of the palp with an internal projection without a strong meso-dorsal spur	*I. auritulus* (Fig. 2.94)
	Article I of the palp with an internal projection with presence of a strong meso-dorsal spur	*I. cornuae* (no figure available)

3.	Coxae without spurs; cornua and auriculae absent; dental formula 2/2	*I. uriae* (Fig. 2.118)
	Coxae with spurs	**4**
4.	Coxae II–III each with two spurs	**5**
	Coxae II–III without two spurs	**6**
5.	Article III of the palp with a strong, pointed, retrograde spur; article II of the palps with a strong, bifurcate, basolateral salient, and a dorsal process rounded; notum with numerous long, stout, blunt setae; auriculae long and slightly directed laterally	*I. longiscutatus* (Fig. 2.96)
	Article I of palp with large anterior and posterior processes, plus a large posterolateral spur on the posterior process, and a minute lateral protuberance near the point of insertion of article II of palp; auriculae small and triangular	*I. neuquenensis* (Fig. 2.104)
6.	Article I of palp with large ventral anterior and posterior processes present	**7**
	Article I of palp with large ventral anterior and posterior processes absent	**9**
7.	Article I of palp ends in a bifurcate formation	*I. sigelos* (Fig. 2.110)
	Article I of palp bluntly ended	**8**
8.	Process in palpal article I reaches the last third of toothed portion of the hypostome	*I. stilesi* (Fig. 2.113)
	Process in palpal article I does not reach the toothed portion of the hypostome	*I. nuttalli* (no figure available)
9.	Hypostome insertion at the level of the apical third of palpal article II; coxa I with two short, triangular spurs, the external spur slightly longer than internal spur	*I. schulzei* (Fig. 2.108)
	Hypostome insertion at the level of the basal third of palpal article II	**10**
10.	Coxa I with two short spurs, subequal in size	*I. loricatus* (Fig. 2.99)
	Coxa I with two spurs, the external spur sharp and longer than the internal spur	*I. luciae* (Fig. 2.102)

MORPHOLOGICAL KEY FOR SPECIES OF *RHIPICEPHALUS*

Morphological Key for Males

1.	Marginal groove absent; anal groove indistinct; dorsal surface covered by numerous, long, pale setae; presence of caudal appendage; dental formula 4/4	*R. microplus* (Fig. 2.119)
	Marginal groove present, incomplete; anal groove conspicuous; dorsal surface without setae; caudal appendage absent; dental formula 3/3	*R. sanguineus* (Fig. 2.122)

Morphological Key for Females

1.	Notum and ventral surface covered by numerous setae; anal groove indistinct; dental formula 4/4; coxa I with two short blunt spurs	*R. microplus* (Fig. 2.120)

| Notum glabrous; anal groove conspicuous; dental formula 3/3; coxa I with two long triangular spurs | *R. sanguineus* (Fig. 2.123) |

Morphological Key for Nymphs

1. Dental formula 3/3; anal groove indistinct; coxa I with two short blunt spurs; ventral process absent; eyes located at the level of the scutal midlength — *R. microplus* (Fig. 2.121)

   Dental formula 2/2; anal groove conspicuous; coxa I with two triangular long spurs; ventral process present; eyes located at the level of the posterior third of the scutum — *R. sanguineus* (Fig. 2.124)

MORPHOLOGICAL KEY FOR GENERA OF ARGASIDAE (ADULTS)

Numbers of figures for Argasidae refer to those numbers used in Chapter 3, "Genera and Species of Argasidae."

1. Lateral suture present — *Argas*

   Lateral suture absent — 2
2. Integument granular (with spines in nymphs); hypostome vestigial — *Otobius*

   Integument mammillated and lacking spines; hypostome not vestigial — *Ornithodoros*

MORPHOLOGICAL KEY FOR GENERA OF ARGASIDAE (LARVAE)

1. Eyes present — *Otobius*

   Eyes absent — 2
2. Palpal segment IV as long or longer than other palpal segments; dorsum with 26–30 pairs of dorsal setae — *Argas*

   Palpal segment IV shorter than palpal segments II and III; dorsum with 13–21 pairs of dorsal setae — *Ornithodoros*

MORPHOLOGICAL KEY FOR SPECIES OF *ARGAS*

Morphological Key for Adults

1. Post palpal setae absent — 2

   Post palpal setae present — 3
2. Presence of subapical dorsal protuberances on tarsus I–IV — *A. neghmei* (Fig. 3.7)

   Presence of subapical dorsal protuberances only on tarsus I — *A. monachus* (Fig. 3.5)
3. Peripheral cells of dorsum with a single large setiferous pit occupying most of surface area — *A. persicus* (no figure available)

   Peripheral cells of dorsum with a single small setiferous pit not occupying the total of surface area — 4
4. Basis capituli with 9–12 pairs of ventral setae — *A. miniatus* (Fig. 3.3)

   Basis capituli with 7–8 pairs of ventral setae — *A. keiransi* (Fig. 3.1)

Morphological Key for Larvae

1.	Presence of a trumpet-shaped sensillum extending from the capsule of Haller's organ into the lumen of the tarsus	2
	Absence of a trumpet-shaped sensillum extending from the capsule of Haller's organ into the lumen of the tarsus	3
2.	Dorsum with 28 pairs of setae; hypostome about three times longer than wide	*A. neghmei* (Fig. 3.8)
	Dorsum with 24 pairs of setae; hypostome just twice longer than wide	*A. monachus* (Fig. 3.6)
3.	Posterolateral setae length 0.065−0.075 mm	*A. miniatus* (Fig. 3.4)
	Posterolateral setae length 0.043−0.050 mm	4
4.	Apex of hypostome with dental formula 3/3 or 4/4	*A. persicus* (Fig. 3.9)
	Apex of hypostome with dental formula 2/2 or 3/3	*A. keiransi* (Fig. 3.2)

MORPHOLOGICAL KEY FOR SPECIES OF *ORNITHODOROS*

Morphological Key for Adults

Adults of *O. lahillei*, *O. microplophi*, *O. peruvianus*, *O. quilinensis*, and *O. xerophylus* are unknown.

1.	Cheeks absent	*O. rostratus* (Fig. 3.22)
	Cheeks present	2
2.	Discs small, faintly differentiated from the mammillae	*O. rudis* (no figure available)
	Discs large, well differentiated from the mammillae	3
3.	Anterior body outline rounded	*O. amblus* (Fig. 3.10)
	Anterior body outline pointed	4
4.	All coxae contiguous	*O. hasei* (no figure available)
	Coxae I and II separated	5
5.	Body size small (female ≅4.5 mm, male ≅3.5 m	*O. mimon* (Fig. 3.16)
	Body size large	6
6.	Micromamillae on tarsus I conspicuous	*O. rioplatensis* (Fig. 3.20)
	Micromamillae on tarsus I inconspicuous	*O. spheniscus* (Fig. 3.24)

Morphological Key for Larvae

1.	Six pairs of sternal setae	*O. microlophi* (Fig. 3.15)
	Three pairs of sternal setae	2
2.	Ventral groove present	*O. rostratus* (Fig. 3.23)
	Ventral grooves absent	3
3.	Postcoxal setae absent	4
	Postcoxal setae present	5
4.	Dorsal shield subtriangular, longer than wide	*O. lahillei* (Fig. 3.14)
	Dorsal shield wider than long	*O. rudis* (no figure available)

5.	Hypostome blunt	6
	Hypostome pointed	8
6.	Dorsal shield pyriform	*O. mimon* (Fig. 3.17)
	Dorsal shield oval	7
7.	Dorsum with 14 pairs of setae, capsule of Haller's organ irregular in shape	*O. quilinensis* (Fig. 3.19)
	Dorsum with 16 pairs of setae, capsule of Haller's organ oval in shape	*O. xerophylus* (Fig. 3.26)
8.	Posteromedian seta present	9
	Posteromedian seta absent	11
9.	Dorsum with 14 pairs of setae	*O. peruvianus* (Fig. 3.18)
	Dorsum with 19–20 pairs of setae	10
10.	Dorsum with 20 pairs of setae, maximum length of hypostome 0.250 mm	*O. rioplatensis* (Fig. 3.21)
	Dorsum with 19 pairs of setae, hypostome maximum length 0.190 mm	*O. hasei* (Fig. 3.12)
11.	Apex of hypostome with dental formula 5/5 or 4/4	*O. amblus*[a] (Fig. 3.11)
	Apex of hypostome with dental formula 4/4 or 3/3	*O. spheniscus*[a] (Fig. 3.25)

[a]*In some specimens these morphological features overlap which makes the larvae of these species indistinguishable.*

Conclusion

Progress to know the situation about ticks for the countries forming the Southern Cone of America has been made in recent years as shown in the previous chapters. Certainly, the contribution on alpha taxonomy and phylogeny of ticks is obvious throughout the recent literature of the species established in the region. Nevertheless several matters should be addressed in the future to obtain a better picture of the tick problem including also alpha and beta taxonomy to solve weak knowledge for genus such as *Ornithodoros* and some complex species such as *Ixodes auritulus* and *Rhipicephalus sanguineus*. Several species of *Ornithodoros* described recently are just known from the larval stage indicating that additional works are needed to find their nymphs and adults as well as their hosts to understand their natural cycles and ecology. The cosmopolitan *I. auritulus* has been recognized as a species complex long ago and it is almost certain that more than one species of this group are established in the Southern Cone of America; more field and laboratory studies are indispensable to define this situation. The *R. sanguineus* species complex represents an important and difficult problem to solve; at least two species with different capacities to transmit dog pathogens are known for the region but to ascribe them to any species will have to wait until an international effort in progress defines with certainty bona fide *R. sanguineus* and related species.

Several studies on tick phylogeny based on molecular data have been carried out in recent years. However, data on mitochondrial 16S rDNA sequences are overrepresented in relation to other mitochondrial (e.g., sequences of 12S rDNA, cytochrome oxidase I and II, and control region or d-loop) and nuclear (e.g., sequences of the inter-transcribed spacer 2 and 18S rDNA) markers. Moreover, the current knowledge of the molecular phylogeny of Argasidae is entirely based on 16S rDNA sequences. Further phylogenetic analyses constructed with other DNA sequences besides those of the 16S rRNA gene are needed to improve our knowledge on the phylogeny and evolution of the ticks species distributed in the Southern Cone of America.

The knowledge on the ecology of several species, mainly for those taxa with economic and sanitary importance, has been improved but it has not been a uniform trend. This situation is of special concern for the case of Paraguay where information for relevant species such as *Amblyomma sculptum* and *A. tonelliae*, among others, is scarce and local studies are necessary to understand their ecologies and ability as vectors of pathogens for humans and animals there. Besides, the ecology of species of sanitary importance as *A. aureolatum* is poorly known for the entire region and this gap should be filled in the near future. With few exceptions the ecology of the species feeding mostly on wild hosts are basically unknown except from their hosts and localities. Probably unique strategies for survival remain unknown. This ignorance should be reversed.

Perhaps the most important advancement in recent years has been the development of diagnostic capacities for pathogens, especially *Rickettsia*, in ticks and their hosts. A wider application of this tool in different areas of the Southern Cone of America will enhance our knowledge of the pathway of *Rickettsia* in nature and may aid to a better control of some rickettsial diseases. It is also expected that studies in progress including other tick-borne organisms causing diseases for human and animals will bring valuable information on this topic in the near future.

One of the most important informative deficiencies for the region is the lack of data about the effect of tick infestation not only on wild animals but also on domestic vertebrates. Some wild hosts are heavily infested by several species of ticks but in our knowledge no studies have been carried out to reveal if this condition results in health impairment. There are data on the effects of some species of ticks on domestic hosts but additional studies should be made to increase our knowledge on this subject especially for native species of ticks, that is, cattle are sometimes severely parasitized by several species of *Amblyomma* but we do not know how much this infestation affects milk or meat production.

In conclusion, we feel that this publication shows that a good effort has been made to increase the knowledge on ticks and associated problems in Argentina, Chile, Paraguay, and Uruguay, but also shows that it is just one step that should be followed by a continuous labor to fill some important knowledge gaps. This includes the necessity to expand the support to deal with this problem as well as inclusion of new technologies to improve the value of the information to be obtained.

Index

Printed in the United States
By Bookmasters